Human Caspases and Neuronal Apoptosis in Neurodegenerative Diseases

Human Caspases and Neuronal Apoptosis in Neurodegenerative Diseases

Anil Gupta

*Professor and Head, Department of Physiology and Biochemistry,
Eklavya Dental College and Hospital,
Rajasthan, India
Formerly Dean of Research, Professor and Head,
Department of Biochemistry, Desh Bhagat University,
Punjab, India*

ACADEMIC PRESS
An imprint of Elsevier

ELSEVIER

Academic Press is an imprint of Elsevier
125 London Wall, London EC2Y 5AS, United Kingdom
525 B Street, Suite 1650, San Diego, CA 92101, United States
50 Hampshire Street, 5th Floor, Cambridge, MA 02139, United States
The Boulevard, Langford Lane, Kidlington, Oxford OX5 1GB, United Kingdom

Notices
Knowledge and best practice in this field are constantly changing. As new research and experience broaden our understanding, changes in research methods, professional practices, or medical treatment may become necessary.

Practitioners and researchers must always rely on their own experience and knowledge in evaluating and using any information, methods, compounds, or experiments described herein. In using such information or methods they should be mindful of their own safety and the safety of others, including parties for whom they have a professional responsibility.

To the fullest extent of the law, neither the Publisher nor the authors, contributors, or editors, assume any liability for any injury and/or damage to persons or property as a matter of products liability, negligence or otherwise, or from any use or operation of any methods, products, instructions, or ideas contained in the material herein.

Library of Congress Cataloging-in-Publication Data
A catalog record for this book is available from the Library of Congress

British Library Cataloguing-in-Publication Data
A catalogue record for this book is available from the British Library

ISBN: 978-0-12-820122-0

For information on all Academic Press publications visit our website at
https://www.elsevier.com/books-and-journals

Publisher: Nikki Levy
Acquisitions Editor: Natalie Farra
Editorial Project Manager: Veronica III Santos
Production Project Manager: Niranjan Bhaskaran
Cover Designer: Alan Studholme

Typeset by TNQ Technologies

I dedicate this book to my parents

Contents

Preface to first edition of book

Persistent alteration in demographic profiles of the world's population including rapid rise in aging population in certain countries has been cited in the global rise in the prevalence and incidence of neurodegenerative diseases. The worldwide prevalence of Alzheimer's disease in 2050 would reach nearly 100 million as per an estimate. The world may struggle with impending global epidemic of Alzheimer's disease. The rising trend in the prevalence of neurodegenerative disease would certainly become the major health hazard especially among the persons in older age group.

Generalized clinical manifestations as neuropsychiatric signs and symptoms, cognitive disturbance, age-linked progression, disability and functional disorders in several neurodegenerative diseases create difficulty in ascribing disease-specific molecular mechanism in the pathology of neurodegenerative diseases.

Rather advancing neuroimaging techniques help us understand the intricate mechanisms and signaling pathways implicated in the separate neurodegenerative disease.

Animal models of neurodegenerative diseases including advanced neuroimaging techniques and postmortem examination of the parts of brain involved in the neurodegenerative diseases in humans have simplified the understanding of the molecular pathology of the frontotemporal dementia, Alzheimer's disease, Parkinson's disease, and amyotrophic lateral sclerosis.

We have much better understanding of the precise role of human caspases in the physiology and pathology of neurons in the central nervous system.

Caspases are the cytosolic proteases controlling the physiology of several signaling pathways involved in proliferation of cells, neuroinflammation, and neuronal apoptosis. Substantial literature in the field of neurology point to involvement of activated caspase-mediated release of cytokines in inducing activation of microglia, neuroinflammation, impaired release of neurotransmitters and neuronal apoptosis.

Present book titled as "*Human Caspases and Neuronal apoptosis in the Neurodegenerative diseases*" is a treatise providing specific details about the human caspases, neuronal apoptosis, signaling pathways, and molecular pathology of the neurodegenerative diseases particularly Parkinson's disease, Alzheimer's disease, and Huntington's disease.

First chapter of the book is titled as *Neurodegenerative Diseases: Neuropathology, Genetics, and Epidemiology.*

It provides in-depth knowledge about the risk factors insinuated in the pathology of neurodegenerative disease (NDD).

Chapter starts with the introduction to NDD and describes the role of transition metals like iron and copper in the pathology of NDD. Mitochondrial dysfunction has significant ill effects on the physiology of neurons in the central nervous system. The chapter describes several factors namely free radicals, oxidative stress, impaired mitochondrial enzymes, impaired mitochondrial dynamics, and distribution of mitochondria in the neurons in brain involved in the mitochondrial abnormalities implicated in the pathology of NDD.

The book elaborates the role of protein misfolding in the pathology of NDD via vivid description of several mutant proteins namely Aβ peptide and Tau protein in Alzheimer's disease, α-Synuclein protein in Parkinson's disease, Lewy bodies in dementia, and Huntingtin protein in the pathology of Huntington's disease. The first chapter of the book provides adequate details about the role of monoamine oxidase production, nitric oxide synthase and nitric oxide production, and formation of highly reactive nitrogen species in the brain regions affected with the neurodegenerative disease.

Furthermore, impaired antioxidant system of the body has harmful effects on the delicate structure of neurons and their normal physiology in the regions of brain. The chapter describes the involvement of dysregulation of ubiquitin proteasome and NADPH oxidase and ROS production in the brain in the NDD.

Next section of the first chapter describes role of genetic in the pathology of NDD via extensive details covering mutations in *APP* gene, gene PSEN1, gene APOE, gene LRRK2, and gene DJ1.

Last section of the first chapter gives concise details about the epidemiology of neurodegenerative diseases.

Second chapter of the book titled as, *"Role of Caspases, Apoptosis, and Additional Factors in Pathology of Alzheimer's disease"* elaborates intricate molecular pathology of Alzheimer's disease involving the role of caspases and neuronal apoptosis. This chapter describes amyloid cascade hypothesis explaining the pathogenesis of Alzheimer' disease with evidences from several sources.

Apoptosis is the programmed cell death. Chapter explains the probable involvement of the apoptosis in the neuronal death in the pathology of Alzheimer's disease considering the in situ detection of DNA fragments, pattern of neuronal loss in brain and up-regulation of expression of proapoptotic factors in the Alzheimer disease.

Chapter provides elaborate details about the role of caspase 3 and 6 in the cause of neuronal apoptosis in Alzheimer's disease. Present book gives detailed account of caspase 3-mediated activity over genes presenilin-1 and 2, beta-site APP-cleaving enzyme (BACE), TAR DNA-binding protein-43, caspase-activated DNase (CAD), rho-associated coiled-coil-containing protein kinase 1 (ROCK1), and mammalian sterile 20-like kinase 1 (MST1).

Activated caspase-6 has its action on the amyloidogenesis and apoptosis in Alzheimer's disease showing its action on the cleavage of amyloid precursor protein, formation of amyloidogenic fragment (Capp6.5), and cleavage of Tau contributing to pathology of Alzheimer's disease.

Chapter is enriched with knowledge concerning role of activated microglia in the pathogenesis of Alzheimer's disease explaining academically and clinically oriented important topics namely microglial TREM2 receptor, microglial LRP1 receptor, microglial Fc γ receptors (FcγRIIb), microglial CD36 transmembrane protein, astrocytic glial α 7 subtype of nAChR (α7nAChRs), astrocytic glial calcium-sensing receptor (CaSR), microglia-Tau interaction, and microglial PU.1 Transcription factor including the role of advanced glycation end product receptor (RAGE) in expressing the ill-effects of AGEs on the pathology of Alzheimer's disease.

Toll-like receptors are integral membrane proteins. These are activated in microglia and astrocytes in brain regions and are circuitously involved in the neuronal death in NDD. Furthermore, present book chapter describes TLR-4 activated MyD88-dependent pathway, and TLR-4 activated TRIF-dependent pathway including pathogenic potential of inflammasome, and P2X purinoreceptors in the initiation and progression of Alzheimer's disease. The NLR family pyrin domain containing 3 (NLRP3) is the predominant element of innate immune system. Chapter discusses about the influence of nuclear factor-κB nuclear translocation, deubiquitylation, and cathepsin B in the activation of NLRP3 inflammasome contributing to Alzheimer's disease.

Last section of the chapter describes the apparent role of chronic psychological stress and effect of microRNAs (miRNAs) on the cognitive impairment in neurodegenerative diseases.

Third chapter of book titled as "*Role of Caspases and Apoptosis in Parkinson's disease*" describes the subject matter extensively. Chapter describes the extrinsic and intrinsic pathways of apoptosis explaining role of caspase 1, 2, 3, and 8 in the activation of neuronal apoptosis in the pathology of Parkinson's disease.

Fourth chapter of book titled as "*Illustrated Etiopathogenesis of Huntington's Disease*" provides minute details about the subject matter. Chapter starts with explanation of structure, and ubiquitous nature of Huntingtin protein and description of pathogenic role of mutated huntingtin, exon 1 HTT protein, correlation between length of expanded polyglutamine tract to Huntington's disease.

Chapter focuses on the fundamental aspect of microglia activation mechanisms explaining implication of nuclear factor kappa B, kynurenine signaling pathway, and autonomous activation process in the progression of Huntington's disease. Chapter describes potential of caspases 1, 2, and 6 in the pathology of Huntington's disease. Role of tumor suppressor protein, PGC-1α gene, and dynamin-related protein 1 in mitochondrial dysfunction contributing to pathology of Huntington's disease is fully elucidated.

Role of advanced glycation end-products in pathology of Huntington's disease is the interesting and comprehensive section of the chapter that explains every minute detail correlating the global increase in proportion of old age population, impaired carbohydrate metabolism, formation of exogenous and endogenous AGES, and their ill effects on the physiology of neurons in brain with plausible contribution to the pathology of Huntington's disease.

Last section of chapter deals with the role of autophagy, mutation of UBB gene, and PARP-1 in the progression of disease.

I hope this book would be helpful in spreading knowledge about the human caspases-mediated molecular mechanisms and signaling pathways involved in the neuronal apoptosis contributing to the pathology of neurodegenerative diseases especially Parkinson's disease, Alzheimer's disease, and Huntington's disease. I have made my sincere and honest efforts to prepare the first edition of book. With humble submission, I welcome criticism, comments, or suggestions for the improvement of forthcoming edition of book.

Acknowledgment

God bestows upon me knowledge and perseverance in writing the book.

My father, *Shree Ved Parkash Gupta*, always inspires me for achieving high endeavor. I am indebted to my father for instilling passion toward study, since my school days.

I am highly thankful to my wife, *Dr. Deepali Gupta*, for persistent motivation to accomplish endeavor. She always stands beside me through thick and thin.

I acknowledge support provided by my daughter, *Deeksha Gupta*, student of M.B.B.S in writing the book.

Furthermore, I owe my sincere gratitude to *Devlin Person* (Former Editorial Project Manager), Elsevier (World's leading publisher of science and health information) for kind cooperation and motivation ushered during the COVID-19 crisis that helped me sustain my efforts in writing the book today.

I am also thankful to *Veronica Santos* who has recently joined as Editorial Project Manager in Elsevier in streamlining the publication of the book.

Entire production team of Elsevier is well-intentioned and is doing the commendable job in translating the thoughts, words, and expression style of author into valuable, readable, comprehensible, and audience friendly format named as BOOK.

Neurodegenerative diseases

Introduction

Neurodegeneration is the slow, progressive irreversible damage to selective groups of neurons in the central nervous system due to metabolic or toxic effects (Dugger and Dickson, 2016).

These represent a heterogeneous class of brain disorders that are manifested by progressive deterioration of the function of neurons progressing to death of neurons. Neurodegenerative diseases involve selective degeneration of neurons(Silberberg et al., 2015).

The human brain is highly complex. It may contain about 100 billion neurons (Kandel et al., 2000). Multiple factors alter brain functions namely cognition, memory, recognition, and motor functions resulting in neurodegenerative diseases. Progressive damage of neurons and neuronal dysfunction are the cardinal features of neurodegenerative diseases (Ahmad et al., 2016). The Neurodegenerative disease is the common term for a broad range of disorders including Alzheimer's disease (AD), Parkinson's disease (PD), and Huntington's disease (HD), and amyotrophic lateral sclerosis.

Each category of neurodegenerative disease is characterized by separate etiopathogenesis and has discrete clinical manifestations.

Based upon a plethora of evidence, it can be inferred that multiple factors can lead to the pathogenesis of neurodegenerative diseases. **Important factors include metals toxicity, mitochondrial dysfunction and altered bioenergetics, free radical formation and oxidative stress, and protein misfolding.** Current evidence need further investigation to understand the pathophysiology of these diseases (Safia et al., 2012).

Risk factors and neuropathology of neurodegenerative diseases
Role of transition metals in neurodegenerative diseases

Transition metals including zinc, copper, iron, and manganese are essential for the metabolic activities in the body. These are actively involved in the normal

Human Caspases and Neuronal Apoptosis in Neurodegenerative Diseases. https://doi.org/10.1016/B978-0-12-820122-0.00004-2

physiological activities in cells. However, altered homeostasis of these metals can be the cause of oxidative stress in tissues (Chen et al., 2018).

Transition metals are redox-active. These can generate reactive oxygen and or nitrogen species via redox cycling reactions. Iron reacts with H_2O_2 and produces hydroxyl radicals, and the reaction is called as Fenton's reaction (Chen et al., 2018).

Surplus accumulation of reacting oxygen and or nitrogen species in the body tissues can impair the intracellular redox state, which is termed as oxidative stress. It manifests as lipid peroxidation, protein denaturation and DNA injury, mitochondrial dysfunction, endoplasmic reticulum stress, cellular inflammation, and/or apoptosis. These molecular and cellular events are implicated in the pathogenesis of neurodegenerative diseases (Chen et al., 2018).

Role of iron in neurodegenerative diseases

Iron is an important transition metal that has a major role in the cellular activities in brain tissues including mitochondrial oxidative phosphorylation, synthesis of myelin, DNA synthesis, synthesis of neurotransmitters, and oxygen transportation (Stankiewicz and Brass, 2009; Ward et al., 2014). Iron is found in oligodendrocytes, neurons, astroglia, and microglial cells in the central nervous system. Iron can be implicated in the synthesis of reactive oxygen species like highly toxic hydroxyl free radicals by the decomposition of H_2O_2 in brain tissues. Thus, iron is potentially toxic to the health of tissues owing to its ability to catalyze Fenton reaction (Halliwell, 2006).

Strict control over the iron hemostasis is essential to prevent iron-induced toxicity in the central nervous tissues and in other body tissues. It is achieved by iron sequestration by iron ligating proteins like ferritin, heme, and neuromelanin (Kruszewski, 2003).

Iron dyshomeostasis in the brain can lead to irreversible damage to neurons in the substantia nigra, nucleus caudate, globus pallidus, putamen, and red nucleus (iron-sensitive brain areas). Excessive iron accumulation in these areas has been involved in the pathogenesis of neurodegenerative diseases including AD, PD, and multiple system atrophy. Moreover, iron-induced damage in the selective brain regions is not completely understood (Ndayisaba et al., 2019).

The PANK2 gene controls the synthesis of the pantothenate kinase 2 enzyme. It has a key role in oxidative phosphorylation activity in the mitochondria. This enzyme controls the synthesis of Coenzyme A in mitochondria that are actively involved in the metabolism of carbohydrates, fats, and amino acids.

Mutations in the PANK2 gene lead to impaired activity of pantothenate kinase 2 enzyme and associated disturbance in the metabolism of biomacromolecules in mitochondria (Zhou et al., 2001). These events result in pantothenate kinase 2-associated neurodegeneration. This disorder is termed as **Hallervorden–Spatz syndrome** (Harper, 1996). It is a hereditary neurodegenerative disorder of the brain in which surplus accumulation of iron takes place in brain tissues, and it is associated with Parkinsonism, dementia, and hypotonia. In this disorder, mitochondrial dysfunction could precipitate dyshomeostasis of iron in the brain tissues.

Iron supplementation in neonatal life could result in iron accumulation in mitochondria and could induce mitochondrial dysfunction in HD mice. These events could lead to neurodegeneration in HD mice (Agrawal et al., 2018). However, it is uncertain whether early-life iron supplementation could be a cause of the induction of mitochondrial dysfunction and subsequent neurodegeneration.

In an Alzheimer's model of *Caenorhabditis elegans*, the gene for mitoferrin-1 protein was depleted in which authors reported a decrease in the progression of Alzheimer's disease. The mitoferrin-1 protein has a key role in mitochondrial iron homeostasis. It acts as a transporter of iron from the intermembrane space to inside the mitochondrial matrix for the formation of Fe−S clusters and heme groups (Hentze et al., 2010). Huang et al. (2018) hypothesized that mitoferrin-1 gene depletion in *Caenorhabditis elegans* prevented accumulation of mitochondrial iron and synthesis of reactive oxygen species (ROS) in mitochondria and as a consequence, the rate of progression of neurodegeneration was reduced in the Alzheimer model of *Caenorhabditis elegans*.

The α-synuclein is the protein that is abundantly found in the presynaptic neuron in the brain. This protein is closely associated with the pathogenesis of Parkinson's disease.

The α-synuclein protein can bind with ferrous and ferric forms of iron (Peng et al., 2010) Phosphorylated α-synuclein is associated with a higher affinity to the ferrous state of iron in brain tissues (Lu et al., 2011).

There exists an interplay between iron overload and α-synuclein in the progression of Parkinson's disease synergistically.

Brain imaging of patients suffering from Alzheimer's disease has forwarded evidence for the coexistence of surplus iron accumulation with Aβ plaques that together could enhance the progression of the disease (Goodman, 1953).

The latest evidence for the coexistence of iron and Aβ plaques in brain tissues comes from studies by Plascencia-Villa et al. (2016) and Everett et al. (2018) in which the presence of iron oxide magnetite nanoparticles was detected in the Aβ plaques.

The Aβ plaque can bind with iron via three histidine amino acid residues and one tyrosine amino acid residue located in the N-terminal site of the peptide, thereby stabilizing iron ions in the plaque (Lane et al., 2018).

Another study by Boopathi and Kolandaivel (2016) and Tahirbegi et al. (2016) confirmed that the association of iron ions with the Aβ plaque resulted in the minimization of helix structure of peptide and lengthening of the β-sheet proportion of the peptide. It could be stated that iron ions could lead to oligomerization of amyloid monomers and aggregation of Aβ peptides.

Furin (related to the **Proprotein convertase** family of proteins that are involved in the activation of other protein molecules). Furin is involved in the conversion of α-secretase and β-secretase from the inactive form to the active form. Further, the translation of furin is regulated by the iron concentration in the tissues. It was hypothesized by Silvestri and Camaschella (2008) that surplus iron in brain tissues downregulates the expression of furin that in turn results in the activation of the β-secretase enzyme that is mainly implicated in the synthesis of Aβ plaque in brain tissues (Ward et al., 2014).

Thus, iron overload in brain tissues is related to increased synthesis and deposition of Aβ plaque in brain tissues in neurodegenerative diseases.

Autophagy—lysosomal pathway is essential for the degradation of cytoplasmic components through the activity of lysosomes. This pathway is closely involved in the release of iron from ferritin inside the cells (Biasiotto et al., 2016).

Ferritin is the cytosolic, nontoxic form of iron storage in the cells. It can store nearly 4500 atoms of iron (Arosio et al., 2009).

Iron is released from the ferritin when its demand arises in the cells. It occurs via nuclear receptor coactivator 4-mediated autophagy-lysosome pathway. In this pathway, NCOA4 attaches to ferritin and brings about the trafficking of ferritin to autophagosomes (Mancias et al. 2014).

The outer membrane of the autophagosome fuses with the lysosome and results in the formation of autolysosome and thus imparting delivery of ferritin to the lysosome. Ferritin is digested with the help of hydrolytic enzymes of lysosomes. The iron is released into the lysosome and, thereafter, it is exported to the cytosol (Mancias et al. 2014).

Impaired NCOA4-mediated ferritinophagy (autophagy of ferritin) disturbs the homeostasis of iron in the cells and could lead to oxidative stress in the brain tissues. Gao et al. (2016) suggested that NCOA4-mediated ferritinophagy might increase the probability of ferroptosis (programmed cell death owing to the accumulation of oxidatively damaged lipids in the cells induced iron toxicity).

Therefore, impaired NCOA4-mediated autophagy of ferritin is involved in the pathogenesis of several neurodegenerative disorders namely Alzheimer's disease, Parkinson's disease, Huntington's disease, and amyotrophic lateral sclerosis (Biasiotto et al., 2016).

Contrarily, Ward et al. (2014) reported the absence of any direct evidence for the involvement of impaired NCOA4-mediated ferritinophagy in the pathogenesis of neurodegenerative diseases. Iron is essential in the brain for activities namely DNA synthesis, transport of oxygen, mitochondrial oxidative phosphorylation, and synthesis of neurotransmitters. Hence, Ward et al. (2014) hypothesized several indirect evidences for the link among iron dyshomeostasis, autophagy, and ferroptosis with the pathogenesis of neurodegenerative diseases.

The accumulation of iron in the brain is a normal process. Iron is bound to ferritin in cytosol and neuromelanin in lysosomes (Zecca et al., 2004).

Moreover, iron dyshomeostasis, abnormal iron accumulation, and impaired ferritinophagy are associated with the pathogenesis and progression of neurodegenerative diseases.

Dixon et al. (2012) studied glutamate treated culture of rat organotypic hippocampal slices. Glutamate treatment of culture represented manifestations of stroke and neurodegenerative disease. Authors reported cell death due to ferroptosis. Authors claimed that ferroptosis and cell death could be prevented through the use of iron chelators (Dixon et al., 2012).

The study by Dixon et al. (2012) provided ferroptosis and iron dyshomeostasis as the primary factors involved in the pathogenesis of neurodegenerative diseases.

Role of copper in neurodegenerative diseases

Copper is the trace metal and is found in the hippocampus, basal ganglia, cerebellum, and synaptic membranes and cerebellar granular neurons (Cox and Moore, 2002). Important enzymes in the brain namely peptidylglycine α-amidating monooxygenase, ceruloplasmin, hephaestin, and dopamine-β-hydoxylase are dependent on the copper as a cofactor for their enzymatic activities (Zucconi et al., 2007).

Surplus accumulation of copper is involved in the pathogenesis of Alzheimer's disease, amyotrophic lateral sclerosis, Menkes' disease, Huntington's disease, and Parkinson's disease and Wilson's disease. Surplus accumulation of copper in brain tissues may be involved in the Haber—Weiss reaction leading to the formation of free radicals in the tissues. As a consequence, copper overload could result in irreversible injury to mitochondria, DNA changes, and injury to neurons in the brain.

A study by Roos et al. (2006) reported increased copper concentration in the cerebrospinal fluid in patients suffering from Alzheimer's disease. Copper ions interact with β-amyloid and enhance the accumulation of amyloid resulting in the pathogenesis of the disorder.

The role of copper ions in the pathogenesis of Alzheimer's disease is still uncertain. Studies by Borchardt et al. (1999) and Kessler et al. (2005) reported deficiency of copper ions is the cause of Alzheimer's disease while studies by Cherny et al. (2001) and Sparks et al. (2006) indicated copper overload is the cause of Alzheimer's disease.

Role of mitochondrial abnormalities in neurodegenerative diseases

Role of free radicals formation in mitochondria

Mitochondria are the powerhouse of cells involved in the synthesis of ATP through the oxidative phosphorylation process occurring on the electron transport chain (ETC) located in the inner membrane of mitochondria. Free radicals are also produced during oxidative phosphorylation. The electrons are passed to oxygen by complex I and complex III, thereby leading to the formation of superoxide radicals. These radicals are catalyzed by manganese superoxide dismutase leading to the formation of $H2O2$ and oxygen. The metabolites of the tricarboxylic acid cycle namely α-ketoglutarate dehydrogenase produce superoxide radicals in the mitochondrial matrix.

These free radicals are transported to cytosol from the mitochondrial matrix through voltage-dependent anion channels. According to an estimate, about 4% of the total O_2 utilized by mitochondria is changed into superoxide radicals (Hansford et al., 1997). Under normal physiological conditions, nearly 10^{10} reactive oxygen species are generated per cell per day (Bonda et al., 2010).

The ROS are removed by body's antioxidant system including enzymes as superoxide dismutase-I, superoxide dismutase-II, glutathione peroxidase, and catalase.

When the concentration of ROS exceeds the capability of the antioxidant system of the body to neutralize, it leads to oxidative stress and oxidative injury to cellular macromolecules. Under oxidative stress, mitochondrial functions are impaired that further leads to the production of excessive reactive oxygen species in the mitochondria (Smith et al., 2000).

Surplus free radicals induce lipid peroxidation, protein denaturation, and oxidation of nucleic acids, thereby causing impairment of mitochondrial structure and or functions. The mitochondrial dysfunction is intimately associated with the pathogenesis of neurodegenerative diseases.

Oxidative stress and mitochondrial dysfunction in neurodegenerative diseases

Environment factors are strongly associated with the pathogenesis of neurodegenerative disorders (Mandemakers et al., 2007). Environmental factors are also implicated in the mechanism of oxidative stress and mitochondrial dysfunction (Van Houten et al., 2006). Environmental factors including pollution, metal toxicity, ionizing radiations, and pesticides directly linked with the surplus production of reactive oxygen species in cells.

These ROS have deleterious effects on the proteins involved in the physiology of ETC in mitochondria, thereby obstructing the activity of ETC (Safia et al., 2012; Mandavilli et al., 2005). There is consequent reduction in production for ATPs and decline in activity of mitochondrial membrane potential leading to death of cells (Safia et al., 2012; Mandavilli et al., 2005).

Oxidative damage is the incipient injury in the pathogenesis of Alzheimer's disease (Nunomura et al., 2001). Studies conducted by Markesbery and Lovell (1998) of postmortem brains of patients who suffered from Alzheimer's disease reported the presence of extensive oxidative damage in the parts of the brain.

Markesbery and Lovell (1998) detected high content of aldehydes named as four-hydroxynonenal and acrolein in the hippocampus of patients suffering from Alzheimer's disease upon postmortem examination. The four-hydroxynonenal and acrolein are the products of lipid peroxidation. Furthermore, Markesbery and Lovell (1998) additionally reported the high content of isoprostanes in the hippocampus of patients suffering from Alzheimer's disease. Isoprostane is the proinflammatory product obtained after the peroxidation of arachidonic acid (Markesbery and Lovell, 1998). Strong evidences have been forwarded by researchers to associate Parkinson's disease with mitochondrial dysfunction (Johri and Beal, 2012).

A study was conducted by Gabbita et al. (1998) related to changes in the nuclear DNA by the oxidative damage in the brains of patients with Alzheimer's disease. In the study, Gabbita et al. (1998) isolated nuclear DNA from cerebellum, frontal, parietal, and temporal lobes from nine patients with AD and from 11 persons as controls. Gabbita et al. (1998) subjected the purine and pyrimidine bases to oxidation, and oxidized products were quantitated through gas chromatography.

Gabbita et al. (1998) reported the presence of higher contents of 5-hydroxyuracil, 5-hydroxycytosine, 8-hydroxyguanine, and 8-hydroxyadenine in the brains of patients with AD in comparison to the brains of controls. Gabbita et al. (1998) concluded that persistent damage to nuclear DNA in brain tissues owing to increased oxidative stress could be closely associated with Alzheimer's disease onset and progression.

Impaired mitochondrial enzymes in neurodegenerative diseases

Deficiency in the activity of NADH dehydrogenase enzyme, complex I in the ETC had been reported in the substantia nigra in patients suffering from Parkinson's disease (Johri and Beal 2012). Further studies reported reduced activity and or absence of NADH dehydrogenase enzyme in mitochondria, lymphocytes, and platelets in the patients suffering from Alzheimer's disease (Johri and Beal, 2012; Beal, 2007).

In Alzheimer's disease, reduction in the concentration of enzymes namely pyruvate dehydrogenase complex, α-ketoglutarate dehydrogenase complex, and cytochrome oxidase have been reported by Gibson et al. (1998).

The deficiency of α-ketoglutarate dehydrogenase complex enzyme in the mitochondria in Alzheimer's disease could be hereditary. However, studies indicated the link between oxidative stress and deficiency of mitochondrial enzymes.

Cytochrome oxidase is the essential terminal enzyme in the **mitochondrial** ETC. Cytochrome oxidase is also termed as complex IV consisting of 13 subunits that mediate the transfer of electrons from ferrocytochrome c to molecular oxygen (Kadenbach and Huttemann 2015).

The Aβ peptides and oligomers were detected in the membranes of mitochondria in the neurons of brain specimens from Alzheimer's disease in postmortem, in the neurons of brain specimens from mice with Alzheimer's disease by Manczak et al. (2006) and Crouch et al. (2005).

Hansson Petersen et al. (2008) studied the transport of Aβ peptides in mitochondria in the neurons in Alzheimer's patients. Authors reported that translocase in the outer membrane was helpful in the uptake of Aβ peptides in mitochondria.

The Aβ peptides exert toxicity in the mitochondria by inhibition of mitochondrial enzymes located on the inner membrane (Wang et al., 2007). Primarily, complex IV named as cytochrome c oxidase that is part of ETC is specifically inhibited by Aβ_{25-35} fragment (Canevari et al., 1999).

The Aβ_{25-35} fragment is devoid of the N-terminal region and metal-binding domain in its structure. It is produced by splitting of Aβ(1−40) peptides. Structurally, it exhibits a β-sheet and β-turn structure (Bond et al., 2003). **The** Aβ_{25-35} fragment can rapidly form aggregates than the complete Aβ peptide. Authors reported that Aβ_{25-35} fragment could inhibit one of the active sites of cytochrome c oxidase in the inner membrane in mitochondria.

Impaired mitochondrial dynamics in neurodegenerative diseases

Impaired mitochondrial dynamics are involved in the pathogenesis of neurodegenerative disorders namely Alzheimer's disease (Chan, 2006).

Mitochondria have the potential for passing through successive cycles of fusion. Separate mitochondria fuse together to form a single cytoplasmic organelle. This physiological activity is opposed by the breakdown of a single mitochondrion into two separate smaller mitochondria (Detmer and Chan, 2007).

The dynamin-related GTPases are involved in these mitochondrial dynamics.

Mitochondrial fission is essential for the rejuvenation of mitochondria and their redistribution inside the synapses. The mitochondrial fusion is vital for the distribution and mitochondrial movement across axons (Chen et al., 2007).

The perfect harmony between mitochondrial fusion and fission is essential for the integrity of mitochondrial structure and functions in neurons in helping the formation of synapses and dendritic spines (Arduino et al., 2011).

The mitofusins (proteins) designated as Mfn1 and Mfn2 are essential for the fusion of outer membranes while Mitochondrial Dynamin Like GTPase labeled as Opa1 is necessary for fusion of outer and inner membranes. Mitochondrial fission needs Dynamin I like protein labeled as Drp1 and mitochondrial fission I protein labeled as Fis1 (Yoon et al., 2003).

Studies done by Manczak et al. (2011a) on the brains of Alzheimer's disease patients reported the downregulation of the Mnf1, Mnf2, and Opa1 genes and upregulation of Fis1 fission gene.

Drp1 activity is impaired in the patients with Alzheimer disease leading to the generation of dysfunctional mitochondria and hence, reduction in energy supply to synapse could occur in Alzheimer's disease in the brain (Barsoum et al., 2006).

Impaired mitochondrial distribution in neurodegenerative diseases

The neurons are highly complex and polarized cells. These comprised the cell body, axon, and dendrites. The dendritic arborization (dendritic branching) is the feature of neurons (Urbanska et al., 2008). Dendritic arborization is the intricate physiological process to generate additional synapses. Neurons have a heterogeneous distribution of mitochondria to fulfill diverse metabolic demands of different portions of neurons in the brain (Hollenbeck and Saxton 2005).

The presynaptic terminals have higher energy demand in comparison to other regions of neurons and thus, have a higher number of mitochondria than other regions of neurons (Li et al., 2004).

The intracellular transport apparatus is helpful in the distribution of mitochondria from the soma to the axons and synaptic terminals. This type of mitochondrial transport in neurons is called as anterograde transport. Mitochondria in the axons and synaptic terminals provide ATP for the transmission of impulse through synapses (Sheng and Cai 2012; Saxton and Hollenbeck, 2012).

The anterograde transport of mitochondria takes place with the help of two types of motor proteins namely dynein and kinesin. These molecular motor proteins require energy via ATP hydrolysis (Hirokawa et al., 2010).

Whereas, the aged or damaged **mitochondria** from the axons are transported back to the soma for their autophagic removal or repair. This type of transport of mitochondria in neurons is called as **retrograde transport** (Sheng and Cai 2012; Saxton and Hollenbeck 2012).

The cytosolic dynein is mainly involved in the retrograde transport of mitochondria in the neurons (Pfister et al., 2006).

Sheng and Cai (2012) concluded that impaired mitochondrial distribution in neurons could be closely related to the pathogenesis of disturbed functions of neurons and neurodegenerative diseases.

A study by Wang et al. (2009) reported the impaired distribution of mitochondria in neurons in the brains of patients with Alzheimer's disease. The authors described the redistribution of mitochondria in the pyramidal neurons in the brains of Alzheimer's disease. Furthermore, on the basis of Immunoblot analysis, Wang et al. (2009) depicted that contents of dynamin I like protein, mitochondrial Dynamin Like GTPase, Mitofusin-1, and Mitofusin-2 were declined while the content of mitochondrial fission I protein was elevated in Alzheimer's disease (Wang et al., 2009). These intracellular proteins are strongly involved in the mitochondrial transport in the neurons. Hence, altered expression of these motor proteins might be involved in the impaired distribution of mitochondria in neurons in the brain of Alzheimer's disease.

The redistribution of mitochondria in the brains of Alzheimer's patients results in a lack of mitochondria in the large axonal region and in the dendritic areas of neurons. This was studied by Pickett et al. (2018) who suggested that amyloid β and tau proteins together are implicated in the deficient distribution of mitochondria in the large axonal region and in the dendritic areas in the brain of patients with Alzheimer's disease. Both amyloid β and τ proteins contribute to synapse dysfunction and degeneration of synapses in Alzheimer's disease.

Stokin et al. (2005) recognized axonal defects in the neurons in the brains of Alzheimer patients in the early stage of the onset of disease as well as the axonal defects were identified by authors in the mouse model of Alzheimer's disease as early as 1 year before the onset of disease in a mouse model.

Axonal defects were manifested in the form of neuronal swellings that were due to excessive accumulation of vesicles, molecular motor proteins, and organelles.

Authors reported that amyloid-β peptides and amyloid plaques in the brains in Alzheimer's disease were associated with a higher frequency of axonal defects and reduced content of kinesin molecular motor protein in the neurons. There is enhanced proteolysis of amyloid precursor protein leading to higher production of amyloid-β peptides and setting a vicious cycle for the pathogenesis of Alzheimer's disease (Stokin et al., 2005; Hollenbeck and Saxton, 2005).

The mutant huntingtin protein is reported in the brains in Huntington's disease. The protein is associated with increased activity of mitochondrial dynamin I like protein that further impairs the normal mitochondrial trafficking and leads to mitochondrial fission (Reddy and Shirendeb, 2011).

De Vos et al. (2007) studied the mitochondrial transport in the neuronal cultures of superoxide dismutase I mutant mice. It was found that anterograde transport of mitochondria was reduced in the superoxide dismutase I mutant mice leading to impaired mitochondrial trafficking that might be involved in the onset of amyotrophic lateral sclerosis.

Mitochondrial trafficking is disturbed in the dopaminergic neurons in the mouse model from Parkinson's disease (Sterkyet al., 2011).

Ishihara et al. (2009) and Wang et al. (2008) worked on the expression of dynamin I-like protein in the fibroblasts in Alzheimer's disease and in the control subjects. There was reduced expression of Drp1 in fibroblasts in Alzheimer's disease coupled with impaired mitochondrial trafficking. Ishihara et al. (2009) and Wang et al. (2008) reported that after restoring contents of dynamin I like protein in fibroblasts to normalcy led to repair of the defective mitochondrial transport in fibroblasts.

Role of protein misfolding in neurodegenerative diseases

Protein folding is the biological activity in which a polypeptide chain undergoes folding to acquire its characteristic and native three-dimensional structure (Alberts et al., 2002). The polypeptide chain exists in the linear form after its synthesis from the ribosomes. It has been suggested that the folding of the polypeptide chain begins cotranslationally as the N-terminal region undergoes folding while the C-terminal region is in the stage of synthesis on the ribosomes.

Misfolded proteins and their aggregates are the important biomarkers for the occurrence of neurodegenerative proteinopathies (disorders in which protein molecules are uncharacteristically aggregated owing to conformational changes). The misfolded proteins induce cellular toxicity and are implicated in impaired cellular proteostasis (reduced protein quality control involved in proteinopathies).

Further, misfolding of proteins leads to an altered state of cell functioning due to cellular toxicity induced by these abnormal protein aggregates. The misfolded proteins expose hydrophobic regions. These hydrophobic surfaces of misfolded proteins in the hydrophilic medium of cells have a tendency to form aggregates (Balch et al., 2008). These misfolded protein aggregates have been reported from the brain tissues of patients suffering from neurodegenerative disorders.

Strong evidences substantiate that the accumulation of misfolded proteins in brain tissues are involved in synaptic dysfunction, apoptosis of neurons, brain injury, and pathogenesis of neurodegenerative disorders. Furthermore, the exact pathogenesis of protein misfolding and neurodegeneration are still unclear (Claudio and Estrada, 2008).

Recent studies provide strong evidence that mutant proteins accumulate in the mitochondrial membranes and exert cellular toxicity. These mutant proteins are involved in the induction of higher oxidative stress and production of ROS in the mitochondria resulting in reduced ATP synthesis leading to cell death. Mutant proteins namely amyloid β in Alzheimer's disease, mutant huntingtin in Huntington's disease, mutant parkin, and mutant α-synuclein in Parkinson's disease are mainly involved in the pathogenesis of neurodegenerative diseases (Beal, 2005; Reddy, 2008).

Role of Aβ peptide in Alzheimer's disease

Surplus accumulation of β-amyloid peptides has been established in the brain tissues in patients with Alzheimer's disease (Tanzi and Bertram, 2005).

The Aβ peptide is produced by proteolysis of amyloid precursor protein (APP). It is composed of nearly 695−770 amino acid residues in humans (Hamley, 2012). The APP is an integral membrane protein that is expressed in several tissues including neuronal synapses in the brain. The APP molecule is involved in anterograde neuronal transport (Turner et al., 2003) and controls the synapse formation and repair (Priller et al., 2006).

The amyloid precursor protein is cleaved by β-secretase and γ-secretase that results in the release of Aβ peptide in brain tissues (Nunan and Small, 2000). The Aβ peptide is nearly composed of 36−43 amino acid residues.

The Aβ monomers undergo aggregation to form several types of protein assemblies namely oligomers, proto-fibrils, and amyloid fibrils. The amyloid fibrils are insoluble and large protein assemblies that are additionally involved in the formation of amyloid plaques that are closely related to the pathogenesis of Alzheimer's disease (Haass et al., 1992).

Two important domains in Aβ peptide are intimately involved in the polymerization of Aβ peptide monomers into insoluble and amyloid fibrils in brain tissues. The amyloid fibrils are neurotoxic both in vitro and in vivo studies (Walsh et al., 1999). The first domain, the C-terminal region with amino acid residues from 32 to 42 and second domain, a hydrophobic region with amino acid residues from 16 to 23, might (Estrada and Soto, 2007) be implicated in the process to raise the content of β-sheet conformation in Aβ peptide and furthermore leading to misfolding of Aβ peptide (Estrada and Soto, 2007). However, In the normal brain, Aβ peptide is composed of an admixture of β-sheet conformation and random coils.

The pathogenesis of Alzheimer's disease is dependent on the Aβ-dependent mechanism as well as Aβ-independent mechanism that triggers the inset of disease. Moreover, there is a paucity of evidences that explains the precise pathophysiology of Alzheimer's disease.

The Aβ-dependent mechanism is also called as amyloid cascade hypothesis or Aβ hypothesis is centered on the accumulation and deposition of fibrillar amyloid in the brain tissues in Alzheimer's disease (Hardy and Allsop, 1991; Selkoe, 1991).

According to Aβ hypothesis, In normal persons, Aβ peptides are rapidly degraded from the tissues. Moreover, in aged persons or genetically predisposed persons, decomposition of Aβ peptides is reduced, and it leads to the accumulation of extracellular Aβ peptides in brain tissues. The Aβ 42 fragment is more hydrophobic than Aβ 40 fragment and higher contents of Aβ 42 over Aβ 40 induce the polymerization and formation of insoluble amyloid fibrils that further undergo aggregation to form senile amyloid plaque in the brain and resulting in neurotoxicity and neurodegeneration (Hardy and Selkoe. 2002).

Study by Price et al. (1998) detected the occurrence of mutations in gene coding APP and gene PS1, and gene PS2 coding proteins presenilin 1 and 2 that control the synthesis of Aβ peptides from the amyloid precursor protein. These mutations are responsible for dominantly inherited familial Alzheimer's disease (Intlekofer and Cotman, 2013).

Familial Alzheimer disease mutations in APP and PS1 genes are responsible for impaired lysosomal-autophagy pathway in neurons in the brain in Alzheimer's disease. In healthy conditions, the lysosomal-autophagy pathway is involved in the degradation of cytoplasmic organelles and proteins through the uptake of cytoplasmic constituents inside autophagosomes that further fuse with lysosomes to form autolysosomes for the degradation of sequestrated substances. Autophagy is vital for the health of neurons in the brain that clears the abnormal proteins in aging and in diseases (Wong and Cuervo, 2010).

It has been established by Lee et al. (2010) presenilin 1 is essential for the acidification of lysosomes and activation of protease enzymes. Mutations in gene PS1 and gene APP are strongly related to the disturbed lysosomal-autophagy pathway in neurons in the brain. Thus, functions of neurons are disturbed leading to a higher incidence of neuronal apoptosis.

In neurons with deficient presenilin 1 protein, vacuolar ATPase (highly conserved family of enzymes in eukaryotes for acidification of organelles) fails to acidify lysosomes. Proteases are not activated inside lysosomes. Hence, the sequestrated cytoplasmic contents are not digested inside the lysosomes and undigested constituents continue to accumulate inside the autolysosomes (Lee et al., 2010).

The disturbed autophagy is involved in the pathogenesis of Alzheimer's disease in which excessive Aβ proteins are accumulated in autolysosomes owing to their deficient degradation. The accumulated Aβ proteins induce neurotoxicity and neurodegeneration in the brain in Alzheimer's disease (Glabe, 2001).

Role of τ-protein in Alzheimer's disease

Tau is the chief microtubule-associated protein in the mature neurons, and it represents the major element in the pathogenesis of Alzheimer's disease. The name "Tau" is derived from "tubulin associated unit." It is largely expressed in the brain (Kolarova et al., 2012).

Microtubules constitute the main protein units of the cytoskeleton. The τ-protein serves to strengthen the microtubules to microtubules attachments (Buee et al., 2000).

Phosphorylation of Tau proteins is essential to perform its functions under normal conditions. However, hyperphosphorylation of τ-protein results in the loss of its physiological functions. Hyperphosphorylation induces the conformational alteration in the structure of τ proteins leading to abnormal aggregation of τ proteins in Alzheimer's disease. The hyperphosphorylation of τ proteins is four times higher in the brains of Alzheimer's patients in comparison to hyperphosphorylation of tau proteins in the brains of control subjects. The hyperphosphorylation of τ proteins that have been reported during hypothermia and general anesthesia (transient hyperphosphorylation) is lesser in the extent that has been reported in the brains of AD patients (abnormal hyperphosphorylation).

The abnormally hyperphosphorylated τ proteins in the AD brain exert their toxicity through the sequestration of normal microtubule-associated protein-I and 2 from microtubules. It induces inhibition and breakage of microtubules (Alonso et al., 1997). The association of abnormally hyperphosphorylated tau proteins with microtubule-associated protein-I or 2 leads to the formation of amorphous aggregates that are without filaments.

Additionally, abnormally hyperphosphorylated τ proteins can sequester normal τ proteins (Alonso et al., 1994) resulting in polymerization and formation of paired helical filaments (PHF) in association with straight filaments that together constitute neurofibrillary tangles in brains (Alonso et al., 1996).

Role of α-synuclein protein in Parkinson's disease

The α-synuclein is the protein linked with presynaptic neurons. It is considered a a biomarker in Parkinson's disease. The α-synuclein is unfolded protein. It exhibits marked localization at the synapses in different regions of areas of the brain (Dauer and Przedborski, 2003). It is localized near presynaptic neurons.

The α-synuclein protein is composed of 140 amino acid residues. The SNCA gene is located on chromosome 4 (4q22.1) and controls the synthesis of α-synuclein protein.

It has three prominent domains. The first domain is the NH2 terminal consisting of amino acid residues from 1 to 60. This domain of α-synuclein contains apolipo-protein lipid-binding motifs and renders α-helix formation ability to α-synuclein.

This domain is amphipathic in nature and is the highly conserved domain. This domain interacts with phospholipids in the membranes. The mutations in the SNCA gene are localized in this domain (Bussell and Eliezer, 2003).

The second domain is the central hydrophobic region consisting of amino acid residues from 61 to 95, and the domain is called as non-Aβ component. This domain provides the β-sheet forming potential to α-synuclein. The non-Aβ component is implicated in protein aggregation.

The third domain is composed of amino acid residues from 96 to 140 and is the carboxyl terminus that is negatively charged. This domain lacks any definite (Burré et al., 2010) structural integrity and is highly acidic in nature. This domain is rich in proline amino acid residues (Burré et al., 2010).

Although, α-synuclein is the unfolded and soluble protein; however, it undergoes aggregation to form insoluble fibrils under diseased states as in Parkinson's disease and dementia (Spillantini et al., 1997). The pathological involvement of α-synuclein in the pathogenesis of diseases is termed as synucleinopathies. The pathological aggregation of α-synuclein could result in oxidative stress, mitochondrial dysfunctions, synaptic dysfunction, impaired lysosomal-autophagy, and disturbed Calcium signaling in the brain tissues (Marvian et al., 2019).

Although, the exact mechanism involved in the pathological aggregation of α-synuclein is uncertain. A study by Sandal et al. (2008) revealed that α-synuclein is found in unfolded state, α-helix, and β sheet conformations that are in equilibrium. But mutations could result in higher content of β conformation that might be related to the pathogenesis of diseases. The familial Parkinson's disease is caused by a mutation in the SNCA gene. To date, five point mutations have been established namely A53T (Polymeropoulos et al., 1997), A30P (Krüger et al., 1998), E46K (Zarranz et al., 2004), H50Q (Appel-Cresswell et al., 2013), and G51D (Lesage et al., 2013).

It might be possible that mutations of α-synuclein may result in amyloid-like fibrils that are associated with Parkinson's disease.

Involving α-synucleinopathy and neuronal degeneration in the pathogenesis of Parkinson's disease, Chung et al. (2009) studied the α-synucleinopathy in the rat model of recombinant adeno-associated virus.

The mutant human α-synuclein (A53T) was delivered into the substantia nigra in the rat model mediated by adeno-associated virus (nonenveloped virus that can serve as a vector) by Chung et al. (2009).

The authors used a small fragment of the human synapsin 1 gene promoter to enhance transgene expression.

After a period of 4 weeks, the authors observed a rise in dopamine metabolite named as 3,4-dihydroxyphenylacetic acid. Authors reported dystrophy of dopaminergic neurons in the striatum (Chung et al., 2009).

The authors observed prominent changes in the content of proteins related to synaptic transmission and axonal transport in the striatum. The levels of rabphilin 3A (membrane trafficking protein responsible for calcium-dependent control of exocytosis in secretory vesicle in neurons) and syntaxin (proteins basically localized to the plasma membrane of the presynaptic neurons) in the striatum were decreased (Chung et al., 2009).

Contents of proteins namely dynamitin, dynein, and dynactin1 (retrograde motor proteins) were raised, whereas contents of proteins designated as KIF1A, KIF1B, KIF2A, and KIF3A (anterograde transport motor proteins) were reduced in the striatum in the rat model (Chung et al., 2009).

Chung et al. (2009) reported alteration in the levels of cytoskeleton proteins namely actin protein and α-and γ-tubulin proteins that were increased and decreased, respectively.

At a period after 8 weeks, Chung et al. (2009) reported a rise in the level of Iba-1 (ionized calcium-binding adaptor molecule **1 that has** actin-bundling property, involved in phagocytosis, membrane ruffling in activated microglia), rise in the level of Interleukin 1 β (cytokine protein that induces cyclooxygenase 2 activity in CNS contributing to inflammatory), and increase in the level of interferon γ, IFN-γ (cytokine protein involved in innate and adaptive immunity, and serves as a basic activator of macrophages, neutrophils, and natural killer cells) and rise in the level of tumor necrosis factor-α (proinflammatory cytokine protein involved in activation of macrophages during acute inflammation) (Chung et al., 2009). The rise in levels of Iba-1, Interleukin 1 β, IFN-γ, and tumor necrosis factor-α were reported in the striatum and not in the substantia nigra (Chung et al., 2009).

Chung et al. (2009) concluded that alterations in levels of cytoskeleton proteins and motor proteins related to axonal transport and synaptic transmission and rise in levels of proinflammatory cytokines and neuro-inflammation preceded the neurodegeneration mediated by α-synuclein in the rat model.

Chung et al. (2009) reported neurodegeneration of dopaminergic neurons at a period of 17 weeks in rat model after adeno-associated virus induced insertion of mutant human α-synuclein in the substantia nigra.

In another studies, calcium ions were implicated in the neuronal degeneration caused by α-synuclein. It is speculated that α-synuclein oligomers can induce changes in the transport ability of voltage-gated receptors located on the plasma membranes of neurons resulting into exaggerated influx of calcium ions in the neurons and neurodegeneration (Leonidas, 2012).

Hettiarachchi et al. (2009) studied the calcium ion signaling in human neuroblastoma (SH-SY5Y) cells modulated by α-synuclein.

Fura-2 is called as aminopolycarboxylic acid. It dye that was used in the study by Hettiarachchi et al. (2009) on human neuroblastoma (SH-SY5Y) cells and Fura-2 has the potential to bind with free cytosolic calcium ions.

Hettiarachchi et al. (2009) transfected the human neuroblastoma (SH-SY5Y) cells with either A53T mutant form (with A53T mutation that is point mutation of the α-synuclein), wild-type α-synuclein, the S129D phosphomimetic mutant (phosphorylation of α-synuclein at serine 129 is seen in Parkinson's disease and α-synucleinopathies) or with control vector (empty). Hettiarachchi et al. (2009) reported an increased influx of calcium ions via voltage-gated Ca^{2+} channels in the human neuroblastoma (SH-SY5Y) cells expressing three types of α-synuclein proteins.

Hettiarachchi et al. (2009) concluded that α-synuclein protein controls the influx of calcium ions in neurons. Moreover, abnormal levels of α-synuclein in neurons enhance the calcium influx leading to dyshomeostasis of calcium ions in neurons and consequently cause the neurodegeneration in Parkinson's disease.

Authors have reported strong interaction between α-synuclein and tubulin protein, thereby influencing the cytoskeletal dynamics in the patients suffering from Parkinson's disease (Alim et al., 2002).

Alim et al. (2002) identified tubulin protein that has the inherent potential to bind with α-synuclein. The authors detected colocalization of tubulin with α-synuclein in Lewy bodies in the brains of humans and rat models. It was tested by authors that tubulin was involved in the onset and progression of α-synuclein fibril formation in in vitro models during physiological conditions. Alim et al. (2002) concluded that interaction between α-synuclein and tubulin might be implicated in the formation of Lewy bodies in Parkinson's disease.

The physiological role of α-synuclein is still not clear; however, Alim et al. (2004) identified the binding site of α-synuclein with protein tubulin and further proved that α-synuclein could catalyze the polymerization of tubulin (purified) into microtubules in cell cultures. Authors claimed that mutant forms of α-synuclein cannot induce the polymerization of tubulin. The emergence of new facts about the physiological role of α-synuclein in the polymerization of tubulin could be useful in understanding the mechanisms of α-synuclein induced neurodegenerative diseases.

Lee et al. (2006) posited that overexpression of α-synuclein could be involved in neurodegenerative diseases like Parkinson's disease. The authors worked upon the effect of aggregated α-synuclein on the microtubules in neurons.

Zhou et al. (2010) identified the binding region of α-synuclein with tubulin protein, and it is located at amino acid residues from 60 to 100 of α-synuclein.

With the help of confocal laser scanning microscopy, Zhou et al. (2010) observed that α-synuclein protein exposure to cultured cells resulted in the inhibition of the formation of microtubule. The authors concluded that α-synuclein is involved in the control of microtubule dynamics in neurons and could be helpful in understanding the physiological and pathological functions of synuclein.

Lee et al. (2006) reported neuronal degeneration and fragmentation of Golgi apparatuses in primary cultures of dorsal root ganglia neurons with overexpression of α-synuclein. The authors observed various spherical aggregations of α-synuclein and tubulin proteins in the primary cultures of dorsal root ganglia neurons. The authors concluded that overexpression of α-synuclein in the neurons might be involved in the impairment of the microtubule network and associated pathology of Parkinson's disease.

Role of lewy bodies in neurodegenerative diseases

Lewy bodies represent abnormal assemblies of protein molecules inside the neurons in the brain and contribute to the pathology of Parkinson's disease and other neurodegenerative diseases.

According to Gibb and Lees (1988), Lewy bodies are the neuronal inclusions and are detected in substantia nigra in all cases of Parkinson's disease. Moreover, the pattern of distribution of Lewy bodies in Parkinson's disease is generalized. These are found in neurons in the dorsal vagal nucleus, locus coeruleus, nucleus basalis of Meynert, and hypothalamus. Additional regions of the brain including cerebral cortex, autonomic ganglia, and thalamus exhibited the presence of Lewy bodies in Parkinson's disease (Gibb and Lees, 1988). Neurodegeneration is the hallmark in most of the regions in which Lewy bodies have been detected in the brain.

Lewy bodies look like spherical aggregation inside the neurons. Lewy bodies are reported in the brain stem and cortex (Jellinger, 2007).

Lewy bodies have 8−30 μm diameter and are made up of nearly 10 nm amyloidogenic fibrils like fibrillary α-synuclein and neurofilaments (Engelender, 2008). Lewy bodies have granular and fibrillar cores surrounded by a halo. A single neuron can have more than one Lewy body (Spillantini et al., 1998).

Two types of Lewy bodies have been described namely classical brainstem Lewy bodies and cortical Lewy bodies. The main morphological difference between the two types of Lewy bodies is that cortical Lewy bodies have diffuse outlines and are generally smaller in diameter without the presence of halo (Spillantini et al., 1998).

The filamentous α-synuclein constitutes the primary component of Lewy bodies (Spillantini et al., 1997). In Parkinson's disease and synucleinopathies, α-synuclein assumes the amyloid-like pattern further undergoes phosphorylation and aggregation. The α-synuclein protein constitutes the halo of Lewy bodies.

Additionally, proteins namely ubiquitinated proteins, neurofilaments, τ, parkin, and heat shock proteins namely Hsp70 and Hsp90 become the structural components of Lewy bodies (McNaught and Olanow, 2006; Licker et al., 2009). More than 70 molecules have been reported from the Lewy bodies (Wakabayashi and Rinsho, 2008).

According to Power et al. (2017), loss of neurons and occurrence of Lewy bodies in definite brain areas are the hallmarks of dementia with Lewy bodies and Parkinson's disease.

Power et al. (2017) studied the association between Lewy bodies and loss of mitochondria in neurons. Authors identified degeneration of microtubule network, mitochondrial dysfunction, and degeneration of nuclei in neurons containing growing Lewy bodies. Authors reported higher content of Lewy bodies in Parkinson's disease that were associated with α-synuclein in the neurons and there was marked nuclear degeneration (Power et al., 2017).

Authors reported dysfunctional mitochondria inside the Lewy bodies in the neurons in dementia with Lewy bodies. Authors asserted cytotoxic effects of Lewy bodies on the mitochondria, microtubule network, and the integrity of nuclei in the neurons in the brain. Degeneration of the microtubule network resulted in reduced axonal transport of mitochondria leading to a reduction in energy levels and neurodegeneration in Parkinson's disease and dementia with Lewy bodies (Power et al., 2017).

Gelb et al. (1999) identified Lewy bodies at the substantia nigra and the locus coeruleus regions of the brain in patients with Parkinson's disease. These regions of the brain have a higher propensity for neuronal loss in Parkinson's disease. Halliday (2013) asserted that colocalization of neuron loss and Lewy bodies in brain regions is the definite hallmark for considering the cytotoxic effect of Lewy bodies that are involved in the pathogenesis of neurodegenerative diseases.

A study by Dickson et al. (2008) identified the presence of Lewy bodies in the brains of asymptomatic persons and reported that the incidence of Lewy bodies in the brain increased with the advancing age of persons. The study related to identification Lewy bodies by Dickson et al. (2008) in brains of asymptomatic persons contrasts the previous study by Gelb et al. (1999) related to documentation of Lewy bodies at the substantia nigra and the locus coeruleus regions of the brain in patients with Parkinson's disease.

Contradictory findings in both the studies resulted in the unconvinced role of Lewy bodies in the brain in neurodegenerative diseases. Whether occurrence of Lewy bodies in neurons constitutes a presymptomatic stage of Parkinson's disease and or represents the physiological characteristics of advancing age in persons?

Studies by Gomez-Tortosa et al. (1999, 2000) concluded lack of significant association among the density of Lewy bodies and onset of Parkinson's disease, its duration, the severity of symptoms, and variation in cognitive functions in the patients.

Another study by authors (Galvin et al., 2006) reported the absence of Lewy bodies in cortical areas in the brains of patients who suffered from Parkinson's disease with dementia.

Based on the improvement in the immunohistochemical staining techniques and cell culture methods, invasive studies were conducted to decode the biological and pathological roles of the components of Lewy bodies.

The α-synuclein and ubiquitin-proteasome complex are the principal components of Lewy bodies. The α-synuclein is a protein in presynaptic neurons.

Abeliovich et al. (2000) reported that α-synuclein-depleted mice possessed normal cell bodies, fibers, and synapses in dopaminergic neurons, intact brain functions, and were fertile. After normal electrical stimulation, dopaminergic axon terminals in the nigrostriatal pathway in α-synuclein-depleted mice showed a standard pattern (Abeliovich et al., 2000) of dopamine release and reuptake.

Moreover, α-synuclein depleted mice showed decreased striatal dopamine release in the absence of impaired locomotive function. The authors concluded that α-synuclein could be involved as a negative regulator of dopamine release under normal conditions.

Another study by Dauer et al. (2002) and Drolet et al. (2004) identified resistance to neuronal death in α-synuclein knockout mice model of Parkinson's disease after exposure to 1-methyl-4-phenyl-1,2,3,6-tetrahydropyridine (neurotoxin that induces inhibition of mitochondrial complex I involved in the environment-induced pathogenesis of PD closely related to the pathology of PD).

Dauer et al. (2002) and Drolet et al. (2004) had designed a subchronic model and a chronic model of wild-type mice and α-synuclein knock-out mice. First, in the subchronic model, α-synuclein knock-out and wild-type mice were exposed to 1-methyl-4-phenyl-1,2,3,6-tetrahydropyridine for five successive days (Dauer et al., 2002; Drolet et al., 2004). Later on, in the chronic model, α-synuclein knock-out mice and wild-type mice were administered two injections of 1-methyl-4-phenyl-1,2,3,6-tetrahydropyridine each week for 5 weeks and probenecid was coadministered (Dauer et al., 2002; Drolet et al., 2004).

Dauer et al. (2002) and Drolet et al. (2004) reported a dose-related reduction in striatal dopamine levels in wild-type mice in the subchronic model, whereas weakened response was reported in the α-synuclein knock-out mice model. In the chronic model, dose-related reduction in striatal dopamine levels with a corresponding loss of vesicular monoamine transporter-2 in the striatum in wild-type mice was reported (Dauer et al., 2002; Drolet et al., 2004). But the weakened loss of striatal dopamine levels with intact vesicular monoamine transporter-2 in the striatum was identified in mice with depleted α-synuclein (Dauer et al., 2002; Drolet et al., 2004).

Dauer et al. (2002) and Drolet et al. (2004) observed that subchronic and chronic exposure to 1-methyl-4-phenyl-1,2,3,6-tetrahydropyridine resulted in elevated DOPAC (3,4-dihydroxyphenylacetic acid) to DA (dopamine) ratio in wild-type mice; however, this effect was absent in mice with depleted α-synuclein. Thus, α-synuclein depleted mice showed a weakened response to the neurotoxic effects of MPTP exposure.

The α-synuclein depleted mice showed significant resistance to the MPTP-dependent neurodegeneration of dopaminergic neurons and release of dopamine.

It can be inferred that either environmental or genetic factors interact with α-synuclein in the neurons and can be implicated in the pathogenesis of Parkinson's disease.

The formation of Lewy bodies in neurons in Parkinson's disease is a cytoprotective response of the body to sequester the abnormal proteins.

McNaught et al. (2002a,b) posited the occurrence of neuronal apoptosis and presence of Lewy bodies in Parkinson's disease and dementia with Lewy bodies.

The study by McNaught et al. (2002a,b) reported intimate relation aggresomes and Lewy bodies. The synthesis of aggresomes is a tightly regulated biological activity near the centrosome of the cell to help compartmentalize the excessively formed misfolded proteins from the other organelles in the cell (Corboy et al., 2005). Aggresomes are cytoprotective in nature and sequester the toxic proteins in the cells.

McNaught et al. (2002a,b) reported γ-tubulin protein (microtubule nucleating factor), and pericentrin protein (helps in centrosome and mitotic spindle formation) that are related to centrosome and aggresomes exhibited aggresomes like the pattern of distribution in the Lewy bodies in Parkinson's disease and dementia with Lewy bodies.

McNaught et al. (2002a,b) identified ubiquitin, proteasomes, ubiquitinated proteins, ubiquitin-activating enzyme (E1), and components of the ubiquitin-proteasome system in the Lewy bodies can ubiquitinate cytotoxic proteins that are further decomposed by the 26S proteasome complex located inside the Lewy bodies.

Bennett et al. (1999) proved that α-synuclein can be decomposed by the ubiquitin-proteasome pathway in an in vitro experiment.

Thus, Lewy bodies might be involved in cytoprotective action by degrading α-synuclein via proteasome complex.

McNaught et al. (2002a,b) identified various aggregations of ubiquitinated proteins inside the cell bodies and axons of neurons in the brains of patients with Parkinson's disease and dementia with Lewy bodies. These protein aggregates might be transformed into Lewy bodies in neurons.

The authors concluded that the formation of Lewy bodies might be comparable with the events involved in the formation of aggresomes when the contents of abnormal proteins rise in the neurons.

Additionally, a study by Schulz-Schaeffer (2012) posited that role of Lewy bodies in neurons could not substantially explain the pathogenesis of neurodegeneration in Parkinson's disease and dementia with Lewy bodies. Furthermore, Olanow et al. (2004) described that occurrence of Lewy bodies might be a physiological cum cytoprotective response of the body to detoxify aggregates of α-synuclein oligomers in the neurons in the brain.

Role of Hhuntingtin protein in Huntington's disease

The huntingtin gene codes for the huntingtin protein (Ross, 2010; Trzesniewska et al., 2004). Huntingtin gene is abbreviated as HD gene or HTT gene. It is situated at the short arm(p) of chromosome 4 at position 16.3 (USNLM, 2020a,b,c; Walker, 2007).

Huntingtin protein has a molecular weight of 350 kDa, and it is made up of 3144 amino acid residues.

Huntingtin protein is universally expressed in body tissues, while its highest levels are reported in brain tissues and testes (Li et al., 1993).

Inside the brain, it is found in neurons and glial cells (Li et al., 1993). The precise role of huntingtin protein is not clear. It might be involved in the physiological functioning of the basal ganglia (Nasir et al., 1995).

Huntingtin protein may occur with cytoplasmic organelles including the endosomes, nucleus, Gogli complex, and endoplasmic reticulum (DiFiglia et al., 1995). Huntington protein is found in synapses, axonal processes, microtubule network, clathrin-coated vesicles, and synaptosomes (DiFiglia et al., 1995). Huntingtin protein might be involved in cell signaling, binding with other proteins, transport of ions, protection against apoptosis.

Huntingtin protein can be involved in the up-regulation of the expression of brain-derived neurotrophic factor (a protein active in cortex, hippocampus, and basal forebrain areas helping in the survival of neurons and normal growth and differentiation of new neurons) (Zuccato et al., 2001).

Furthermore, huntingtin protein is directly associated with the microtubule network and vesicles (Hoffner et al., 2002). Thus, it has a potential role in the mitochondrial transport in neurons. The huntingtin protein interacts with microtubule motor proteins, dynactin, kinesin, and dynein. This interaction is mediated by Huntingtin-associated protein 1. Hence, it can be assumed that the huntingtin protein can be involved in anterograde and retrograde axon transports (Engelender et al., 1997; Li et al., 1998; Rong et al., 2006).

Huntingtin protein interacts with dynamin (is a GTPase that acts in clathrin-dependent endocytosis and vesicular trafficking) and clathrin (is a protein helpful in the formation of clathrin-coated vesicles). Huntingtin protein might be involved in endocytosis and vesicular trafficking (Velier et al., 1998; Elias et al., 2015).

Huntington's disease represents trinucleotide repeat disorder and is also called as microsatellite expansion disease (Walker, 2007).

This is a type of mutation in which lengthening of trinucleotide repeats takes place beyond the threshold when the repeats become abnormal and unstable (Orr and Zoghbi, 2007). The HTT gene contains cytosine-adenine-guanine (CAG), a sequence of three nitrogenous bases that is repeated several times and is termed as trinucleotide repeat(Walker, 2007).

The sequence CAG is the codon for glutamine amino acids. The trinucleotide repeats result in the formation of a chain of polyglutamine tract (poly Q tract). The domain of the gene that contains the trinucleotide repeats is called as poly Q domain (Katsuno et al., 2008).

A study by the author (Walker, 2007) identified that lesser than 36 repeated resides of glutamine amino acid in the poly Q domain of gene HTT leads to the synthesis of the huntingtin protein. The wild-type HTT gene has 35 CAG repeats.

Author (Walker, 2007) posited that 36 repeated residues of glutamine amino acid or higher could result in the synthesis of an altered protein with different properties. This altered protein is termed as mutant huntingtin protein (mHTT) (Walker, 2007). The mutation lengthens the CAG repeats to ≥ 40 resulting in an expression of Huntington's disease with full penetrance, while the lengthening of CAG repeats between 36−39 glutamine residues leads to variation in the expression of Huntington disease (Myers, 2004). The mutant huntingtin protein might be implicated in the accelerated neuronal apoptosis in the brain and could be implicated in the pathology of Huntington's disease.

The number of CAG repeats in the HTT gene is highly unstable and could lead to variation in the number of CAG repeats in parents and offspring (Duyao et al., 1993). The difference in the number of CAG repeats has been reported in brain and sperm cells of the same patient and the phenomenon is called as mosaicism (presence of two or more genetically different sets of cells in the same person) (Telenius et al., 1994).

Inside brain tissues, mutant huntingtin protein undergoes misfolding that further assembles into aggregates that might have neurotoxic potential in Huntington's disease. The rate of mutant huntingtin protein aggregation is directly related to the rate of expansion of poly Q (Scherzinger et al., 1999).

The ubiquitin-proteasomal degradation network helps to degrade the misfolded proteins. Higher contents of misfolded poly Q Htt in the brain exhaust the ubiquitin-proteasomal degradation network and it leads to aggregation of the mutant protein (Bence et al., 2001).

Mutant huntingtin protein has the ability to interacts and coaggregate cytosolic AMP response element-binding protein (CREB) and downregulates the expression of various cellular proteins in the cells (Steffan et al., 2000).

Mutant huntingtin protein promotes various abnormal protein–protein interactions leading to neuronal apoptosis and dysfunction in the cortex, striatum, and other parts of the brain (Imarisio et al., 2008).

Lengthening CAG repeats coding for polyglutamine residues in the huntingtin are implicated in Huntington's disease (Cooper et al., 1998). The N-terminal fragment of huntingtin protein undergoes aggregation inside neurons in patients with Huntington's disease (Cooper et al., 1998). Aggregation of N-terminal fragments inside the cytoplasm leads to dystrophic neuritis (pathological neuronal processes containing anomalous sprouting and **dystrophic** expansion), while inside the nucleus, results in the formation of intranuclear inclusion bodies in the neurons (Cooper et al., 1998).

Full-length huntingtin constructs were inserted into neuroblastoma cells with normal and expanded CAG repeats (Cooper et al., 1998). Diffused cytoplasmic localization of the proteins was reported in the neuroblastoma cells (Cooper et al., 1998). Contrarily, truncated N-terminal fragments were inserted into the neuroblastoma cells. Aggregation of proteins was noted with expanded CAG repeats. Cooper et al. (1998) concluded that truncated N-terminal fragments of protein huntingtin with expanded CAG repeats have the potential to aggregate in cell culture and aggregation has cytotoxicity.

Gutekunst et al. (1999) detected neuropil aggregates (small huntingtin protein aggregates localized in axons or dendrites) are abundantly found in different areas of the brain before the clinical manifestation of Huntington disease.

In presymptomatic cases of the disease, neuropil aggregates are mainly found in cortical layers V and VI (connect the cerebral **cortex** with subcortical areas) (Gutekunst et al., 1999). As the disease progresses, the number of neuropil aggregates declines in the deeper cortical layers and it increases in the cortical layer III. In the disease, the striatum is the highly affected area of the brain (Gutekunst et al., 1999). The occurrence of neuropil aggregates in the striatum increases as the disease progresses. Neuropil aggregates have been reported from the dendritic spines (small membranous protrusion dendrite of the neuron) and dendrites that can be involved in the dendritic pathology(loss of dendritic spine density related to decline in mental faculties) associated with Huntington's disease (Gutekunst et al., 1999).

Expanded polyglutamine repeats in mutant Huntingtin protein might have a role in the impaired gene expression and pathogenesis of HD.

Luthi-Carter et al. (2000) experimented on the R6/2 mouse model that is the most frequent transgenic animal model used for Huntington's disease. This mouse model has truncated N-terminal mutant huntingtin with expanded CAG repeat ($\sim$125 repeats) (Menalled and Chesselet, 2002) inside the huntingtin gene exon 1. The R6/2 mouse model exhibits symptoms resembling Huntington's disease.

Luthi-Carter et al. (2000) conducted DNA microarray analysis R6/2 mouse and reported fewer expressed mRNAs. The authors reported a decline in gene expression in another mouse model (N171-82Q). Authors concluded that mutant huntingtin protein directly or indirectly declines the gene expression of a specific group of genes entailed in signaling pathways for neuron functions in the striatum.

The mutant huntingtin protein interacts with transcription factors namely TP53 (important transcription protein, prevents against tumor formation, called guardian of genome), cAMP response element-binding protein (CREB protein controls gene expression importantly involved in dopaminergic neurons), and CREB-binding protein (transcription activator). These transcription factors are entailed in mitosis and cell survival (Steffan et al., 2000). Authors reported that the TP53 gene has a high binding affinity with expanded CAG repeat and later resulted in an expression of the TP53 gene.

Role of reactive oxygen species and oxidative stress in neurodegenerative diseases

Excessive production of reactive oxygen species and cumulative oxidative stress are incriminated in the etiopathogenesis of several neurodegenerative diseases (Gandhi and Abramov, 2012). The oxidative stress could induce damage to cellular macromolecules, damage to DNA, mitochondrial dysfunction, and endoplasmic reticulum stress that are either collectively or separately accelerate pathophysiology of neurodegenerative disorders (Patten et al., 2010; Ienco et al., 2011).

Reactive oxygen species are referred as the class of highly reactive molecules produced from oxygen that have mostly short life span and retain high reactivity potential owing to the presence of unpaired valence electrons (Bolisetty and Jaimes, 2013). Reactive oxygen species include hydroxyl radicals ($\cdot$OH), superoxide ions (O_2^-), and nonradicals as hydrogen peroxide (H_2O_2).

Endogenous production of ROS in brain
Mitochondrial ROS production
The chief endogenous source of ROS is the mitochondria in body tissues including the brain. Oxidative phosphorylation is the essential biological activity concerned with the synthesis of ATP in mitochondria that is the principal source of energy for neurons in the brain (Voet et al., 2008). The reactive oxygen species (superoxide) are produced as a by-product during the activity in mitochondria (Voet et al., 2008).

Superoxide has a short life span and undergoes dismutation into H_2O_2 and O_2. The reaction is catalyzed by mitochondrial superoxide dismutase-2 that is a manganese-dependent enzyme. Otherwise, superoxide has a role in the regulation of signaling pathways in the brain (Halliwell, 1992). The high metabolic rate of the brain requires a higher number of ATP molecules that consequently release a large amount of ROS as a by-product. However, the balance between the amount of production of ROS and the rate of clearance of ROS by an antioxidant system of the body determines the deleterious impact of ROS in the brain (Halliwell, 1992).

NADH-coenzyme Q oxidoreductase represents the Complex I in the ETC that is mainly involved in the production of superoxide free radicals in the brain. Complex I catalyzes the transport of electron NADH to coenzyme Q and premature leak of an electron to oxygen takes place resulting in the formation of superoxide (Zoccarato et al., 2007).

Zoccarato et al. (2007) reported that complex I is mainly involved in the mitochondrial production of H_2O_2 in the brain during oxidation of NADH and oxidation of succinate (FADH2). Authors concluded that succinate-dependent production of H_2O_2 occurs during electron downflow in Complex I. Succinate concentration modulates the rate of production of H_2O_2 via influencing hydroquinone to quinone ratio.

Monoamine oxidase ROS production

Monoamine oxidases (flavin-containing amine oxidoreductases) are located on the outer membrane of mitochondria. These catalyze the oxidative deamination of monoamines leading to the production of H_2O_2 as a by-product (Voet et al., 2008).

These are probably involved in the oxidative stress in neurons in the brain. Two isoforms of monoamine oxidase (MAO) have been reported namely MAO-A and MAO-B that are entailed in the modulation of the redox potential of neurons and glia in the brain (Voet et al., 2008).

Nitric oxide synthase, nitric oxide, and reactive nitrogen species

Nitric oxide synthase is the flavin-enriched calcium ions/calmodulin-dependent enzyme. It can modulate the tone of blood vessels, peristalsis, insulin secretion, neural development, and angiogenesis. NOS needs NADPH, tetrahydrobiopterin as coenzymes, and O_2 as a cofactor for its activity. It is mainly entailed in the synthesis of nitric oxide utilizing L-arginine. The nitric oxide has the potential to react with superoxide to generate highly reactive nitrogen species as active peroxynitrite (Radi et al., 1991). Nitric oxide is intimately involved in cell signaling, synaptic plasticity, and synaptic transmission (Schuman and Madison, 1991).

NADPH oxidase and ROS production

NADPH oxidase is the membrane-bound enzyme complex. It faces toward the extracellular space. It is composed of six subunits. The first subunit is rho-GTPase namely Rac 1 or Rac 2, and the remaining five subunits are phagocytic oxidases abbreviated as Phox subunits (Babior, 2004; Lambeth, 2004). These are designated as $gp91^{phox}$, $p47^{phox}$, $p22^{phox}$, $p67^{phox}$, and $p40^{phox}$.

Under the condition of inactive NADPH oxidase, gp91*phox* and p22*phox* remain bound to the membrane whereas rho-GTPase, p47phox, p67phox, and p40phox are located in the cytosol (Babior, 2004; Lambeth, 2004).

The rho-GTPase, p47phox, p67phox, and p40phox subunits translocate to the membrane and form a complex with the gp91*phox* and p22*phox* subunits. This event results in the activation of NADPH oxidase that in turn catalyzes the oxidation of NADPH into NADP+ and the loss of an electron. It couples with O_2 to form superoxide radicals (Sumimoto et al., 2005).

Serrano et al. (2003) utilizing various antibodies to stain sections of the mouse brain, identified extensive distribution of subunits of NADPH oxidase namely $p47^{phox}$, $p40^{phox}$, $p22^{phox}$, $p67^{phox}$, and $gp91^{phox}$ in neurons in all regions of brain of mouse.

Oxidative stress and neurodegenerative diseases

Oxidative stress is a state of imbalance between the production ROS in the body tissues and the ability of the body's antioxidant system to detoxify these reactive intermediates. Serrano et al. (2003) posited that H_2O_2 is also essential for the normal physiological neuron functions including neuronal differentiation, synaptic plasticity, and synaptic transmission in the brain. Additionally, Serrano et al. (2003) reported neurotoxicity of H2O2 that is released after the NADPH oxidase activity in brain tissues. H_2O_2 induced neurotoxicity has a role in the pathogenesis of Parkinson's disease, ischemic stroke, and possibly Alzheimer's diseases.

Utilizing immunohistochemical techniques for staining subunits of NADPH oxidase in the brain of adult mouse, Serrano et al. (2003) identified staining for cytosolic subunits namely p40(phox), p47(phox), and p67(phox), while membrane-bound subunits p22(phox) and gp91(phox) in the neurons in all regions of the mouse brain, localized particularly in cortex, hippocampus, striatum, amygdala, and thalamus. The authors concluded that the production of H_2O_2 in brain regions is involved in normal as well as in the pathophysiology of the disease.

A study by Markesbery and Lovell (1998) described the role of oxidative stress in the pathophysiology of Alzheimer's disease. ROS-induced lipid peroxidation and reduced levels of polyunsaturated fatty acids have been mentioned in the brain tissues in Alzheimer's disease.

Markesbery and Lovell (1998) reported the neurotoxicity of four-hydroxynonenal (aldehyde) that is produced in response to the lipid peroxidation in tissue culture studies and in vivo studies. The authors detected raised concentration of four-hydroxynonenal in the fluid in the ventricles in patients with Alzheimer's disease. Authors reported elevated levels of free four-hydroxynonenal in various regions of the brain in patients with Alzheimer's disease in comparison to age-matched controls. The authors concluded that four-hydroxynonenal is implicated in the development of neurodegeneration.

Smith et al. (1991) conducted a study on the autopsy of frontal lobes and occipital lobes from brain samples of 16 patients with Alzheimer's disease, eight controls with age-matching, and five controls in young ages. The authors analyzed carbonyl (protein oxidation products) and activities of glutamine synthetase and creatine kinase in the brain samples. Smith et al. (1991) reported a rise in carbonyl contents in the brain samples augmented with age and which were almost double in the frontal lobes than in the occipital lobes. Furthermore, Smith et al. (1991) identified reduced activities of creatine kinase and glutamine synthetase in the brains in age patients with Alzheimer's disease and age-matched controls in comparison to young controls. The reduced enzyme activities were pronounced in the frontal lobes than occipital lobes.

Smith et al. (1991) concluded that carbonyl contents accumulate in the brain regions in Alzheimer disease and in advancing age with the decline in oxidation sensitive enzyme activities.

Stability of genome is essential for cellular integrity and cellular activities. Oxidative stress can induce damage to the genome leading to ischemic injury of cells and degenerative changes in the affected cells.

The guanine base lesion, 7,8-dihydro-8-oxoguanine (8-oxoG), is the most common oxidation product of DNA. According to one estimate, around 100,000 8-oxoG lesions can be induced per day in the DNA per cell (Lindahl and Barnes, 2000).

This guanine base lesion is the biomarker of oxidative stress and it accumulates at the $5'$-end of guanine strings in the DNA helix, RNA, and guanine quadruplexes (produced from two separate **guanine-**rich strands) (Barnes and Lindahl, 2004; Lindahl and Barnes, 2000).

The oxidation product of DNA, 8-OxoG has mutagenic potential. It can undergo base pairing with adenine instead of cytosine leading to G:C to T:A transversion (point mutation in DNA where either adenine or guanine is changed with thymine or cytosine or vice versa) in replication of DNA (Grollman, and Moriya 1993).

The guanine base lesion (8-oxoG) and its open-ring product, 2,6-diamino-4-hydroxy-5-formamidopyrimidine are mainly recognized and repaired with the help of enzyme, 8-oxoguanine glycosylase (Dizdaroglu et al., 2008) through the Base excision repair pathway (OGG1-initiated base excision repair) (Krokan and Bjoras 2013; Izumi et al., 2003).

The 8-oxoguanine glycosylase I-base excision repair (OGG1-BER) might be incriminated in oxidative stress-induced chronic inflammation and cell death. The guanine base prominently undergoes rapid oxidative damage induced by ROS owing to its lowest redox potential in comparison to other nitrogenous bases in DNA molecule(Margolin et al., 2006).

A study by Burrows and Muller (1998) described the high reactivity of hydroxyl free radicals (•OH) that interact with guanine base in DNA.

Earlier it was assumed that OH• free radical reactivity with DNA molecule resulted in the addition of OH• free radical to the C=C bonds in the bases in DNA molecule (Kumar et al., 2011).

A recent study by Kumar et al. (2011) identified that the main interaction of OH• free radical with deoxy-guanosine leads to direct removal of a hydrogen atom from the amine group instead of adding OH• free radical to the C=C bonds in DNA molecule (Kumar et al., 2011).

There is the formation of reducing neutral radical that in turn induces the electron transfer to molecular oxygen resulting in the formation of 7,8-dihydro-8-oxoguanine (8-oxoG) (Belanger et al., 2016).

The DNA base lesion (8-oxoG) and 2,6-diamino-4-hydroxy-5-formamido pyrimidine (open-ring product of 8-oxoG) is recognized and excised and repaired by 8-oxoguanine DNA glycosylase 1 (OGG1) that is the functional homolog of *Escherichia coli* protein MutM (Formamidopyrimidine-DNA glycosylase) (Wallace, 2013; Izumi et al., 2003).

The 8-oxoguanine DNA glycosylase 1 exhibits apurinic/apyrimidinic lyase potential, and it induces nicks in the phosphodiester bond 3′ to (apurinic/apyrimidinic) sites in the DNA strand. This activity involves a β-elimination reaction that is proceeded by a delta-elimination reaction Bailly and Verly (1988) and Dizdaroglu et al. (2008).

During base excision repair by OGG1, the 7,8-dihydro-8-oxoguanine (8-oxoG) base is excised from the DNA molecule and it is released into the cytoplasm. The 8-oxoG binds with high affinity to cytosolic OGG1 and induces conformational changes in the structure of OGG1 that in turn interacts and activates small GTPases (enzymes that hydrolyze GTP into GDP) (Belanger et al., 2016). Elevated levels of the 8-oxoG in cytoplasm and DNA molecule constitute biomarkers of oxidative stress and might be implicated in chronic inflammation, aging process, and neurodegeneration.

In another work (Wang et al., 2003), authors studied the role of apoptosis-inducing factor and poly(ADP-ribose) polymerase (PARP)-mediated injury in the mouse model of Parkinson's disease. DNA single-stranded breaks represent a discontinuity in the strand of DNA molecule that can arise due to ROS-induced oxidative injury of DNA base, intermediate products of DNA base excision repair (8-oxoG), and oxidized nucleotides (Caldecott, 2008).

Superoxide free radicals, peroxynitrite, and DNA breaks activate the PARP that is a highly conserved protein of molecular weight, 116 kDa (Hassa and Hottiger, 2008).

This protein is involved in regulating gene transcription, DNA repair, genomic stability, control of cell cycle, and apoptosis (Hassa and Hottiger, 2008).

Regarding the involvement of oxidative damage of the DNA base and activation of PARP-I in the pathogenesis of neurodegenerative diseases, there are two pathways.

The PARP-1 serves as a sensor of DNA damage in the cell. It has a high affinity for binding with DNA single-stranded breaks. After binding of activated PARP-1 with damaged DNA base, it induces the cleavage of NAD^+ resulting in the formation of nicotinamide and ADP-ribose. Further, PARP-1 induces the polymerization of ADP-ribose into ADP-ribose polymers on the proteins namely histones and PARP-1 itself.

Under normal physiological condition, ADP-ribose polymers (poly(ADP-ribosylation)) has a role in the DNA repair and genomic stability. Moreover, under oxidative stress conditions, overexpression of PARP-1 and resultant synthesis of ADP-ribose polymers might be insinuated in the cellular ATP depletion, dysfunction of mitochondrial, and apoptosis of cells including neurodegeneration (Virág and Szabó, 2002).

Love et al. (1999) posited that oxidative damage to DNA, overexpression of poly(ADP-ribose) polymerase (PARP), and depletion of ATP in neurons are linked with Alzheimer's disease.

Love et al. (1999) studied the autopsy materials of frontal and temporal lobes obtained from 20 patients with Alzheimer's disease and 10 control subjects. Love et al. (1999) examined the autopsy materials for the PARP and poly(ADP-ribose).

Love et al. (1999) identified poly(ADP-ribose) and PARP-immuno-labeled cells in the brains of all patients with Alzheimer's disease in a much higher proportion than in control subjects. The authors concluded that PARP activity is highly raised in brains in patients with Alzheimer's disease and pharmacotherapeutics aimed at inhibition of PARP might be helpful in slowing the disease progression.

Soos et al. (2004) utilized immunohistochemical techniques to study the levels of PARP in the brain of patients with Parkinson's disease.

Authors identified a rise in PARP-holding nuclei in the dopaminergic neurons in the substantia nigra in Parkinson's disease as well as in diffuse Lewy body disease (Soos et al., 2004) when compared with dopaminergic neurons in the brains of patients with other neurodegenerative diseases and control subjects.

Soos et al. (2004) reported the presence of nuclear factor kappa B (NF-kappa B) in the nuclei of dopaminergic neurons in the brains of patients with Parkinson's disease and diffuse Lewy body disease (Soos et al., 2004).

Soos et al. (2004) concluded that oxidative damage of DNA, overactivation of PARP, and nuclear translocation of NF-kappa B might be implicated in inducing lesions in the neurons in the substantia nigra leading to the onset of neurodegenerative diseases.

Dysregulation of ubiquitin-proteasome system in neurodegenerative diseases

The dysregulated ubiquitin-proteasome system is insinuated in the pathology of neurodegenerative diseases. The presence of depositions made up of ubiquitylated proteins has been reported in the involved neurons in neurodegenerative diseases. Several abnormal proteins namely polyglutamine proteins, α-synuclein, and β-amyloid peptide have been identified in association with the compromised functioning of the ubiquitin-proteasome system.

Activity of ubiquitin-proteasome system

The ubiquitin-proteasome system is predominantly responsible for scavenging the abnormal and misfolded proteins from the cytoplasm and the nucleus. The ubiquitin-proteasome system employs ubiquitination of substrate followed by proteasomal degradation (Hershko and Ciechanover, 1998).

The ubiquitination of substrate requires ubiquitin-activating enzyme (E1 enzyme that catalyzes the initial step in the **ubiquitination of substrate**), **Ubiquitin**-conjugating **enzymes** (E2) and **ubiquitin ligase** (E3) collectively bring about ubiquitination of substrate via covalent bonding of ubiquitin (intensively conserved polypeptide of 76-amino acid residues in eukaryotes) to substrate protein in which glycine at the carboxy-terminal of ubiquitin is conjugated with the free amino-terminal of substrate protein (Pickart, 2001).

Repeated ubiquitination of substrate protein leads to the synthesis of discrete ubiquitin polymers that initiate various signals for biodegradation of abnormal misfolded proteins tagged with ubiquitin by the activity of proteasome (multiprotein complex with proteolytic activity).

The proteolytic sites in the proteasome are located in the interior surface of the 20S core particle (Bedford et al., 2010). The proteasome catalyzes the unfolding of tagged proteins and spit them into short peptides. These peptides are further processed into constituent amino acids by cellular aminopeptidases (Bhattacharyya et al., 2014).

Dysregulation of ubiquitin-proteasome system in Alzheimer's disease

The ubiquitin polypeptides have been identified in the neurofibrillary tangles and amyloid β plaques in the cortex section of patients with Alzheimer's disease (Cole and Timiras, 1987).

PHF are closely associated with the pathogenesis of Alzheimer's disease and related tauopathies. These are the principal structural components of neurofibrillary tangles in Alzheimer's disease. The tau proteins are the chief antigenic component of PHF. Various authors worked to determine the structure of PHF. These have predominantly admixture of random coil and β-sheet conformations (Barghorn et al., 2004).

PHF exhibited insolubility in sodium dodecyl sulfate and urea, thus the contents of PHF including the colocalization of ubiquitin protein could not be established through the use of gel electrophoresis (Mori et al., 1987).

Later on, Mori et al. (1987) treated the PHF with the concentrated formic acid and lysyl endopeptidase enzyme, and resultant PHF products were further subjected to reversed-phase high-performance liquid chromatography.

The authors reported the presence of two fragments in the digested PHF products that belonged to ubiquitin derivatives. Authors concluded the occurrence of ubiquitin associated with aggregates of abnormal and misfolded proteins in various regions of the brain depending upon the neurodegenerative diseases (Mori et al., 1987).

Hyo-Jin et al. (2009) employed an enhanced green fluorescent protein system (d2EGFP) (a **protein** in the jellyfish Aequorea Victoria that emits **green fluorescence** on exposure to light). The authors reported the relation between the attenuated activity of the 26S proteasome with the increase in the intracellular deposits of Amyloid β in the perinuclear area.

Hyo-Jin et al. (2009) with the help of immunofluorescence assays reported the elevated levels of 20S proteasome α-subunits in the aggregates around perinuclear regions in human neuroglioma H4 (H4) and human embryonic kidney (HEK293) cell lines.

Authors reported elevated levels of apoptosis regulator Bax protein (involved in the release of protein from the perimitochondrial space that activates Caspases involved in apoptosis) and PARP-1 cleavage fragments (PARP-1 cleavage fragments are biomarkers of the proteolytic activity involved in the apoptosis of cells) in the human neuroglioma H4 (H4).

Hyo-Jin et al. (2009) concluded that intracellular deposits of Amyloid β in the perinuclear area induce mitochondrial-dependent neuronal death through raised levels of apoptosis regulator Bax protein coupled with inhibition of proteasome.

It can be inferred that ubiquitin-conjugated amyloid β aggregates in the brain of patients with Alzheimer's disease might be involved in the suppression of proteasome activity and activation of mitochondria-dependent apoptosis of cells and phagocytosis of cytoplasmic inclusions.

Authors identified surplus levels of ubiquitin in amyloid β plaques and neurofibrillary tangles in the brains of patients with AD (Perry et al., 1987). Authors further commented that polypeptide ubiquitin is attached covalently with the senile plaques and neurofibrillary tangles in the brains in AD.

van Leeuwen et al. (1998) detected frameshift mutation in the mRNA involving deletion of two nucleotide bases that resulted in the synthesis of mutant ubiquitin protein designated as UBB+1 that contains a 19-amino acid C-terminal extension. Possibly, mutant ubiquitin protein is implicated in the inhibition of proteasomal degradation of misfolded protein tagged with ubiquitin in the neurons.

Lam et al. (2000); **UBB^{+1} is ubiquitylated and appears to be a protein with dual properties; it acts as a ubiquitin-fusion-degradation substrate for the proteasome and also acts as a specific inhibitor of the proteasome**.

Paula van et al. (2007) identified that mutant ubiquitin, UBB^{+1}, collects in the brain regions in diseases namely Alzheimer's disease, tauopathies, and polyglutamine diseases and thus acts as a biomarker for the dysfunction of proteasomal degradation in these diseases.

Paula van et al. (2007) posited that the transition of the substrate to inhibitor activity of UBB^{+1} mutant protein can be explained on the basis of expression and dose of UBB^{+1} in the tissues. Paula van et al. (2007) employed different expression levels of UBB^{+1} protein in cell cultures and reported that UBB^{+1} mutant protein exhibited an inhibitory effect on the proteasomal degradation ability at its high dose and escaped its degradation by the proteasomal system. It resulted in the accumulation of UBB^{+1} mutant protein that subsequently was associated with reversible functional impairment of the ubiquitin-proteasome system (Paula van et al., 2007).

Dysregulation of ubiquitin-proteasome system in Parkinson's disease
The occurrence of ubiquitin-tagged α-synuclein in Lewy bodies has been reported in the substantia nigra in the brain in patients with Parkinson's disease (Kuzuhara et al., 1988).

McNaught and Jenner (2001) posited that dysfunction of proteasomal degradation of abnormal proteins in the substantia nigra was insinuated in the pathogenesis of Parkinson's disease.

Later on, McNaught et al. (2002a,b) confirmed the occurrence of structural as well as functional impairment in the 26S (2000-kDa) and 20S (700-kDa) proteasomes in the substantia nigra that was involved in the formation of Lewy bodies and subsequent degeneration of dopaminergic neurons in the Parkinson's disease.

McNaught et al. (2002a,b) worked on the rat models of Parkinson disease to test the posited hypothesis related to impairment in the 26S (2000 kDa) and 20S (700 kDa) proteasomes. Authors infused lactacystin (selective proteasome inhibitor) on the single side into the substantia nigra pars compacta of rats. The infused rats developed bradykinesia, curved posture, and tilting of head on the contralateral side (opposite to that received infusion of lactacystin).

Further, McNaught et al. (2002a,b) administer apomorphine (nonselective agonist of dopamine that activates D_2-like and D_1-like receptors) to lactacystin-infused rats that led to a reversal of the behavioral irregularities and induced contralateral rotations.

The authors identified Lactacystin-induced accumulation of α-synuclein and death of dopaminergic neurons. The authors concluded that an impaired ubiquitin-proteasome system to degrade α-synuclein resulted in manifestations like Parkinson's disease in rat models.

Dysregulation of ubiquitin-proteasome system in Huntington disease

The polyQ repeats consisting of more than 40 amino acid residues are located within the amino-terminal of huntingtin protein in patients with Huntington's disease. The presence of mutant huntingtin protein tagged with ubiquitin protein has been identified by Difiglia et al. (1997) in the neurons as the intranuclear inclusion bodies in the cortex and striatum of patients with HD.

Venkatraman et al. (2004) identified that the proteasomal degradation system failed to cleave polyQ repeats consisting of more than 25 amino-acid residues that subsequently accumulate in the neurons leading to neurodegeneration in HD.

Kisselev et al. (1999) described that proteasome degradation produces 3−9 amino-acid residues long peptides. The polyQ sequences with more than 25 amino acids length might be unable to diffuse through the narrow α-pores resulting in blockage of proteasomal degradation system and in turn lead to their accumulation in the neurons in Huntington's disease.

Bennett et al. (2005) refuted the proteasomal blockage hypothesis by the long stretch of polyQ repeats in their in vitro study through the use of synthetic polyglutamine aggregates.

The dysregulation of UPS activity in Huntington's disease associated with mutant polyQ repeats is controversial. Jana et al. (2001) posited a reduction in the proteasome degradation activity, while Bett et al. (2006) described an increase in the proteasome degradation activity and Ding et al. (2002) claimed unaltered proteasome degradation activity in Huntington's disease.

Genetics of neurodegenerative diseases

Genetics of Alzheimer's disease

Alzheimer's disease is the most prevalent form of age-dependent neurodegenerative dementia. It is one of the serious health problems in developed countries. The prevalence of dementia is around 35 million affected persons worldwide (Barber, 2012). Alzheimer's disease has a multifactorial origin with nearly 70% contribution from genetic factors while around 30% of environmental factors participating in the pathogenesis of AD (Dorszewska et al., 2016).

Based on the age predilection for the onset of Alzheimer's disease, it can be sub-categorized as early-onset Alzheimer's disease (EOAD) and late-onset Alzheimer's disease (LOAD).

Role of genetics in early-onset familial Alzheimer's disease (EO-FAD)

The disease begins before the age of 60 or 65 years. The early-onset familial AD is hereditary, and the symptoms start early in life (Tanzi, 2012). The term "familial" refers to the disease state when it has tendency to occur more often in family members than is expected by chance alone and or a disease condition that occurs in at least two generations of a family. The prevalence of early-onset familial AD is nearly 1%—5% of total cases of Alzheimer's disease (Thomas, 2008). The mutation of one of the three genes is insinuated in the pathogenesis of early-onset familial AD. The early-onset familial AD exhibits a peculiar classic Mendelian pattern of inheritance as an autosomal-dominant inheritance (Tanzi, 2012).

Role of mutations in *APP* gene (amyloid precursor protein) in (EO-FAD)

The APP gene encodes the amyloid precursor protein in humans. APP is located on chromosome 21q21.3. The gene consists of 19 exons and covers approximately 240 kilobases of DNA(Tharp and Sarkar, 2013).

The APP is highly conserved in humans. It undergoes maturation in Golgi and endoplasmic reticulum. It consists of amino acid residues between 365 and 770 (Barber, 2012). The amyloid precursor protein has eight isoforms out of which isoforms APP695, APP751, and APP770 are the prominent isoforms (Barber, 2012). The isoform APP751 and APP770 contain exon 7 while isoform APP695 lacks exon 7. Isoform APP695 is prominently expressed in neuronal tissues (Barber, 2012).

The cleavage of AP Protein by α-secretase creates a fragment that is not associated with plaques or Alzheimer's. However, cleavage by β-, and γ-secretases creates β-amyloid, which is encoded by exons 16 and 17 and is 39—42 amino acids in length. It is the β-amyloid fragment that forms the extracellular plaques that are the hallmarks of Alzheimers disease.

Lustbader et al. demonstrated that β-amyloid interacted with amyloid β-binding alcohol dehydrogenase (ABAD) to induce mitochondrial toxicity in Alzheimer's disease patients and in transgenic mice (Lustbader et al., 2004).

A study by Suzuki et al. (1994) identified three mutations at amino acid residue 717 namely V717F, V717G, and V717I out of several mutations involving APP gene resulted in a higher probability of synthesis of β-amyloid fragments with longer lengths than the β-amyloid fragments with shorter lengths. Suzuki et al. (1994) utilized human neuroblastoma (M17) cells and transfected the cells with artificially constructed DNA segments expressing wild-type β APP and mutant β APP717.

Suzuki et al. (1994) for the sake of comparison between lengths of wild-type β APP and mutant β APP717, authors utilized two methods that furnished data that APP717 mutations resulted in nearly 1.5—1.9 times rise in the percentage of synthesis of longer β-amyloid fragments. Authors further posited that longer β-amyloid fragments have a higher tendency to assemble into insoluble amyloid fibrils rapidly than shorter β-amyloid fragments.

Yamatsuji et al. (1996) reported that missense mutations in the amino acid residue 695 of the amyloid precursor protein, designated as APP695, are responsible for substitution of valine at 642 positions with phenylalanine, isoleucine, or glycine resulting in the synthesis of mutant amyloid precursor protein that was implicated in the nucleosomal DNA fragmentation in the neuronal cells in culture. The DNA fragmentation was mediated by GTP-coupled proteins (G proteins).

Murrell et al. (2000) identified a novel mutation at amino acid residue 717 in the gene APP that is designated as V717L that is associated with the history of dementia in the family beginning in the age group between 25 years and late 30s.

Murrell et al. (2000) reported single nucleotide substitution in 1 allele where guanine was replaced with cytosine. It led to the substitution of amino acid residue valine with Leucine at codon 717 in gene APP.

Chartier-Harlin et al. (1991) reported that APP gene mutation at codon 717 was found to co-segregate with early-onset familial Alzheimer's disease in a single family. Further, this mutation was reported in another five out of nearly 100 families affected by Alzheimer's disease. Chartier-Harlin et al. (1991) utilizing direct sequencing and analysis of exon 17 confirmed a substitution of amino acid residue valine with glycine at codon 717 of APP gene.

Mullan et al. (1992) after analysis of exon 16 identified mutations at codons 670 and 671 in APP 770 transcript that cosegregate with the early-onset familial AD in two large families from Sweden. The double mutations at codon 670 and 671 involved transversions of base pair G to T, and A to C that further led to the substitution of amino acid residue, Lys to Asn and substitution of amino acid residue, Met to Leu at codons 670 and 671 in APP 770 transcript. The mutations occurred at the -NH2 terminal of β-amyloid fragment and might be pathogenic.

Nilsberth et al. (2001) reported mutation in 693 codon inside Aβ segment of the amyloid precursor protein. There was the substitution of glutamic acid with glycine and labeled as E693G. Authors named it as Arctic mutation due to the location of the family to northern Sweden. The concentrations of Aβ42 and Aβ40 were reduced in the plasma of the family members with mutations in comparison to the healthy members.

Nilsberth et al. (2001) identified that the concentration of protein Aβ42 was declined in the culture containing cells transfected with APPE693G. Nilsberth et al. (2001) concluded that Arctic mutation coupled with a high tendency for the formation of Aβ protofibrils are associated with the pathogenesis of AD.

Role of mutations in gene PSEN1 (presenilin 1) in (EO-FAD)

The **gene PSEN1** is located on chromosome 14 (Schellenberg et al., 1992). It encodes the Presenilin 1 protein that is a transmembrane protein and represents the important catalytic domain of γ-secretase enzyme and determinant of its enzymatic activity (Schellenberg et al., 1992). The γ-secretase enzyme splits the amyloid precursor protein at various points resulting in the formation of Aβ fragments with varying lengths (Lee et al., 2002). The Aβ 40 (40 amino acid residues long) and Aβ 42 (42 amino acid residues long) are linked with the pathogenesis of Alzheimer's disease. Nevertheless, the Aβ 42 fragment has been reported to be associated with a higher tendency to assemble into Aβ protofibrils that are incriminated in the pathology of AD (Lee et al., 2002).

According to an estimate, around 150 mutations have been reported in the *PSEN1* gene that is implicated in the etiology of early-onset Alzheimer's disease (USNLM, 2020a,b,c).

Queralt et al. (2002) reported a novel mutation, V89L in the PSEN1 gene in a family with established early-onset familial Alzheimer's disease. The affected members (two) in the family presented with significant behavior changes.

Snider et al. (2005) described kindred affected with early-onset familial AD at an age less than 40 years. Novel mutation, S170F, in the gene PSEN1 was linked with the disorder. Three members in the family had dementia at a mean age of 27 years (Snider et al., 2005). The proband suffered from seizures, rigidity, and myoclonus. The Lewy bodies were detected in proband in limbic areas, brainstem, and neocortex. The mutation, S170F in exon 6 of the gene PSEN1 segregates with disease in proband (Snider et al., 2005).

Samura et al. (2006) studied the A β amyloidosis and tauopathy in the double transgenic mouse (APP-PS).

The double transgenic mouse was generated through crossing APP mutant line, Tg2576, with a mutant line (presenilin-1 L286V). The Tg2576 model is a thoroughly characterized and extensively used mouse model for Alzheimer's disease. The mouse expresses a mutant form of the amyloid precursor protein (APP 695) containing the Swedish mutation (KM670/671NL) (Hsiao et al., 1996).

Samura et al. (2006) reported the occurrence of A β amyloid deposits at 8 weeks in the double transgenic mouse, while τ deposits were reported at 4.5 months including a significant increase in abnormal neuronal processes (dystrophic neuritis) surrounding A β amyloid deposits. Samura et al. (2006) identified phosphorylated τ accumulation was increased in the brains in the double transgenic mouse, APP-PS in comparison to Tg2576 mice.

Samura et al. (2006) described the occurrence of twisted tubules simulating paired helical filament at 16-month of the double transgenic mouse, APP-PS. The authors concluded that mutant presenilin-1 protein promoted the accumulation of A β deposits, tauopathy, and fibril formation.

Citron et al. (1997) claimed that mutations in gene presenilin I led to the rise in the ratio of Aβ 42 fragments to Aβ 40 fragments, moreover, the total contents of Aβ fragments remained undisturbed. The higher contents of Aβ 42 fragments were linked to a higher propensity to the pathogenesis of early-onset familial AD.

Van Giau et al. (2019) posited that the presenilin-1 gene is mainly involved in the cause of early-onset Alzheimer's disease. Van Giau et al. (2019) reported that single nucleotide substitution from G > T in presenilin-1 gene resulted in the substitution of amino acid residue tryptophan to cysteine. The mutation Trp165Cys is located in the transmembrane-III (TM-III) domain that is highly conserved in between PSEN1 to PSEN2 (Van Giau et al., 2019).

The affected person presented with loss of memory at the age of 53 years. He had a family history including his mother and one brother affected with the symptoms of dementia (Van Giau et al., 2019). Further, Van Giau et al. (2019) confirmed that MRI of brain regions depicted the mild level of atrophy of parietal lobes and hippocampus. Deposition of amyloid β was identified in the parietal, frontal, temporal lobes, and precuneus.

Enzymatic activity of PSEN1 protein and PSEN2 protein is necessary for the synthesis of β-amyloid 42 fragment from the cleavage of the amyloid precursor protein. Duff et al. (1996) reported that transgenic mice with overexpression of mutant presenilin-1 resulted in a rise in contents of β-amyloid 42 fragments, while transgenic mice with overexpression of wild-type presenilin-1 had failed to show an increase in the amyloid 42 fragments (Duff et al., 1996).

Duff et al. (1996) concluded that mutations in gene PSEN1 led to harmful gain of function by increasing the synthesis of β-amyloid 42 fragments that are insinuated in the pathology of early-onset familial AD.

Role of mutations in gene PSEN2 (presenilin 2) in (EO-FAD)

The mutations in the PSEN2 gene are extremely less in comparison to mutations in PSEN1.

Globally, fewer than 40 mutations in PSEN2 have been identified (Barber, 2012; Cai et al., 2015; Thomas, 2018).

Gene PSEN2 was established to be the causative factor in early-onset familial AD in 1995 (Levy-Lahad et al., 1995). The Gene PSEN2 is localized on chromosome I consisting of 12 exons. Exon 1 and exon 2 have untranslated domains.

Gene PSEN2 encodes the PSEN2 protein. It is a transmembrane protein made up of 448 amino acid residues. The PSEN1 and PSEN 2 proteins are homologous, representing a similarity of 67% (Rademakers et al., 2003; Cai et al., 2015).

The protein, PSEN2 is highly protein that undergoes endoproteolysis inside the large cytoplasmic loop domain and forms a physiological active heterodimer (Cai et al., 2015).

In protein PSEN2, aspartyl residue, D263, and aspartyl residue, D366 are located in transmembrane domain-VI and VII represent the catalytic sites of γ-secretase enzyme(Cai et al., 2015).

Citron et al. (1997) described that mutation in gene PSEN2 resulted in a rise in γ-secretase enzymatic activity. The fact was authenticated by utilizing mouse models and cell-based studies that showed that PSEN2 mutations led to a higher synthesis of Aβ42 fragments that are the biomarkers in the brains of patients with early-onset familial AD.

Studies by Vito et al. (1996) and Wolozin et al. (1996) hypothesized that the PSEN2 gene is implicated in apoptosis that was authenticated in the studies involving HEK293 human cells and murine neurons. It was identified that wild-type and mutant N141I-PSEN2 induced apoptosis in the HEK293 human cells and murine neurons that were induced by the p53-dependent pathway.

The point mutation (single nucleotide substitution) was reported in gene PSEN2 at codon 62. It led to single nucleotide substitution (CGC to TGC) that in turn resulted in the substitution of amino acid arginine with cysteine. The mutation was reported from the Asian population by Van Giau et al. (2019). The mutation was reported in a patient with dementia. Symptoms namely personality change and impairment of memory were reported at 49 years old.

Ezquerra et al. (2003) described a novel mutation in gene *PSEN2* linked to early-onset familial Alzheimer's disease. The proband presented with dementia starting at age 45 years. Proband's one parent at age 64 years and one of the grandparents at age 60 years had dementia.

Ezquerra et al. (2003) identified a missense mutation in gene PSEN 2 at codon 430 that led to the substitution of amino acid threonine with methionine. The mutation was found in affected persons. This mutation was independent of the *APOE* genotype.

Jayadev et al. (2010) reported a female patient of age 59 years with slowly progressive memory loss without a history of dementia in her family. Sequencing and analysis of APP gene and PSEN1 gene was normal. Further analysis (Jayadev et al., 2010) of the PSEN2 gene explained the occurrence of novel frame-shift mutations in exon 5 in the PSEN2 gene. It involved two base pair deletion of GA at positions 342 and 343 in exon 5 (Jayadev et al., 2010) resulting in the introduction of premature termination codon in exon 5 and consequently loss of gene expression.

Role of genetics in late-onset Alzheimer's disease

The late-onset Alzheimer's disease (LOAD) begins after the age of 65 years. Etiologically, the late-onset Alzheimer's disease has a heterogeneous pattern including genetic components as well as environmental factors. However, major genetic contributions in the pathogenesis of late-onset Alzheimer's disease remain elusive. A study by Gatz et al. (2006) identified significant genetic involvement in the cause of late-onset Alzheimer disease in a population-based study of Swedish twins, where the inheritance rate of 79% of late-onset Alzheimer's disease was reported.

Role of mutation in gene APOE in (LOAD)

The gene *APOE* is located on chromosome 19. The *APOE* gene is made up of 4 exons and 3 introns with a total of 3597 nucleotide pairs. The gene *APOE* transcription is activated by liver X receptor, retinoid X receptor, and PPAR- γ (Hoek et al., 2008).

The gene *APOE* exhibits polymorphism (Singh et al., 2006). There are three main allelic forms of the APOE gene namely *ε2 (epsilon 2)*, *ε3 (epsilon 3)*, and *ε4 (epsilon 4)* (Eisenberg et al., 2010). The allele *APOE-ε2* consists of cysteine at positions 112 and 158, allele *APOE-ε3* consists of cysteine at position 112 and arginine at position 158, and allele *APOE-ε4* consists of arginine at positions 112 and 158 (Rall et al., 1982; Zuo et al., 2006).

The gene *APOE* encodes the APOE protein that is made up of 299 amino acid residues. It contains several amphipathic α-helices in its structure (Eichner et al., 2002).

The amino terminal has amino acid residues from 1 to 167, and this domain constitutes a bundle of four α-helices (antiparallel) (Eichner et al., 2002; Phillips, 2014).

The carboxy terminal contains amino acid residues 206 to 299 and constitutes a bundle of three α-helices (Phillips, 2014).

The four-helices bundle of N-terminal links with three α-helices bundle of C-terminal via formation of hydrogen bonds and salt bridges. The C-terminal possesses LDL-receptor binding domain (Phillips, 2014; Eichner et al., 2002).

The allele, APOE-ε2 has a worldwide frequency of 8.4% (Liu et al., 2013), and it encodes APOE2 apolipoprotein. The APOE2 apolipoprotein is linked to a higher prevalence of atherosclerosis. It is identified that persons exhibiting genotype, apolipoprotein E2/E2 together with mutations in the gene LDL-R (LDL receptor gene) are responsible for type III hyperlipoproteinemia (Breslow et al., 1982; Civeira et al., 1996).

The allele, *APOE-ε3*, has a worldwide frequency of 77.9% (Liu et al., 2013) and it encodes APOE3 apolipoprotein. This genotype is considered as neutral.

The allele, APOE-ε4, has a worldwide frequency of 13.7% (Liu et al., 2013), and it encodes APOE4 apolipoprotein. This apolipoprotein is involved in the pathogenesis of late-onset Alzheimer disease (Corder et al., 1993), multiple sclerosis (Chapman et al., 2001), ischemic cerebrovascular disease (McCarron et al., 1999), and impaired cognitive functions (Deary et al., 2002) and in the etiology of COVID-19 (Chia-Ling et al., 2020).

Corder et al. (1993) posited that allele, APOE- ε4 exhibits genetic involvement with the late-onset familial AD andlate-onset sporadic Alzheimer's disease. With the increase in the dose of allele APOE-ε4, an increase in the prevalence of Alzheimer's disease from 20% to 90% (Corder et al., 1993) was reported. There was a decline in the mean age for the beginning of Alzheimer's disease from 84 to 68 years in the presence of an increasing number of allele APOE-ε4 (Corder et al., 1993). The study was conducted in 42 families by Corder et al. (1993). The authors concluded that the dose of the APOE-ε 4 gene is prominently a risk component in the late-onset AD.

Tamaoka (1994) reported a protective effect of allele APOE-ε2 against the beginning of late-onset AD. In vitro study by Tamaoka (1994) showed that apolipoprotein E4 exhibited a high affinity to bind with synthetic amyloid β peptide that is the major structural component of senile plaque. The apolipoprotein E4 formed a complex with synthetic amyloid β peptide in the in vitro study (Tamaoka, 1994). In the same study by Tamaoka (1994), it was identified that apolipoprotein E3 exhibited a higher affinity to bind with microtubule-associated protein τ that constitutes the main constituent of neurofibrillary tangles. Thus, the authors concluded that APOE-E3 might be involved in the prevention of abnormal phosphorylation of τ and reducing the formation of neurofibrillary tangles.

Gomez-Isla et al. (1999) hypothesized the influence of genotype, APOE-ε 4 upon the clinical course and clinical features of Alzheimer's disease. Gomez-Isla et al. (1996) conducted a prospective longitudinal study involving the age of beginning of AD, duration, and rate of progression of AD in 359 patients with APOE-ε 4 genotype.

Gomez-Isla et al. (1996) reported a strong association between the occurrence of genotype APOE-ε 4 in patients with the earlier onset of Alzheimer's disease in patients, moreover, without any association in between genotype and rate of progression of dementia in patients.

Gomez-Isla et al. (1996) identified the amyloid β deposits were higher in patients with APOE-ε 4 genotype, however, the presence of neurofibrillary tangles were not related with the APOE-ε 4 genotype in patients.

Authors concluded that APOE-ε 4 genotype was selectively linked with clinical and pathological features in Alzheimer's disease.

Tiraboschi et al. (2004) described that APOE-ε4 gene is strongly linked with the increased formation and deposition of β-amyloid senile plaques and neuritic plaques in Alzheimer's disease. Tiraboschi et al. (2004) studied the autopsy materials of 296 patients with Alzheimer's disease. Authors identified the APOE-ε4 genotype from the blood and autopsy brain tissue. Tiraboschi et al. (2004) identified that AD patients with two APOE-ε4 alleles exhibited higher contents of neuritic plaques and neurofibrillary tangles in all neocortical areas in the brains in comparison to patients with either single or absence of ε4 alleles.

Rebeck et al. (1993) reported higher prevalence of APOE-ε 4 allele in patients with late onset familial Alzheimer disease. Further, Rebeck et al. (1993) reported 62% prevalence of occurrence of APOE-ε 4 allele in patients with sporadic Alzheimer disease in comparison to 20% prevalence of APOE-ε 4 allele in controls.

Genetics of Parkinson disease

The etiopathogenesis of Parkinson disease is multifactorial including influences from environmental factors as well as from the genetic components. Moreover, the precise pathogenesis of PD is still incomprehensible. The familial PD has strong genetic influence on its pathogenesis. Study by Bonifati (2014) reported the role of 14 genes in the Mendelian inheritance of Parkinson's disease.

Role of mutations in gene LRRK2

The gene *LRRK2* is the most frequently mutated gene involved in the pathogenesis of Parkinson's disease with autosomal-dominant pattern of inheritance (Di Fonzo et al., 2006). Mutation in gene LRRK2 is responsible for pathophysiology of sporadic as well as familial Parkinson disease (Kumari and Tan, 2009).

The gene LRRK2 is located on chromosome 12p11.2−q13.1. It encodes large protein called as Dardarin, which has multiple domains as armadillo repeats domain, an ankyrin repeat domain, leucine-rich repeat domain, GTPase domain (also called ROC), COR (C-terminal of Roc) domain, kinase domain, and WD40 domain. The LRRK2 protein is located in cytosol and remains attached to outer mitochondrial membrane (Gilks et al., 2005; Shen, 2004).

Mutation G2019S in the gene LRRK2 is responsible for nearly 3%−6% of cases of sporadic and 1%−2% cases of familial Parkinson's disease (Gilks et al., 2005; Shen, 2004).

Study by Kumari and Tan (2009) augments the fact that mutations in leucine-rich repeat kinase 2 gene is linked to familial and sporadic cases of Parkinson's disease.

But, the prevalence of both types of cases in the study by Kumari and Tan (2009) is slightly higher (between 5% and 13% of familial and between 1% and 5% of sporadic cases of Parkinson's disease) than the prevalence shown in the previous study by (Gilks et al., 2005; Shen, 2004) (between 1% and 2% of familial and between 3% and 6% of sporadic cases of Parkinson disease).

It was first time that mutation G2019S in gene LRRK2 was identified and established as the genetic component of sporadic form of PD (Gilks et al., 2005).

The mutation G2019S is frequent, pathogenic, and cosegregates with Parkinson's disease in families (Di Fonzo et al., 2006).

Mutations in the LRRK2 gene is the predominant genetic factor in the development of Parkinson's disease, and around 100 mutations in LRRK2 gene have been reported to increase the risk for the development of PD. Study by Kumari and Tan (2009) posited that mutation G2019S is located in the kinase domain in gene LRRK2. This mutation is identified across all the six ethnic groups with maximum prevalence in North African Arabs and Ashkenazi Jews.

Another estimate by Kumari and Tan (2009) commented upon higher prevalence of mutation G2019S in the southern European countries than northern European countries as per the report based on meta-analysis of pooled data from 24 populations across the world.

The penetrance of mutation G2019S in the southern European countries is reported to be nearly 75% at 79 years old (Kumari and Tan, 2009).

The mutation G2019S in gene LRRK2 is responsible for substitution of glycine at codon 2019 by serine amino acid reside resulting into increased kinase activity in the kinase domain of LRRK2 is involved in the onset of familial Parkinson's disease (Gilks et al., 2005).

The gene LRRK2 encodes multidomain protein comprised 2527-amino acid residues belonging to Leucine-rich repeat (sequence of 20−30 leucine amino acid residues, held together to form Solenoid protein domain called as leucine rich repeat domain) kinase family.

The pathogenic mutations in LRRK2 gene are localized in the central region of gene that is comprised of GTPase and kinase domains (catalytic domains). Alteration in the enzymatic activity of either kinase domain or GTPase domain in the LRRK2 leads to expression of pathogenic manifestations in the PD (Cookson, 2010; Mata et al., 2006). The mutation G2019S in the kinase domain in LRRK2 gene is responsible for the occurrence of around 40% of prevalence of familial Parkinson's disease based on ethnicity (Farrer, 2006).

The G2019S mutation in the kinase domain of LRRK2 gene induces substitution of glycine at codon 2019 by serine, thus enhancing the enzymatic activity of kinase domain of LRRK2. The activities, auto-phosphorylation of substrate as well as phosphorylation of substrate in the kinase domain are increased (Lee et al., 2010).

According to a study by West et al. (2007), rise in auto-phosphorylation and phosphorylation activities of kinase domain renders neurotoxicity upon the mutated LRRK2.

Recent study by Xiong et al. (2017) related to G2019S mutation in LRRK2 gene, and its neurotoxic potential was conducted on transgenic mice expressing LRRK2 gene in nondopaminergic neurons. Authors identified that LRRK2 protein with its inactive kinase domain (mutation D1994E) exhibited absence of neurotoxicity, thus authors concluded that hyperactivity of kinase domain is critical in the expression of neurotoxicity related to LRRK2.

Pereira et al. (2014) worked upon the yeast model expressing wild-type LRRK2 and mutant LRRK2 for the assessment of the role of LRRK2 protein in the oxidative stress. Pereira et al. (2014) reported that expression of wild-type LRRK2 in yeast resulted in development of resistance to harmful effects of H2O2 generated during mitochondrial induced oxidative phsophorylation, while this potential was not reported in yeast expressing LRRK2 with mutation G2019S and R1441C.

Authors concluded that full length LRRK2 provides cellular protection against the oxidative stress produced during activity of mitochondria.

Increased activity of the catalytic domains in mutant LRRK2 is responsible for neurotoxicity, additionally in a study by Sheng et al. (2010), it was identified that WD40 domain of LRRK2 gene confers neurotoxicity to the protein. Sheng et al. (2010) treated zebrafish LRRK2 (zLRRK2) by morpholinos (nucleotide analogs that bind with short sequences, nearly 25 nucleotides at the transcription start site). The activity resulted in blockage of translation of zLRRK2 protein. Sheng et al. (2010) reported intense developmental defects including growth retardation and neurodegeneration leading to high embryonic lethality.

Further, Sheng et al. (2010) induced deletion of WD40 domain in zLRRK2 gene through activity of morpholinos. Authors found Parkinsonism-like manifestations as loss of dopaminergic neurons in diencephalon and associated locomotion defects. Authors reported that locomotion defects were prevented by administration of L-dopa that supplemented the deficiency of dopamine in the brain regions. Authors concluded that zLRRK2 gene is ortholog of hLRRK2 gene. Deletion of domain WD40 in gene zLRRK2 could be used s study model for PD (Sheng et al., 2010).

In vitro studies reported that mutation G2019S in gene LRRK2 induced increased enzymatic activity in the kinase domain leading to increased phosphorylation of mitogen-activated protein kinase (MAPK) and stimulation of pathway for neuronal death.

Chen et al. (2012) generated transgenic mice of 12−16 months old with mutation G2019S in gene LRRK2. Authors identified increased kinase activity in LRRK2 protein causing overphosphorylation of MAPK kinase 4 (MKK4) at amino acid residue, Serine at codon 257. This led to upregulation of phosphorylated phospho-MKK4(Ser257) in the substantia nigra in brain in transgenic mice with mutation G2019S in LRRK2 gene.

The activated MKK4(Ser257) in turn phosphorylates and activates downstream JNK(Thr183/Tyr185) and phospho-c-Jun(Ser63) in the substantia nigra in brain in transgenic mice with mutation G2019S in LRRK2 gene (Chen et al., 2012).

Expression of proapoptotic Bim (**proapoptotic** BCL2-family protein) and FasL (homotrimeric type II transmembrane protein on cytotoxic T-cell) was increased. Further, synthesis of caspase-8, caspase-9, and caspase-3 was reported in substantia nigra in brain of transgenic mice (Chen et al., 2012).

Chen et al. (2012) concluded that mutation G2019S in gene LRRK2 stimulates MKK4-JNK-c-Jun pathway in the substantia nigra leading to degeneration of dopaminergic neurons in the substantia nigra PD transgenic mice.

Mutation G2019S is mapped to kinase domain of LRRK2 gene, and it exhibits a gain-of-function and enhancement in the kinase activity of LRRK2 (Chen et al., 2012; Hsu et al., 2010).

There is close homology between the kinase domain of LRRK2 and mixed lineage kinases (MLKs) belonging to a large family of MAPKKKs (mitogen-activated protein kinase kinase) (Mata et al., 2006).

The mixed lineage kinases phosphorylate and activate the downstream MAPK kinases. In turn, phosphorylated MAPK kinases catalyze phosphorylation and activation of c-Jun N-terminal kinases, especially **c-Jun N-terminal kinase 3** (JNK3) (Wang et al., 2004).

MAPK kinases namely MKK4 and MKK7 are actively involved in the phosphorylation of threonine and tyrosine residues inside the Thr-Pro-Tyr motif located in kinase subdomain VIII of **c-**Jun N-terminal kinase 3 (Ip and Davis, 1998).

In vitro as well as in vivo studies identified potential of JNK pathways in the neuronal degeneration and pathogenesis of Alzheimer's disease. Activation of JNK3 pathway induced Aβ production and formation of neurofibrillary tangles in brain (Antoniou et al., 2011). Additionally, activation of MKK4-JNK pathway is involved in the degeneration of dopaminergic neurons in substantia nigra in brain in Parkinson's disease(Burke, 2007).

It can be concluded that mutation G2019S in LRRK2 gene induces gain-of-function and associated enhanced kinase activity of LRRK2. In turn, overphosphorylation of MAPK kinases, MKK4 and MKK7, leads to activation of **c-Jun N-terminal kinase 3**, which is insinuated in the activation of neuronal apoptotic pathway in the substantia nigra implicated in the pathogenesis of Parkinson's disease.

Studies were conducted by Brecht et al. (2005) and Ries et al. (2008) on the animal models of Parkinson's disease, which demonstrated that phosphorylated and activated mitogen-activated protein kinases namely MKK4 and MKK7 induced activation of **c-Jun N-terminal kinase 3** pathways are implicated in the neurodegeneration of dopaminergic neurons in the substantia nigra.

Brecht et al. (2005) and Ries et al. (2008) studied the effect of homozygous jnk2/3 double null mutations on the neuronal apoptosis of dopaminergic neurons in a transgenic model of mice. The double jnk2/3 null mutations resulted in either nontranscription of mRNA or synthesis of nonfunctional jnk2 and jnk3 proteins. Authors injected 6-hydroxydopamine in the striatum of homozygous jnk2/3 double null mutations mice.

Brecht et al. (2005) and Ries et al. (2008) reported complete cancellation of neuronal degeneration of dopaminergic neurons in transgenic mice that indicated that absence of jnk2 and jnk3 proteins in homozygous jnk2/3 double null mutations mice models prevented the neurotoxic effects of 6-hydroxydopamine on dopaminergic neurons.

This effect can be further supplemented by another study by Choi et al. (1999). The 6-hydroxydopamine is the specific neurotoxin which is employed in the evaluation of neuronal apoptosis in models of Parkinson's disease. There is increased phosphorylation of c-Jun-NH2-terminal kinase in neurons exposed to 6-hydroxydopamine leading to jnk2 and jnk3 activation and onset of neuronal apoptotic pathways in dopaminergic neurons.

Role of mutations in gene SNCA (PARK1/4)

The gene *SNCA* is located on chromosome 4q21.3-q22 and encodes α-synuclein protein. It is chiefly expressed in brain on the presynaptic terminals (Medline Plus 2020; NCBI, 2020). *SNCA* gene mutations are linked to early-onset Parkinson's disease and nearly 30 mutations in the gene *SNCA* have been reported (Medline Plus, 2020; Iwai et al., 1995).

Mutation in gene SNCA is the important genetic factor involved in the pathogenesis of early onset of familial Parkinson disease. Five missense mutations have been reported in the gene SNCA namely H50Q (Appel-Cresswell et al., 2013), E46K (Zarranz et al., 2004), G51D (Lesageet al. 2013), A30P (Krüger et al., 1998), and A53T (Polymeropoulos et al., 1997).

Krüger et al. (1998) reported missense mutation in SNCA gene leading to one amino acid substitution by another. In mutation A30P in SNCA gene, there is substitution of alanine with proline amino acid residue in the α-synuclein protein. The altered α-synuclein protein has higher tendency to misfolding and aggregate formation that is linked to pathology of Parkinson's disease.

Zarranz et al. (2004) identified a Spanish family affected with autosomal dominant inheritance of Parkinson's disease and dementia.

Authors described autopsy based reports citing the presence of Lewy bodies with atrophy of the substantia nigra in the cortical and subcortical areas of brains of patients with Parkinson's disease. Authors confirmed immunoreactive potential of Lewy bodies to ubiquitin and α-synuclein in cortical and subcortical areas (Zarranz et al., 2004).

Zarranz et al. (2004) performed DNA sequencing and analyzed the SNCA gene for the presence of mutation E46K in family members affected with dementia and Parkinson's disease.

The novel E46K led to substitution of glutamic acid (dicarboxylic amino acid) with lysine (basic amino acid) in Zarranz et al. (2004) the α-synuclein protein. The mutation resulted into severe change in the function of protein in affected patients. Authors concluded that dementia with presence of Lewy bodies in brain regions is linked to mutation of E46K mutation in α-synuclein (Zarranz et al., 2004).

Appel-Cresswell et al. (2013) identified and explained the occurrence of a novel mutation, H50Q in the SNCA gene that is related to pathogenesis of Parkinson's disease. It is the novel missense mutation mapped in the exon 4 in gene SNCA. The mutation H50Q in the SNCA gene resulted in the substitution of histidine with glutamine amino acid residue at codon 50 in cases affected with familial PD and dementia. The mutation H50Q led to increased rate of aggregate formation of α-synuclein and altered fibrillization potential of α-synuclein.

In vitro study by Conway et al. (1998) was performed to compare the α-Syn aggregation of wild type and mutant, H50Q α-synuclein. Conway et al. (1998) matched the kinetic properties and the extent of fibrillization of both proteins at 37°C under constant agitation over the concentration range of 5—45 μm (Conway et al., 1998). Authors reported higher fibrillization potential of mutated H50Q α-synuclein than wild-type α-synuclein at had slightly higher fibril formation rates compared with the WT protein at 37°C under constant agitation over the concentration range of 5—45 μm.

Lesageet al. (2013) reported novel G51D mutation in gene SNCA, which cosegregated with the disease. The mutation G51D manifested in the form of early onset of Parkinson's disease, rapid progression of disease, moderate response to levodopa, and reported death of affected patients within a few years. Lesage et al. (2013) identified occurrence of diffuse cytoplasmic inclusions rich in phospho-α-synuclein in superficial layers of the cerebral cortex and entorhinal cortex.

Lesage et al. (2013) concluded that mutation G51D resulted in slower rate of oligomerization of α-synuclein protein than wild-type protein; however, fibrils of mutated α-synuclein were more toxic than wild type protein.

Polymeropoulos et al., 1997) reported a missense point mutation A53T in the SNCA gene leading to the substitution of one amino acid with another in the α-synuclein protein. There was the substitution of alanine with threonine at 53rd position in the protein owing to the substitution of nucleotide base guanine with adenine at position 209 in the gene SNCA.

The mutation A53T is associated with autosomal dominant pattern of inheritance and early-onset form of PD that was initially reported in Italian families and Greek families. Alanine and threonine have almost similar chemical structures; however, the substitution of alanine with threonine have profound effects on the properties of α-synuclein. The mutation A53T might be responsible for stabilization of β-sheets that are implicated in the formation of α-synuclein oligomers and fibrils (Russel, and Eliezer 2001).

Role of mutations in gene DJ1

The gene DJ1 is also referred as gen *PARK7*. The gene is located at chromosome 1p36.23. The gene DJ1 contains eight exons (Nagakubo et al., 1997; Rizzu et al., 2004).

This gene encodes human protein deglycase DJ-1 belonging to C56 peptidase family. DJ-1 is made up of 189 amino acid residues organized into nine α-helices and seven β-strands that exists as a dimer (Nagakubo et al., 1997; Rizzu et al., 2004).

The DJ-1 protein is expressed in all cells including neurons and glia cells in brain (Nagakubo et al., 1997; Rizzu et al., 2004).

The human DJ-1 protein is multifunctional exhibiting its role in regulation of transcription (Takahashi et al., 2001), serves as antioxidant (Yanagida et al., 2009), and acts as redox-dependent chaperone that prevents α-synuclein aggregate formation (Shendelman et al., 2004).

Surplus accumulation of reactive oxygen species and the resultant oxidative stress in reactive astrocytes lead to overexpression of human DJ-1 protein (Yanagida et al., 2009). The human DJ1 protein in reactive astrocytes serves as antioxidant and helps to scavenge the free radicals and minimize the oxidative injury in the astrocytes (Yanagida et al., 2009).

Abou-Sleiman et al. (2003) reported Mutations DJ-1 (PARK7) in two consanguineous families affected with young-onset Parkinson's disease. Authors sequenced and analyzed complete reading frame of the gene DJ-1 in 185 unrelated patients of young-onset Parkinson's disease and in 190 patients of Parkinson's disease with proven diagnosis. Authors reported single homozygous missense mutation and single heterozygous missense mutation in two patients of young-onset Parkinson's disease. Authors concluded that occurrence of pathogenic mutations in DJ-1 gene in young-onset Parkinson's disease with a prevalence of around 1% and further commented that mutations in DJ-1 gene are rarely involved in the onset of young-onset Parkinson's disease. Lorraine et al. (2004) posited that genetic contribution and frequency of mutations in DJ-1 gene in the pathogenesis of early-onset Parkinson's disease is largely unknown. Authors selected a cohort of 89 patients with early-onset Parkinson's disease at mean age of 41.5 ± 7.2 y (Lorraine et al., 2004) at the time of onset of manifestations.

Lorraine et al. (2004) performed gene sequencing and analysis of DJ-1 gene. Missense mutation, A104T, was reported in exon 5 in patient with early-onset Parkinson's disease belonging to Asian ethnicity.

Lorraine et al. (2004) identified variations in DJ1 gene sequence covering three polymorphisms in the 5′ noncoding region of exon 1A/1B, 1 polymorphism in the coding region of exon 5, and 2 polymorphisms in introns IVS1 and IVS5. Lorraine et al. (2004) concluded that DJ-1 mutations are of rare occurrence in the familial early-onset Parkinson's disease and sporadic early-onset Parkinson's disease.

Genetics of Huntington's disease

Mutations in the **HTT** gene (Huntingtin gene) cause Huntington's disease. The HTT gene is positioned on chromosome 4 at its short arm (p) at position 16.3, from nucleotide base pair 3,074,510 to 3,243,960 (USNLM, 2020a,b,c). The HTT gene contains a codon (CAG) at the 5-prime end of HTT gene. It codes for glutamine. The trinucleotide sequence is repeated several times in the HTT gene, and the region is termed as "Trinucleotide repeat." Under normal physiological condition, the CAG repeat count below 27 repeats is considered normal (USNLM, 2020a,b,c).

Huntington's disease exhibits autosomal dominant inheritance. The mutant HTT gene is transmitted from parents to offspring. This leads to increase in size of trinucleotide repeat that is related to onset of clinical features of disease. This feature in the field of genetics is called as "anticipation" (repeated inheritance of a disease makes the symptoms more severe in the next generation with the early onset of disease) (Polito et al., 1996).

The mutation responsible for increase in the CAG repeat count in the coding domain of exon 1 in the HTT gene is implicated in the pathogenesis of Huntington's disease (Goldberg et al., 1994).

The patients suffering from the adult-onset type of Huntington disease generally have CAG repeats between 40 and 50 in the HTT gene; whereas, the patients affected with juvenile form of the Huntington's disease have more than 60 CAG repeats in the HTT gene.

Expansion of the CAG repeat results into rise in polyglutamine that in turn causes synthesis of abnormal intracellular aggregates in the cytoplasm and nucleus that are responsible for cellular dyshomeostasis leading to cell death (Bates 2003).

Semaka et al. (2006) posited that the abnormal expansion of CAG repeat in HTT gene is the basis of disease. The expansion of length of trinucleotide CAG repeat more than 35 repeats might be linked to clinical features of Huntington disease. Moreover, persons having length of trinucleotide CAG repeat with count less than 36 repeats remain affected with the disease. Semaka et al. (2006) reported the occurrence of "intermediate allele" that have CAG repeat count below the abnormal range of CAG repeat; moreover, the "intermediate allele" has the potential to undergoes expansion beyond 35 CAG repeats in a single generation. The children belonging to carriers of "intermediate alleles" have the lowest risk for developing features of HD. The "intermediate allele" is labeled with CAG repeat count ranging between 27 and 35 repeats (Semaka et al., 2006).

Aronin et al. (1995) detected mutant protein in cortical synaptosomes from the brains of patients with both adult onset HD and more severe, juvenile onset form of HD. It showed that expansion of CAG repeat in HTT gene might be involved in the synthesis of mutant protein that was transported along with wild-type protein to the nerve endings.

The size of mutant huntingtin protein in relation to the wild-type huntingtin protein was higher due to expansion of CAG repeats in the brain of HD patients (Aronin et al., 1995). The mutant huntingtin protein was reported from gray and white matter of brains without any difference in expression of mutant protein in affected regions of the brains in HD patients (Aronin et al., 1995). Aronin et al. (1995) concluded that mutation in HTT gene and expansion of CAG repeat might be responsible for a gain of function in the pathology of Huntington's disease.

Huang et al. (1998) described that expansion of CAG repeats beyond a threshold level of nearly 38 repeats renders new property on the protein huntingtin leading to cell- and region-specific neurodegeneration.

Huang et al. (1998) identified a property of self-initiated fibrillar aggregation from amino terminal fragment of mutant huntingtin in the in vitro study. Authors confirmed this property of aggregation of mutant huntingtin in the in vitro study comparable to the mutant huntingtin aggregates involved in the pathogenesis of Huntington's disease.

Huang et al. (1998) reported that mutant huntingtin amino terminus fragment had the ability to coaggregate with normal polyglutamine tract proteins, thus impairing the normal neuronal functions by imparting toxic effects on the normal polyglutamine tract proteins.

Epidemiology of neurodegenerative diseases
Global prevalence of Parkinson's disease

Parkinson''s disease was first described by Dr. James Parkinson in 1817 as a "shaking palsy."

A change in the epidemiological pattern of neurological disorders like Parkinson's disease and parkinsonism, late onset multiple sclerosis and amyotrophic lateral sclerosis has been reported by Rocca (2017, 2018) in high income group countries like North America and Western Europe in the past the 3 decades. There is comparative declining prevalence of stroke and dementia.

Dorsey et al. (2016) from the Global Burden of Diseases (2016) Parkinson's Disease Collaborators described the indirectly estimated alterations in the worldwide prevalence, morbidity and death rates associated with Parkinson's disease in the period between 1996 and 2016.

Dorsey et al. (2016) depicted that prevalence of Parkinson disease has increased from 2.5 million patients in 1990 to 6.1 million patients in 2016. Thereby, global burden of Parkinson's disease has risen to more than two-fold over the past 26 years starting from 1990 to 2016. The authors hypothesized that increase in prevalence of Parkin disease is partially attributed to rise in life expectancy and partially to factors arising from environmental pollution and social changes in the population (Dorsey et al. 2016).

Dorsey et al. (2016) commented that doubling of prevalence of Parkinson disease over the past 26 years can sustain its prevalence in the next 30 years. Dorsey et al. (2016) concluded that if the changing trend in the rise of Parkinson's disease continues in future, there will be more than 12 million Parkinson's disease patients worldwide by the end of 2050 (Dorsey et al., 2016).

Global prevalence of Alzheimer's disease

Alzheimer's disease and vascular dementia are the two most common forms of dementia across the world.

Alzheimer's disease is the most frequent type of dementia constituting about 50%—75% of the total population with dementia. Prevalence of Alzheimer's disease is higher in older age group of population (Braak and Tredici, 2012).

The population-based studies conducted in Europe reported 6.4% prevalence of dementia in population over the age of 65 years, while the 4.4% prevalence is depicted for Alzheimer's disease (Lobo et al., 2000).

The prevalence of 9.7% of Alzheimer's disease was reported in the United States in the population of age more than 70 years (Plassman et al., 2007).

Global prevalence of 3.9% of dementia was reported in the population in the age group of more than 60 years. The regional prevalence of dementia was 1.6% in Africa, 4.0% in China and Western Pacific regions, while its prevalence rises to 4.6% in Latin America, 5.4% in Western Europe, and 6.4% in North America (Ferri et al., 2005).

The prevalence of dementia is expected to increase twofold after a span of 20 years, while the age-related prevalence of Alzheimer's disease is expected to increase by twofold after every 5 years in the age group of 65 years (Lobo et al., 2000; Scazufca et al., 2008).

In the developed countries, the prevalence of 10% of dementia was reported in the population in the age group above 65 years, while more than 33% of the population in the age group above 85 years is inflicted with dementia-related symptoms (von Strauss et al., 1999).

Global prevalence of Huntington's disease

Huntington's disease is the most devastating neurological disorder manifested as chorea, dementia, and other changes in behavior.

The global prevalence of Huntington's disease as based on meta-analysis of 13 studies was 2.71 per 100,000 persons (Pringsheim et al., 2012).

Further, as per the meta-analysis of 11 studies that were conducted in Europe, North American, and Australia reported, the overall prevalence of Huntington's disease was 5.70 per 100,000 persons (Pringsheim et al., 2012).

Additionally, as per meta-analysis of three studies conducted in Asia reported, the overall prevalence of Huntington's disease was 0.40 per 100,000 persons (Pringsheim et al., 2012).

Meta-regression depicted significantly lower prevalence of Huntington's disease in Asia than European, North American, and Australian (Pringsheim et al., 2012).

Age predilection of neurodegenerative diseases

As per the estimate of United Nations (2015), the world population of age 60 years and older is reported to increase more than twofold in the coming 35 years. It is estimated to reach a population of 2.1 billion, mostly coming from the developing countries of the world. There is a drastic improvement in life expectancy. Moreover, the prevalence of neurodegenerative diseases in the aged population is becoming a major problem for the older population (Wyss-Coray, 2016).

Accumulation of abnormal proteins, inclusion bodies, and occurrence of impaired lysosomes in the aging brain are the common findings. Neurologists are uncertain about their involvement in the pathogenesis of neurodegeneration in the brain.

The occurrence of age-dependent assemblies of abnormal proteins and inclusion bodies in the aging brain represents impaired proteostasis (controlling the biosynthesis, folding, trafficking, and biodegradation of proteins inside and outside the cells).

According to Brunk and Terman (2002) organelles like mitochondria and lysosomes in neurons are highly susceptible to age-dependent changes in structure and functions. There is enhanced tendency to accumulation of nondegradable, and polymeric material named as Lipofuscin in the lysosomes (Brunk and Terman, 2002). Possibly, persistent oxidative stress results in oxidation of phagocytized materials and constituents of mitochondria. These damaged structures cannot be removed from the cells. Mitochondria are enlarged that in turn limits the mitochondrial turnover (Brunk and Terman, 2002).

There is decline in the synthesis of ATP molecules and increase in the production of reactive oxygen species in the cells that further enhance the oxidative damage (Brunk and Terman, 2002).

Lipofuscin containing lysosomes become functionally disturbed. As the Lipofuscin is nondegradable, its accumulation continues and leads to decline in degradative activity of lysosomes by inhibiting lysosomal enzymes from directing to autophagosomes (double-membrane vesicles containing cellular garbage to be degraded by autophagy). Well-established nondegradable material is Lipofuscin in the lysosome that is documented biomarker of aging in neurons (Brunk and Terman, 2002).

In aging microglia, myelin debris has been reported to accumulate where it forms insoluble and nondegradable lysosomal inclusions (Safaiyan et al., 2016).

Cellular stress is implicated in the accumulation of stress granules in the neurons. Stress granules are clusters of RNA and protein that could have role in the pathogenesis of Alzheimer's disease and amyotrophic lateral sclerosis and frontotemporal dementia (Ash et al., 2014).

The immune factors and chemokines such as CCL11, CCL2, CCL12, and CCL19 could be involved in the low-grade inflammation associated with aging. This age-dependent inflammation is named as inflammageing (Franceschi and Campisi, 2014).

These inflammatory chemokines in aging brain might arise from the senescent microglia and astrocyte (Coppé et al., 2008).

The process of inflammation has been established to be linked with neurodegeneration for several years (Lucin and Wyss-Coray, 2009). Further, administration of nonsteroidal antiinflammatory drugs has been in practice for many years and was linked with decreased risk of Alzheimer's disease (Vlad et al., 2008).

Conclusion

The pathophysiological characteristics of neurodegenerative diseases are established with the help of observation of behavioral alterations, and reporting by the clinician, which are further subjected to confirmation by clinical examination and diagnostic imaging.

Moreover, the established characteristics might be the epiphenomena of neurodegenerative diseases and additional incomprehensible causal relationships of these diseases remain obscure and untapped.

Understanding of etiopathogenesis of neurodegenerative diseases and adopting a precise therapeutic approach for their management are still evolving. It is true that animal models contribute substantive evidences related to etiopathogenesis of these diseases, furthermore, the human neurodegenerative diseases exhibit heterogeneity in etiology, pathophysiology, and clinical manifestations, while the animal models simulate a fraction of the complex pathophysiology.

Owing to complexities involved in the brain disorders, advances in diagnostic imaging, genotypic, and phenotypic data and the help from the computational methods can be beneficial in understanding the casual relationships in these diseases.

Stem cells technology can reduce the need for the use of animal models in the study of pathophysiology of neurodegenerative diseases.

Pluripotent stem cells have vast potential for self-renewal and can undergo differentiation into all types of cells, including neural cell types. Stem cell technology can be employed to generate human model of neurodegeneration that would be helpful in understanding chemical and molecular events that triggers the clinical manifestations of these diseases.

References

Abeliovich, A., Schmitz, Y., Farinas, I., et al., 2000. Mice lacking alpha-synuclein display functional deficits in the nigrostriatal dopamine system. Neuron 25, 239–252.

Abou-Sleiman, P.M., Healy, D.G., Quinn, N., Lees, A.J., Wood, N.W., 2003. The role of pathogenic DJ-1 mutations in Parkinson's disease. Ann. Neurol. 54 (3), 283–286.

Agrawal, S., Fox, J., Thyagarajan, B., Fox, J.H., 2018. Brain mitochondrial iron accumulates in Huntington's disease, mediates mitochondrial dysfunction, and can be removed pharmacologically. Free Radic. Biol. Med. 120, 317–329.

Ahmad, K., Baig, M.H., Guptac, G.K., Kamal, M.A., Pathaka, N., Choi, I., November 2016. Identification of common therapeutic targets for selected neurodegenerative disorders: an in silico approach. J. Comput. Sci. 17 (Part 1), 292–306.

Ahmed, Z., Asi, Y.T., Lees, A.J., Revesz, T., Holton, J.L., 2013. Identification and quantification of oligodendrocyte precursor cells in multiple system atrophy, progressive supranuclear palsy and Parkinson's disease. Brain Pathol. 23, 263–273.

Alberts, B., Johnson, A., Lewis, J., Raff, M., Roberts, K., Walters, P., 2002. The shape and structure of proteins. In: Molecular Biology of the Cell, fourth ed. Garland Science, New York and London.

Alim, M.A., Hossain, M.S., Arima, K., Takeda, K., Izumiyama, Y., Nakamura, M., Kaji, H., Shinoda, T., Hisanaga, S., Ueda, K., 2002. Tubulin seeds alpha-synuclein fibril formation. J. Biol. Chem. 277 (3), 2112–2117.

Alonso, A.C., Zaidi, T., Grundke-Iqbal, I., Iqbal, K., 1994. Role of abnormally phosphorylated tau in the breakdown of microtubules in Alzheimer disease. Proc. Natl. Acad. Sci. U. S. A. 91 (12), 5562–5566.

Alonso, A.C., Grundke-Iqbal, I., Iqbal, K., 1996. Alzheimer's disease hyperphosphorylated tau sequesters normal tau into tangles of filaments and disassembles microtubules. Nat. Med. 2 (7), 783–787.

Alonso, A.D., Grundke-Iqbal, I., Barra, H.S., Iqbal, K., 1997. Abnormal phosphorylation of tau and the mechanism of Alzheimer neurofibrillary degeneration: sequestration of microtubule-associated proteins 1 and 2 and the disassembly of microtubules by the abnormal tau. Proc. Natl. Acad. Sci. U. S. A. 94 (1), 298−303.

Antoniou, X., Falconi, M., Di Marino, D., Borsello, T., 2011. JNK3 as a therapeutic target for neurodegenerative diseases. J. Alzheim. Dis. 24, 633−642.

Appel-Cresswell, S., Vilarino-Guell, C., Encarnacion, M., Sherman, H., Yu, I., Shah, B., Weir, D., Thompson, C., Szu-Tu, C., Trinh, J., Aasly, J.O., Rajput, A., Rajput, A.H., Jon Stoessl, A., Farrer, M.J., 2013. Alpha-synuclein p.H50Q, a novel pathogenic mutation for Parkinson's disease. Mov. Disord. 28 (6), 811−813.

Arduíno, D.M., Esteves, A.R., Cardoso, S.M., 2011. Mitochondrial fusion/fission, transport and autophagy in Parkinson's disease: when mitochondria get nasty. Parkinsons Dis. 2011, 767230.

Aronin, N., Chase, K., Young, C., Sapp, E., Schwarz, C., Matta, N., Kornreich, R., Landwehrmeyer, B., Bird, E., Beal, M.F., et al., 1995. CAG expansion affects the expression of mutant Huntingtin in the Huntington's disease brain. Neuron 15 (5), 1193−1201.

Arosio, P., Ingrassia, R., Cavadini, P., 2009. Ferritins: a family of molecules for iron storage, antioxidation and more. Biochim. Biophys. Acta Gen. Subj. 1790, 589−599.

Ash, P.E., Vanderweyde, T.E., Youmans, K.L., Apicco, D.J., Wolozin, B., 2014. Pathological stress granules in Alzheimer's disease. Brain Res. 1584, 52−58.

Babior, B.M., 2004. NADPH oxidase. Curr. Opin. Immunol. 16 (1), 42−47.

Bailly, V., Verly, W.G., 1988. Possible roles of beta-elimination and delta-elimination reactions in the repair of DNA containing AP (apurinic/apyrimidinic) sites in mammalian cells. Biochem. J. 253 (2), 553−559.

Balch, W.E., Morimoto, R.I., Dillin, A., Kelly, J.W., 2008. Adapting proteostasis for disease intervention. Science 319 (5865), 916−919.

Barber, R.C., 2012. The genetics of Alzheimer's disease. Scientifica 2012, 246210, 14 pages.

Barghorn, S., Davies, P., Mandelkow, E., 2004. Tau paired helical filaments from Alzheimer's disease brain and assembled in vitro are based on β-structure in the core domain. Biochemistry 43 (6), 1694−1703.

Barnes, D.E., Lindahl, T., 2004. Repair and genetic consequences of endogenous DNA base damage in mammalian cells. Annu. Rev. Genet. 38, 445−476.

Barsoum, M.J., Yuan, H., Gerencser, A.A., Liot, G., Kushnareva, Y., Gräber, S., Kovacs, I., Lee, W.D., Waggoner, J., Cui, J., White, A.D., Bossy, B., Martinou, J.C., Youle, R.J., Lipton, S.A., Ellisman, M.H., Perkins, G.A., Bossy-Wetzel, E., 2006. Nitric oxide-induced mitochondrial fission is regulated by dynamin-related GTPases in neurons. EMBO J. 25 (16), 3900−3911.

Bates, G., 2003. Huntingtin aggregation and toxicity in Huntington's disease. Lancet 361, 1642−1644.

Beal, M.F., 2005. Mitochondria take center stage in aging and neurodegeneration. Ann. Neurol. 58, 495−505.

Beal, M.F., 2007. Mitochondria and neurodegeneration. Novartis Found. Symp. 287, 183−192 discussion 192-6.

Bedford, L., Paine, S., Sheppard, P.W., Mayer, R.J., Roelofs, J., 2010. Assembly, structure, and function of the 26S proteasome. Trends Cell Biol. 20 (7), 391−401.

Belanger, K.A.K., Ameredes, B.T., Boldogh, I., Aguilera-Aguirre, L., 2016. The potential role of 8-oxoguanine DNA glycosylase-driven DNA base excision repair in exercise-induced asthma. Mediat. Inflamm. 2016.

Bence, N.F., Sampat, R.M., Kopito, R.R., 2001. Impairment of the ubiquitin-proteasome system by protein aggregation. Science 292 (5521), 1552−1555.

Bennett, M.C., Bishop, J.F., Leng, Y., et al., 1999. Degradation of alpha-synuclein by proteasome. J. Biol. Chem. 274, 33855−33858.

Bennett, E.J., Bence, N.F., Jayakumar, R., Kopito, R.R., 2005. Global impairment of the ubiquitin-proteasome system by nuclear or cytoplasmic protein aggregates precedes inclusion body formation. Mol. Cell 17 (3), 351−365.

Bett, J.S., Goellner, G.M., Woodman, B., Pratt, G., Rechsteiner, M., Bates, G.P., 2006. Proteasome impairment does not contribute to pathogenesis in R6/2 Huntington's disease mice: exclusion of proteasome activator REGgamma as a therapeutic target. Hum. Mol. Genet. 15 (1), 33−44.

Bhattacharyya, S., Yu, H., Mim, C., Matouschek, A., 2014. Regulated protein turnover: snapshots of the proteasome in action. Nat. Rev. Mol. Cell Biol. 15 (2), 122−133.

Biasiotto, G., Di Lorenzo, D., Archetti, S., Zanella, I., 2016. Iron and neurodegeneration: is ferritinophagy the link? Mol. Neurobiol. 53 (8), 5542−5574.

Bolisetty, S., Jaimes, E.A., 2013. Mitochondria and reactive oxygen species: physiology and pathophysiology. Int. J. Mol. Sci. 14 (3), 6306−6344.

Bond, J.P., Deverin, S.P., Inouye, H., El-Agnaf, O.M.A., Teeter, M.M., Kirschner, D.A., 2003. Assemblies of Alzheimer's peptide Aβ25−35 and Aβ31−35: reverse-turn conformation and side-chain interactions revealed by X-ray diffraction. J. Struct. Biol. 141, 156−170.

Bonda, D.J., Wang, X., Perry, G., Nunomura, A., Tabaton, M., Zhu, X., et al., 2010. Oxidative stress in Alzheimer disease: a possibility for prevention. Neuropharmacology 59 (4−5), 290−294.

Bonifati, V., 2014. Genetics of Parkinson's disease—state of the art, 2013. Park. Relat. Disord. 20 (1), S23−S28.

Boopathi, S., Kolandaivel, P., 2016. Fe(2+) binding on amyloid beta-peptide promotes aggregation. Proteins 84, 1257−1274.

Borchardt, T., Camakaris, J., Cappai, R., Masters, C.L., Beyreuther, K., Multhaup, G., 1999. Copper inhibits beta-amyloid production and stimulates the non-amyloidogenic pathway of amyloid-precursor-protein secretion. Biochem. J. 344 (Pt 2), 461−467.

Braak, H., Del Tredici, K., 2012. Where, when, and in what form does sporadic Alzheimer disease begin? Curr. Opin. Neurol. 25 (6), 708−714.

Brecht, S., Kirchhof, R., Chromik, A., Willesen, M., Nicolaus, T., Raivich, G., et al., 2005. Specific pathophysiological functions of JNK isoforms in the brain. Eur. J. Neurosci. 21, 363−377.

Breslow, J.L., Zannis, V.I., SanGiacomo, T.R., Third, J.L., Tracy, T., Glueck, C.J., 1982. Studies of familial type III hyperlipoproteinemia using as a genetic marker the apoE phenotype E2/2. J. Lipid Res. 23 (8), 1224−1235.

Brunk, U.T., Terman, A., 2002. The mitochondrial-lysosomal axis theory of aging: accumulation of damaged mitochondria as a result of imperfect autophagocytosis. Eur. J. Biochem. 269 (8), 1996−2002.

Buee, L., Bussiere, T., Buee-Scherrer, V., Delacourte, A., Hof, P.R., 2000. Tau protein isoforms, phosphorylation and role in neurodegenerative disorders. Brain Res. Rev. 33, 95−130.16.

Burke, R.E., 2007. Inhibition of mitogen-activated protein kinase and stimulation of Akt kinase signaling pathways: two approaches with therapeutic potential in the treatment of neurodegenerative disease. Pharmacol. Ther. 114, 261−277.

Burré, J., Sharma, M., Tsetsenis, T., Buchman, V., Etherton, M.R., Südhof, T.C., 2010. Alpha-synuclein promotes SNARE-complex assembly in vivo and in vitro. Science 329 (5999), 1663−1667.

Burrows, C.J., Muller, J.G., 1998. Oxidative nucleobase modifications leading to strand scission. Chem. Rev. 98 (3), 1109—1152.

Bussell, R., Eliezer, D., 2003. A structural and functional role for 11-mer repeats in alpha-synuclein and other exchangeable lipid binding proteins. J. Mol. Biol. 329 (4), 763—778.

Cai, Y., An, S.S.A., Kim, S.Y., 2015. Mutations in presenilin 2 and its implications in Alzheimer's disease and other dementia-associated disorders. Clin. Interv. Aging 10, 1163—1172.

Caldecott, K.W., 2008. Single-strand break repair and genetic disease. Nat. Rev. Genet. 9 (8), 619—631.

Canevari, L., Clark, J.B., Bates, T.E., 1999. β-Amyloid fragment 25-35 selectively decreases complex IV activity in isolated mitochondria. FEBS Lett. 457 (1), 131—134.

Chan, D.C., 2006. Mitochondria: dynamic organelles in disease, aging, and development. Cell 125 (7), 1241—1252.

Chapman, J., Vinokurov, S., Achiron, A., Karussis, D.M., Mitosek-Szewczyk, K., Birnbaum, M., Michaelson, D.M., Korczyn, A.D., 2001. APOE genotype is a major predictor of long-term progression of disability in MS. Neurology 56 (3), 312—316.

Chartier-Harlin, M.C., Crawford, F., Houlden, H., Warren, A., Hughes, D., Fidani, L., Goate, A., Rossor, M., Roques, P., Hardy, J., et al., 1991. Early-onset Alzheimer's disease caused by mutations at codon 717 of the beta-amyloid precursor protein gene. Nature 353 (6347), 844—846.

Chen, H., McCaffery, J.M., Chan, D.C., 2007. Mitochondrial fusion protects against neurodegeneration in the cerebellum. Cell 130 (3), 548—562.

Chen, C.Y., Weng, Y.H., Chien, K.Y., Lin, K.J., Yeh, T.H., Cheng, Y.P., Lu, C.S., Wang, H.L., 2012. (G2019S) LRRK2 activates MKK4-JNK pathway and causes degeneration of SN dopaminergic neurons in a transgenic mouse model of PD. Cell Death Differ. 19 (10), 1623—1633.

Chen, P., Bornhorst, J., Neely, M.D., Avila, D.S., 2018. Mechanisms and disease pathogenesis underlying metal-induced oxidative stress. Oxid. Med. Cell. Longevity 2018.

Chia-Ling, K., Pilling, L.C., Atkins, J.L., Masoli, J.A.H., Delgado, J., Kuchel, G.A., Melzer, D., 2020. APOE e4 genotype predicts severe COVID-19 in the UK Biobank community cohort. J. Gerontol. glaa131.

Choi, W.S., Yoon, S.Y., Oh, T.H., Choi, E.J., O'Malley, K.L., Oh, Y.J., 1999. Two distinct mechanisms are involved in 6-hydroxydopamine- and MPP+-induced dopaminergic neuronal cell death: role of caspases, ROS, and JNK. J. Neurosci. Res. 57 (1), 86—94.

Chung, C.Y., Koprich, J.B., Siddiqi, H., Isacson, O., 2009. Dynamic changes in presynaptic and axonal transport proteins combined with striatal neuroinflammation precede dopaminergic neuronal loss in a rat model of AAV alpha-synucleinopathy. J. Neurosci. 29 (11), 3365—3373.

Citron, M., Westaway, D., Xia, W., Carlson, G., Diehl, T., Levesque, G., Johnson-Wood, K., Lee, M., Seubert, P., Davis, A., Kholodenko, D., Motter, R., Sherrington, R., Perry, B., Yao, H., Strome, R., Lieberburg, I., Rommens, J., Kim, S., Schenk, D., Fraser, P., St George Hyslop, P., Selkoe, D.J., 1997. Mutant presenilins of Alzheimer's disease increase production of 42-residue amyloid beta-protein in both transfected cells and transgenic mice. Nat. Med. 3 (1), 67—72.

Civeira, F., Pocoví, M., Cenarro, A., Casao, E., Vilella, E., Joven, J., González, J., Garcia-Otín, A.L., Ordovás, J.M., 1996. Apo E variants in patients with type III hyperlipoproteinemia. Atherosclerosis 127 (2), 273—282.

Claudio, S., Estrada, L.D., 2008. Protein misfolding and neurodegeneration. Neurol. Rev. 65 (2), 184—189.

Cole, G.M., Timiras, P.S., 1987. Ubiquitin-protein conjugates in Alzheimer's lesions. Neurosci. Lett. 79, 207—212.

Conway, K.A., Harper, J.D., Lansbury, P.T., 1998. Accelerated in vitro fibril formation by a mutant alpha-synuclein linked to early-onset Parkinson disease. Nat. Med. 4 (11), 1318—1320.

Cookson, M.R., 2010. The role of leucine-rich repeat kinase 2 (LRRK2) in Parkinson's disease. Nat. Rev. Neurosci. 11, 791—797.

Cooper, J.K., Schilling, G., Peters, M.F., Herring, W.J., Sharp, A.H., Kaminsky, Z., Masone, J., Khan, F.A., Delanoy, M., Borchelt, D.R., Dawson, V.L., Dawson, T.M., Ross, C.A., 1998. Truncated N-terminal fragments of huntingtin with expanded glutamine repeats form nuclear and cytoplasmic aggregates in cell culture. Hum. Mol. Genet. 7 (5), 783—790.

Coppé, J.P., Patil, C.K., Rodier, F., Sun, Y., Muñoz, D.P., Goldstein, J., Nelson, P.S., Desprez, P.Y., Campisi, J., 2008. Senescence-associated secretory phenotypes reveal cell-nonautonomous functions of oncogenic RAS and the p53 tumor suppressor. PLoS Biol. 6 (12), 2853—2868.

Corboy, M.J., Thomas, P.J., Wigley, W.C., 2005. Aggresome formation. Methods Mol. Biol. 301, 305—327.

Corder, E.H., Saunders, A.M., Strittmatter, W.J., Schmechel, D.E., Gaskell, P.C., Small, G.W., Roses, A.D., Haines, J.L., Pericak-Vance, M.A., 1993. Gene dose of apolipoprotein E type 4 allele and the risk of Alzheimer's disease in late onset families. Science 261 (5123), 921—923.

Cox, D.W., Moore, S.D.P., 2002. Copper transporting P-type ATPases and human disease. J. Bioenerg. Biomembr. 34, 333—338.

Crouch, P.J., Blake, R., Duce, J.A., Ciccotosto, G.D., Li, Q.X., Barnham, K.J., Curtain, C.C., Cherny, R.A., Cappai, R., Dyrks, T., Masters, C.L., Trounce, I.A., 2005. Copper-dependent inhibition of human cytochrome c oxidase by a dimeric conformer of amyloid-beta1-42. J. Neurosci. 25 (3), 672—679.

Dauer, W., Przedborski, S., 2003. Parkinson's diseases: mechanisms and models. Neuron 39, 889—909.

Dauer, W., Kholodilov, N., Vila, M., et al., 2002. Resistance of alphasynuclein null mice to the parkinsonian neurotoxin MPTP. Proc. Natl. Acad. Sci. U.S.A. 99, 14524—14529.

De Vos, K.J., Chapman, A.L., Tennant, M.E., Manser, C., Tudor, E.L., Lau, K.F., et al., 2007. Familial amyotrophic lateral sclerosis-linked SOD1 mutants perturb fast axonal transport to reduce axonal mitochondria content. Hum. Mol. Genet. 16 (22), 2720—2728.

Deary, I.J., Whiteman, M.C., Pattie, A., Starr, J.M., Hayward, C., Wright, A.F., Carothers, A., Whalley, L.J., 2002. Cognitive change and the APOE epsilon 4 allele. Nature 418 (6901), 932.

Detmer, S.A., Chan, D.C., 2007. Functions and dysfunctions of mitochondrial dynamics. Nat. Rev. Mol. Cell Biol. 8 (11), 870—879.

Di Fonzo, A., Tassorelli, C., De Mari, M., et al., 2006. Comprehensive analysis of the LRRK2 gene in sixty families with Parkinson's disease. Eur. J. Hum. Genet. 14, 322—331.

Dickson, D.W., Fujishiro, H., DelleDonne, A., Menke, J., Ahmed, Z., Klos, K.J., Josephs, K.A., Frigerio, R., Burnett, M., Parisi, J.E., et al., 2008. Evidence that incidental Lewy body disease is pre-symptomatic Parkinson's disease. Acta Neuropathol. 115, 437—444.

DiFiglia, M., Sapp, E., Chase, K., Schwarz, C., Meloni, A., Young, C., Martin, E., Vonsattel, J.P., Carraway, R., Reeves, S.A., Boyce, F.M., Aronin, N., 1995. Neuron 14, 1075.

Difiglia, M., Sapp, E., Chase, K.O., Davies, S.W., Bates, G.P., Vonsattel, J.P., et al., 1997. Aggregation of huntingtin in neuronal intranuclear inclusions and dystrophic neurites in brain. Science 277, 1990−1993.

Ding, Q., Lewis, J.J., Strum, K.M., Dimayuga, E., Bruce-Keller, A.J., Dunn, J.C., Keller, J.N., 2002. Polyglutamine expansion, protein aggregation, proteasome activity, and neural survival. J. Biol. Chem. 277 (16), 13935−13942.

Dixon, S.J., Lemberg, K.M., Lamprecht, M.R., Skouta, R., Zaitsev, E.M., Gleason, C.E., et al., 2012. Ferroptosis?: an iron-dependent form of nonapoptotic cell death. Cell 149, 1060−1072.

Dizdaroglu, M., Kirkali, G., Jaruga, P., 2008. Formamidopyrimidines in DNA: mechanisms of formation, repair, and biological effects. Free Radic. Biol. Med. 45, 1610−1621.

Dorsey, E.R., Elbaz, A., Nichols, E., Abd-Allah, F., Abdelalim, A., Adsuar, J.C., Ansha, M.G., Brayne, C., Choi, J.-Y.J., et al., 2016. Global, regional, and national burden of Parkinson's disease, 1990−2016: a systematic analysis for the Global Burden of Disease Study 2016. Lancet Neurol. 2018.

Dorszewska, J., Prendecki, M., Oczkowska, A., Dezor, M., Kozubski, W., 2016. Molecular basis of familial and sporadic Alzheimer's disease. Curr. Alzheimer Res. 13 (9), 952−963.

Drolet, R.E., Behrouz, B., Lookingland, K.J., et al., 2004. Mice lacking alpha-synuclein have an attenuated loss of striatal dopamine following prolonged chronic MPTP administration. Neurotoxicology 25, 761−769.

Duff, K., Eckman, C., Zehr, C., et al., 1996. Increased amyloid-β42(43) in brains of mice expressing mutant presenilin 1. Nature 383 (6602), 710−713.

Dugger, B.N., Dickson, D.W., 2016. Pathology of neurodegenerative diseases. Cold Spring Harb. Perspect. Biol. 9, a028035.

Duyao, M., Ambrose, C., Myers, R., Novelletto, A., Persichetti, F., Frontali, M., Folstein, S., Ross, C., Franz, M., Abbott, M., 1993. Trinucleotide repeat length instability and age of onset in Huntington's disease. Nat. Genet. 4 (4), 387−392.

Eichner, J.E., Dunn, S.T., Perveen, G., Thompson, D.M., Stewart, K.E., Stroehla, B.C., 2002. Apolipoprotein E polymorphism and cardiovascular disease: a HuGE review. Am. J. Epidemiol. 155 (6), 487−495.

Eisenberg, D.T., Kuzawa, C.W., Hayes, M.G., 2010. Worldwide allele frequencies of the human apolipoprotein E gene: climate, local adaptations, and evolutionary history. Am. J. Phys. Anthropol. 143 (1), 100−111.

Elias, S., McGuire, J.R., Yu, H., Humbert, S., 2015. Huntingtin is required for epithelial polarity through RAB11A-mediated apical trafficking of PAR3-aPKC. PLoS Biol. 13 (5), e1002142. https://doi.org/10.1371/journal.pbio.1002142.

Engelender, S., 2008. Ubiquitination of alpha-synuclein and autophagy in Parkinson's disease. Autophagy 4 (3), 372−374.

Engelender, S., Sharp, A.H., Colomer, V., Tokito, M.K., Lanahan, A., Worley, P., Holzbaur, E.L., Ross, C.A., 1997. Hum. Mol. Genet. 6, 2205.

Estrada, L.D., Soto, C., 2007. Disrupting β-amyloid aggregation for Alzheimer diseases treatment. Curr. Top. Med. Chem. 7, 115−126.

Everett, J., Collingwood, J.F., Tjendana-Tjhin, V., Brooks, J., Lermyte, F., Plascencia-Villa, G., et al., 2018. Nanoscale synchrotron X-ray speciation of iron and calcium compounds in amyloid plaque cores from Alzheimer's disease subjects. Nanoscale 10, 11782−11796.

Ezquerra, M., Lleó, A., Castellví, M., et al., 2003. A novel mutation in the PSEN2 gene (T430M) associated with variable expression in a family with early-onset Alzheimer disease. Arch. Neurol. 60 (8), 1149−1151.

Farrer, M.J., 2006. Genetics of Parkinson disease: paradigm shifts and future prospects. Nat. Rev. Genet. 7 (4), 306–318.

Ferri, C.P., Prince, M., Brayne, C., Brodaty, H., Fratiglioni, L., Ganguli, M., Hall, K., Hasegawa, K., Hendrie, H., Huang, Y., Jorm, A., Mathers, C., Menezes, P.R., Rimmer, E., Scazufca, M., 2005. Global prevalence of dementia: a Delphi consensus study. Alzheimer's disease international. Lancet 366 (9503), 2112–2117.

Franceschi, C., Campisi, J., 2014. Chronic inflammation (inflammaging) and its potential contribution to age-associated diseases. J. Gerontol. A Biol. Sci. Med. Sci. 69 (Suppl. 1(3)), S4–S9.

Gabbita, S.P., Lovell, M.A., Markesbery, W.R., 1998. Increased nuclear DNA oxidation in the brain in Alzheimer's disease. J. Neurochem. 71 (5), 2034–2040.

Galvin, J.E., Pollack, J., Morris, J.C., 2006. Clinical phenotype of Parkinson disease dementia. Neurology 67, 1605–1611.

Gandhi, S., Abramov, A.Y., 2012. Mechanism of oxidative stress in neurodegeneration. Oxid. Med. Cell. Longevity 2012, 428010 [Apolipoprotein E4 and late-onset Alzheimer's disease].

Gao, M., Monian, P., Pan, Q., Zhang, W., Xiang, J., Jiang, X., 2016. Ferroptosis is an autophagic cell death process. Cell Res. 26, 1021–1032.

Gatz, M., Reynolds, C.A., Fratiglioni, L., et al., 2006. Role of genes and environments for explaining Alzheimer disease. Arch. Gen. Psychiatr. 63 (2), 168–174.

Gelb, D.J., Oliver, E., Gilman, S., 1999. Diagnostic criteria for Parkinson disease. Arch. Neurol. 56, 33–39.

Gibb, W.R.G., Lees, A.J., 1988. The relevance of the Lewy body to the pathogenesis of idiopathic Parkinson's disease. J. Neurol. Neurosurg. Psychiatr. 51, 745–752.

Gibson, G.E., Sheu, K.F., Blass, J.P., 1998. Abnormalities of mitochondrial enzymes in Alzheimer disease. J. Neural. Transm. 105 (8–9), 855–870.

Gilks, W.P., Abou-Sleiman, P.M., Gandhi, S., Jain, S., Singleton, A., Lees, A.J., Shaw, K., Bhatia, K.P., Bonifati, V., Quinn, N.P., Lynch, J., Healy, D.G., Holton, J.L., Revesz, T., Wood, N.W., 2005. A common LRRK2 mutation in idiopathic Parkinson's disease. Lancet 365 (9457), 415–416.

Glabe, C., 2001. Intracellular mechanisms of amyloid accumulation and pathogenesis in Alzheimer's disease. J. Mol. Neurosci. 17, 137–145.

Goldberg, Y.P., Telenius, H., Hayden, M.R., 1994. The molecular genetics of Huntington's disease. Curr. Opin. Neurol. 7 (4), 325–332.

Gomez-Isla, T., West, H.L., Rebeck, G.W., Harr, S.D., Growdon, J.H., Locascio, J.J., Perls, T.T., Lipsitz, L.A., Hyman, B.T., 1996. Clinical and pathological correlates of apolipoprotein E epsilon 4 in Alzheimer's disease. Ann. Neurol. 39 (1), 62–70.

Gomez-Isla, T., Growdon, W.B., McNamara, M., Newell, K., Gomez-Tortosa, E., Hedley-Whyte, E.T., Hyman, B.T., 1999. Clinicopathologic correlates in temporal cortex in dementia with Lewy bodies. Neurology 53, 2003–2009.

Gomez-Tortosa, E., Newell, K., Irizarry, M.C., Albert, M., Growdon, J.H., Hyman, B.T., 1999. Clinical and quantitative pathologic correlates of dementia with Lewy bodies. Neurology 53, 1284–1291.

Gomez-Tortosa, E., Irizarry, M.C., Gomez-Isla, T., Hyman, B.T., 2000. Clinical and neuropathological correlates of dementia with Lewy bodies. Ann. N. Y. Acad. Sci. 920, 9–15.

Goodman, L., 1953. Alzheimer's disease; a clinico-pathologic analysis of twenty-three cases with a theory on pathogenesis. J. Nerv. Ment. Dis. 118, 97–130.

Grollman, A.P., Moriya, M., 1993. Mutagenesis by 8-oxoguanine: an enemy within. Trends Genet. 9, 246–249.

Gutekunst, C.A., Li, S.H., Yi, H., Mulroy, J.S., Kuemmerle, S., Jones, R., Rye, D., Ferrante, R.J., Hersch, S.M., Li, X.J., 1999. Nuclear and neuropil aggregates in Huntington's disease: relationship to neuropathology. J. Neurosci. 19 (7), 2522–2534.

Haass, C., Schlossmacher, M.G., Hung, A.Y., Vigopelfrey, C., Mellon, A., Ostaszewski, B.L., et al., 1992. Amyloid beta-peptide is produced by culturedcells during normal metabolism. Nature 359, 322–325.

Halliday, G., June 17, 2013. Is the Lewy body telling us anything useful about the pathogenesis of Parkinson's disease?—to understand how the study of the Lewy body gives a profound insight into the pathogenesis of Parkinson's disease. In: Proceedings of the 17th International Congress of Parkinson's Disease and Movement Disorders; Sydney, Australia.

Halliwell, B., 1992. Reactive oxygen species and the central nervous system. J. Neurochem. 59 (5), 1609–1623.

Halliwell, B., 2006. Oxidative stress and neurodegeneration: where are we now? J. Neurochem. 97, 1634–1658.

Hamley, I.W., 2012. The amyloid beta peptide: a chemist's perspective. Role in Alzheimer's and fibrillization. Chem. Rev. 112 (10), 5147–5192.

Hansford, R.G., Hogue, B.A., Mildaziene, V., 1997. Dependence of H_2O_2 formation by rat heart mitochondria on substrate availability and donor age. J. Bioenerg. Biomembr. 29 (1), 89–95.

Hardy, J., Allsop, D., 1991. Amyloid deposition as the central event in the aetiology of Alzheimer's disease. Trends Pharmacol. Sci. 12 (10), 383–388.

Hardy, J., Selkoe, D.J., 2002. The amyloid hypothesis of Alzheimer's disease: progress and problems on the road to therapeutics. Science 297 (5580), 353–356.

Harper, P.S., 1996. Naming of syndromes and unethical activities: the case of Hallervorden and Spatz. Lancet 348 (9036), 1224–1225.

Hassa, P.O., Hottiger, M.O., 2008. The diverse biological roles of mammalian PARPS, a small but powerful family of poly-ADP-ribose polymerases. Front. Biosci. 13, 3046–3082.

Hansson Petersen, C.A., Alikhani, N., Behbahani, H., Wiehager, B., Pavlov, P.F., Alafuzoff, I., Leinonen, V., Ito, A., Winblad, B., Glaser, E., Ankarcrona, M., 2008. The amyloid beta-peptide is imported into mitochondria via the TOM import machinery and localized to mitochondrial cristae. Proc. Natl. Acad. Sci. U. S. A. 105 (35), 13145–13150.

Hentze, M.W., Muckenthaler, M.U., Galy, B., Camaschella, C., 2010. Two to tango: regulation of mammalian iron metabolism. Cell 142 (1), 24–38.

Hershko, A., Ciechanover, A., 1998. The ubiquitin system. Annu. Rev. Biochem. 67, 425–479.

Hettiarachchi, N.T., Parker, A., Dallas, M.L., Pennington, K., Hung, C.C., Pearson, H.A., Boyle, J.P., Robinson, P., Peers, C., 2009. alpha-Synuclein modulation of Ca^{2+} signaling in human neuroblastoma (SH-SY5Y) cells. J. Neurochem. 111 (5), 1192–1201.

Hirokawa, N., Niwa, S., Tanaka, Y., 2010. Molecular motors in neurons: transport mechanisms and roles in brain function, development, and disease. Neuron 68, 610–638.

Hoek, K.S., Schlegel, N.C., Eichhoff, O.M., Widmer, D.S., Praetorius, C., Einarsson, S.O., Valgeirsdottir, S., Bergsteinsdottir, K., Schepsky, A., Dummer, R., Steingrimsson, E., 2008. Novel MITF targets identified using a two-step DNA microarray strategy. Pigm. Cell Melanoma Res. 21 (6), 665–676.

Hoffner, G., Kahlem, P., Djian, P., 2002. Perinuclear localization of huntingtin as a consequence of its binding to microtubules through an interaction with beta-tubulin: relevance to Huntington's disease. J. Cell Sci. 115 (Pt 5), 941–948.

Hollenbeck, P.J., Saxton, W.M., 2005. The axonal transport of mitochondria. J. Cell Sci. 118 (Pt 23), 5411—5419.

Hsiao, K., Chapman, P., Nilsen, S., Eckman, C., Harigaya, Y., Younkin, S., Yang, F., Cole, G., 1996. Correlative memory deficits, Abeta elevation, and amyloid plaques in transgenic mice. Science 274 (5284), 99—102.

Hsu, C.H., Chan, D., Wolozin, B., 2010. LRRK2 and the stress response: interaction with MKKs and JNK-interacting proteins. Neurodegener. Dis. 7 (1—3), 68—75.

Huang, C.C., Faber, P.W., Persichetti, F., Mittal, V., Vonsattel, J.P., MacDonald, M.E., Gusella, J.F., 1998. Amyloid formation by mutant huntingtin: threshold, progressivity and recruitment of normal polyglutamine proteins. Somat. Cell Mol. Genet. 24 (4), 217—233.

Hyo-Jin, P., Sang-Soo, K., Seongman, K., Hyangshuk, R., 2009. Intracellular Abeta and C99 aggregates induce mitochondria-dependent cell death in human neuroglioma H4 cells through recruitment of the 20S proteasome subunits. Brain Res. 1273, 1—8.

Ienco, E.C., LoGerfo, A., Carlesi, C., et al., 2011. Oxidative stress treatment for clinical trials in neurodegenerative diseases. J. Alzheimer's Dis. 24 (Suppl. 2), 111—126.

Imarisio, S., Carmichael, J., Korolchuk, V., Chen, C.W., Saiki, S., Rose, C., Krishna, G., Davies, J.E., Ttofi, E., Underwood, B.R., Rubinsztein, D.C., 2008. Huntington's disease: from pathology and genetics to potential therapies. Biochem. J. 412 (2), 191—209.

Intlekofer, K.A., Cotman, C.W., 2013. Exercise counteracts declining hippocampal function in aging and Alzheimer's disease. Neurobiol. Dis. 57, 47—55.

Ip, Y.T., Davis, R.J., 1998. Signal transduction by the c-Jun N-terminal kinase (JNK)–from inflammation to development. Curr. Opin. Cell Biol. 10 (2), 205—219.

Ishihara, N., Nomura, M., Jofuku, A., Kato, H., Suzuki, S.O., Masuda, K., et al., 2009. Mitochondrial fission factor Drp1 is essential for embryonic development and synapse formation in mice. Nat. Cell Biol. 11 (8), 958—966.

Iwai, A., Masliah, E., Yoshimoto, M., Ge, N., Flanagan, L., de Silva, H.A., et al., 1995. The precursor protein of non-A beta component of Alzheimer's disease amyloid is a presynaptic protein of the central nervous system. Neuron 14 (2), 467—475.

Izumi, T., Wiederhold, L.R., Roy, G., Roy, R., Jaiswal, A., Bhakat, K.K., Mitra, S., Hazra, T.K., 2003. Mammalian DNA base excision repair proteins: their interactions and role in repair of oxidative DNA damage. Toxicology 193, 43—65.

Jana, N.R., Zemskov, E.A., Wang, G., Nukina, N., 2001. Altered proteasomal function due to the expression of polyglutamine-expanded truncated N-terminal huntingtin induces apoptosis by caspase activation through mitochondrial cytochrome c release. Hum. Mol. Genet. 10 (10), 1049—1059.

Jayadev, S., Leverenz, J.B., Steinbart, E., Stahl, J., Klunk, W., Yu, C.-E., Bird, T.D., 2010. Alzheimer's disease phenotypes and genotypes associated with mutations in presenilin 2. Brain 133, 1143—1154.

Jellinger, K.A., 2007. More frequent Lewy bodies but less frequent Alzheimer-type lesions in multiple system atrophy as compared to age-matched control brains. Acta Neuropathol. 114 (3), 299—303.

Johri, A., Beal, F.M., 2012. Mitochondrial dysfunction in neurodegenerative diseases. J. Pharmacol. Exp. Therapeut. 342 (3), 619—630.

Kadenbach, B., Huttemann, M., 2015. The subunit composition and function of mammalian cytochrome c oxidase. Mitochondrion 24, 64—76.

Kandel, E.R., Schwartz, J.H., Jessell, T.M., 2000. Principles of Neural Science, fourth ed. McGraw-Hill, New York, pp. 19—20.

Katsuno, M., Banno, H., Suzuki, K., Takeuchi, Y., Kawashima, M., Tanaka, F., Adachi, H., Sobue, G., 2008. Molecular genetics and biomarkers of polyglutamine diseases. Curr. Mol. Med. 8 (3), 221–234.

Kessler, H., Pajonk, F.G., Supprian, T., Falkai, P., Multhaup, G., Bayer, T.A., 2005. The role of copper in the pathophysiology of Alzheimer's disease. Nervenarzt 76, 581–585.

Kisselev, A.F., Akopian, T.N., Woo, K.M., Goldberg, A.L., 1999. The sizes of peptides generated from protein by mammalian 26 and 20 S proteasomes. Implications for understanding the degradative mechanism and antigen presentation. J. Biol. Chem. 274 (6), 3363–3371.

Kolarova, M., Garcia-Sierra, F., Bartos, A., Ricny, J., Ripova, D., 2012. Structure and pathology of tau protein in Alzheimer disease. Int. J. Alzheimer's Dis. 41.

Krokan, H.E., Bjoras, M., 2013. Base excision repair. Cold Spring Harb. Perspect. Biol. 5, a012583.

Krüger, R., Kuhn, W., Müller, T., Woitalla, D., Graeber, M., Kösel, S., Przuntek, H., Epplen, J.T., Schöls, L., Riess, O., 1998. Ala30Pro mutation in the gene encoding alpha-synuclein in Parkinson's disease. Nat. Genet. 18 (2), 106–108.

Kruszewski, M., 2003. Labile iron pool: the main determinant of cellular response to oxidative stress. Mutat. Res. 531, 81–92.

Kumar, A., Pottiboyina, V., Sevilla, M.D., 2011. Hydroxyl radical (OH•) reaction with guanine in an aqueous environment: a DFT study. J. Phys. Chem. B 115 (50), 15129–15137.

Kumari, U., Tan, E.K., 2009. LRRK2 in Parkinson's disease: genetic and clinical studies from patients. FEBS J. 276 (22), 6455–6463.

Kuzuhara, S., Mori, H., Izumiyama, N., Yoshimura, M., Ihara, Y., 1988. Lewy bodies are ubiquitinated. A light and electron microscopic immunocytochemical study. Acta Neuropathol. 75, 345–353.

Lam, Y.A., Pickart, C.M., Alban, A., Landon, M., Jamieson, C., Ramage, R., Mayer, R.J., Layfield, R., 2000. Inhibition of the ubiquitin-proteasome system in Alzheimer's disease. Proc. Natl. Acad. Sci. U.S.A. 97, 9902–9906.

Lambeth, J.D., 2004. NOX enzymes and the biology of reactive oxygen. Nat. Rev. Immunol. 4 (3), 181–189.

Lane, D.J.R., Ayton, S., Bush, A.I., 2018. Iron and Alzheimer's disease: an update on emerging mechanisms. J. Alzheimer's Dis. 64, S379–S395.

Lee, S.F., Shah, S., Li, H., Yu, C., Han, W., Yu, G., 2002. Mammalian APH-1 interacts with presenilin and nicastrin and is required for intramembrane proteolysis of amyloid-β precursor protein and Notch. J. Biol. Chem. 277 (47), 45013–45019.

Lee, H.J., Khoshaghideh, F., Lee, S., Lee, S.J., 2006. Impairment of microtubule-dependent trafficking by overexpression of alpha-synuclein. Eur. J. Neurosci. 24 (11), 3153–3162.

Lee, B.D., Shin, J.-H., VanKampen, J., Petrucelli, L., West, A.B., Ko, H.S., Lee, Y., Maguire-Zeiss, K.A., Bowers, W.J., Federoff, H.J., et al., 2010. Inhibitors of leucine rich repeat kinase 2 (LRRK2) protect against LRRK2-models of Parkinson's disease. Nat. Med. 16, 998–1000.

Leonidas, S., 2012. α-Synuclein in Parkinson's disease. Cold Spring Harb. Perspect. Med. 2 (2), a009399.

Lesage, S., Anheim, M., Letournel, F., Bousset, L., Honoré, A., Rozas, N., Pieri, L., Madiona, K., Dürr, A., Melki, R., Verny, C., Brice, A., French Parkinson's Disease Genetics Study Group, 2013. G51D α-synuclein mutation causes a novel parkinsonian-pyramidal syndrome. Ann. Neurol. 73 (4), 459–471.

Levy-Lahad, E., Wasco, W., Poorkaj, P., Romano, D.M., Oshima, J., Pettingell, W.H., Yu, C.E., Jondro, P.D., Schmidt, S.D., Wang, K., 1995. Candidate gene for the chromosome 1 familial Alzheimer's disease locus. Science 269 (5226), 973—977.

Li, S.H., Schilling, G., Young 3rd, W.S., Li, X.J., Margolis, R.L., Stine, O.C., Wagster, M.V., Abbott, M.H., Franz, M.L., Ranen, N.G., Folstein, S.E., Hedreen, J.C., Ross, C.A., 1993. Neuron 11, 985.

Li, S.H., Gutekunst, C.-A., Hersch, S.M., Li, X.J., 1998. J. Neurosci. 18, 1261.

Li, Z., Okamoto, K., Hayashi, Y., Sheng, M., 2004. The importance of dendritic mitochondria in the morphogenesis and plasticity of spines and synapses. Cell 119, 873—887.

Licker, V., Kövari, E., Hochstrasser, D.F., Burkhard, P.R., 2009. Proteomics in human Parkinson's diseases research. J. Proteomics 73, 10—29.

Lindahl, T., Barnes, D.E., 2000. Repair of endogenous DNA damage. Cold Spring Harbor Symp. Quant. Biol. 65, 127—133.

Liu, C.C., Liu, C.C., Kanekiyo, T., Xu, H., Bu, G., 2013. Apolipoprotein E and Alzheimer disease: risk, mechanisms and therapy. Nat. Rev. Neurol. 9 (2), 106—118.

Lobo, A., Launer, L.J., Fratiglioni, L., Andersen, K., Di Carlo, A., Breteler, M.M., Copeland, J.R., Dartigues, J.F., Jagger, C., Martinez-Lage, J., Soininen, H., Hofman, A., 2000. Prevalence of dementia and major subtypes in Europe: a collaborative study of population-based cohorts. Neurology 54, S4—S9.

Lorraine, N.C., Afridi, S., Mejia-Santana, H., Harris, J., Louis, E.D., Cote, L.J., Andrews, H., Singleton, A., Wavrant De-Vrieze, F., Hardy, J., Mayeux, R., Fahn, S., Waters, C., Ford, B., Frucht, S., Ottman, R., Marder, K., 2004. Analysis of an early-onset Parkinson's disease cohort for DJ-1 mutations. Mov. Disord. 19 (7), 796—800.

Love, S., Barber, R., Wilcock, G.K., 1999. Increased poly(ADP-ribosyl)ation of nuclear proteins in Alzheimer's disease. Brain 122, 247—253.

Lu, Y., Prudent, M., Fauvet, B., Lashuel, H.A., Girault, H.H., 2011. Phosphorylation of α-synuclein at Y125 and S129 alters its metal binding properties: implications for understanding the role of α-synuclein in the pathogenesis of Parkinson's disease and related disorders. ACS Chem. Neurosci. 2, 667—675.

Lucin, K.M., Wyss-Coray, T., 2009. Immune activation in brain aging and neurodegeneration: too much or too little? Neuron 64 (1), 110—122.

Lustbader, J.W., Cirilli, M., Lin, C., et al., 2004. ABAD directly links Aβ to mitochondrial toxicity in Alzheimer's disease. Science 304 (5669), 448—452.

Luthi-Carter, R., Strand, A., Peters, N.L., Solano, S.M., Hollingsworth, Z.R., Menon, A.S., Frey, A.S., Spektor, B.S., Penney, E.B., Schilling, G., Ross, C.A., Borchelt, D.R., Tapscott, S.J., Young, A.B., Cha, J.H., Olson, J.M., 2000. Decreased expression of striatal signaling genes in a mouse model of Huntington's disease. Hum. Mol. Genet. 9 (9), 1259—1271.

Mancias, J.D., Wang, X., Gygi, S.P., Harper, J.W., Kimmelman, A.C., 2014. Quantitative proteomics identifies NCOA4 as the cargo receptor mediating ferritinophagy. Nature 509 (7498), 105—109.

Manczak, M., Anekonda, T.S., Henson, E., Park, B.S., Quinn, J., Reddy, P.H., 2006. Mitochondria are a direct site of A beta accumulation in Alzheimer's disease neurons: implications for free radical generation and oxidative damage in disease progression. Hum. Mol. Genet. 15 (9), 1437—1449.

Manczak, M., Calkins, M.J., Reddy, P.H., 2011. Impaired mitochondrial dynamics and abnormal interaction of amyloid beta with mitochondrial protein Drp1 in neurons from patients with Alzheimer's disease: implications for neuronal damage. Hum. Mol. Genet. 20 (13), 2495—2509.

Mandavilli, B.S., Boldogh, I., Van Houten, B., 2005. 3-Nitropropionic acid-induced hydrogen peroxide, mitochondrial DNA damage, and cell death are attenuated by Bcl-2 overexpression in PC12 cells. Mol. Brain Res. 133 (2), 215−223.

Mandemakers, W., Morais, V.A., De Strooper, B., 2007. A cell biological perspective on mitochondrial dysfunction in Parkinson disease and other neurodegenerative diseases. J. Cell Sci. 120 (10), 1707−1716.

Margolin, Y., Cloutier, J.-F., Shafirovich, V., Geacintov, N.E., Dedon, P.C., 2006. Paradoxical hotspots for guanine oxidation by a chemical mediator of inflammation. Nat. Chem. Biol. 2 (7), 365−366.

Markesbery, W.R., Lovell, M.A., 1998. Four -hydroxynonenal, a product of lipid peroxidation, is increased in the brain in Alzheimer's disease. Neurobiol. Aging 19 (1), 33−36.

Marvian, A.T., Koss, D.J., Aliakbari, F., Morshedi, D., Outeiro, T.F., 2019. In vitro models of synucleinopathies: informing on molecular mechanisms and protective strategies. J. Neurochem. 150 (5), 535−565.

Mata, I.F., Wedemeyer, W.J., Farrer, M.J., Taylor, J.P., Gallo, K.A., 2006. LRRK2 in Parkinson's disease: protein domains and functional insights. Trends Neurosci. 29, 286−293.

McCarron, M.O., Delong, D., Alberts, M.J., 1999. APOE genotype as a risk factor for ischemic cerebrovascular disease: a meta-analysis. Neurology 53 (6), 1308−1311.

McNaught, K.S.P., Olanow, C.W., 2006. Protein aggregation in the pathogenesis of familial and sporadic Parkinson's diseases. Neurobiol. Aging 27, 530−545.

McNaught, K.S.P., Björklund, L.M., Belizaire, R., Isacson, O., Jenner, P., Olanow, C.W., 2002. Proteasome inhibition causes nigral degeneration with inclusion bodies in rats. Neuroreport 13 (11), 1437−1441.

McNaught, K.S., Shashidharan, P., Perl, D.P., et al., 2002. Aggresomerelated biogenesis of Lewy bodies. Eur. J. Neurosci. 16, 2136−2148.

Medline Plus, U.S. National Library of Medicine, 2020. SNCA Gene. Available at: https://medlineplus.gov/genetics/gene/snca/#conditions.

Menalled, L.B., Chesselet, M.F., 2002. Mouse models of Huntington's disease. Trends Pharmacol. Sci. 23 (32−39).

Mori, H., Kondo, J., Ihara, Y., 1987. Ubiquitin is a component of paired helical filaments in Alzheimer's disease. Science 235, 1641−1644.

Mullan, M., Crawford, F., Axelman, K., Houlden, H., Lilius, L., Winblad, B., Lannfelt, L., 1992. A pathogenic mutation for probable Alzheimer's disease in the APP gene at the N-terminus of beta-amyloid. Nat. Genet. 1 (5), 345−347.

Murrell, J.R., Hake, A.M., Quaid, K.A., Farlow, M.R., Ghetti, B., 2000. Early-onset Alzheimer disease caused by a new mutation (V717L) in the amyloid precursor protein gene. Arch. Neurol. 57 (6), 885−887.

Myers, R.H., 2004. Huntington's disease genetics. NeuroRx 1 (2), 255−262.

Nagakubo, D., Taira, T., Kitaura, H., et al., 1997. DJ-1, a novel oncogene which transforms mouse NIH3T3 cells in cooperation with ras. Biochem. Biophys. Res. Commun. 231 (2), 509−513.

Nasir, J., Floresco, S.B., O'Kusky, J.R., Diewert, V.M., Richman, J.M., Zeisler, J., Borowski, A., Marth, J.D., Phillips, A.G., Hayden, M.R., 1995. Targeted disruption of the Huntington's disease gene results in embryonic lethality and behavioral and morphological changes in heterozygotes. Cell 81 (5), 811−823.

National Center for Biotechnology Information, 2020. SNCA Gene. U.S. National Library of Medicine 8600 Rockville Pike, Bethesda MD, 20894 USA. Available at: https://www.ncbi.nlm.nih.gov/gene/6622.

Ndayisaba, A., Kaindlstorfer, C., Wenning, G.K., 2019. Iron in neurodegeneration - cause or consequence? Front. Neurosci. 13, 180. https://doi.org/10.3389/fnins.2019.00180.

Nilsberth, C., Westlind-Danielsson, A., Eckman, C.B., Condron, M.M., Axelman, K., Forsell, C., et al., 2001. The 'Arctic' APP mutation (E693G) causes Alzheimer's disease by enhanced Abeta protofibril formation. Nat. Neurosci. 4 (9), 887–893.

Nunan, J., Small, D.H., 2000. Regulation of APP cleavage by alpha-, beta- and gamma-secretases. FEBS Lett. 483, 6–10.

Nunomura, A., Perry, G., Aliev, G., Hirai, K., Takeda, A., Balraj, E.K., et al., 2001. Oxidative damage is the earliest event in Alzheimer disease. J. Neuropathol. Exp. Neurol. 60 (8), 759–767.

Olanow, C.W., Perl, D.P., DeMartino, G.N., McNaught, K.S., 2004. Lewy-body formation is an aggresome-related process: a hypothesis. Lancet Neurol. 3, 496–503.

Orr, H.T., Zoghbi, H.Y., 2007. Trinucleotide repeat disorders. Annu. Rev. Neurosci. 30 (1), 575–621.

Patten, D.A., Germain, M., Kelly, M.A., Slack, R.S., 2010. Reactive oxygen species: stuck in the middle of neurodegeneration. J. Alzheimer's Dis. 20 (Suppl. 2), S357–S367.

Paula van, T., Femke, M.S. de V., Schuurman, K.G., Dantuma, N.P., Fischer, D.F., van Leeuwen, F.W., Hol, E.M., 2007. Dose-dependent inhibition of proteasome activity by a mutant ubiquitin associated with neurodegenerative disease. J. Cell Sci. 120, 1615–1623.

Peng, Y., Wang, C., Xu, H.H., Liu, Y.N., Zhou, F., 2010. Binding of alpha-synuclein with Fe(III) and with Fe(II) and biological implications of the resultant complexes. J. Inorg. Biochem. 104, 365–370.

Pereira, C., Miguel Martins, L., Saraiva, L., 2014. LRRK2, but not pathogenic mutants, protects against H2O 2 stress depending on mitochondrial function and endocytosis in a yeast model. Biochim. Biophys. Acta Gen. Subj. 1840, 2025–2031.

Perry, G., Friedman, R., Shaw, G., Chau, V., 1987. Ubiquitin is detected in neurofibrillary tangles and senile plaque neurites of Alzheimer disease brains. Proc. Natl. Acad. Sci. U. S. A. 84 (9), 3033–3036.

Pfister, K.K., Shah, P.R., Hummerich, H., Russ, A., Cotton, J., Annuar, A.A., King, S.M., Fisher, E.M., 2006. Genetic analysis of the cytoplasmic dynein subunit families. PLoS Genet. 2 (1), e1. https://doi.org/10.1371/journal.pgen.0020001.

Phillips, M.C., 2014. Apolipoprotein E isoforms and lipoprotein metabolism. IUBMB Life 66 (9), 616–623.

Pickart, C.M., 2001. Mechanisms underlying ubiquitination. Annu. Rev. Biochem. 70, 503–533.

Pickett, E.K., Rose, J., McCrory, C., McKenzie, C.A., King, D., Smith, C., et al., 2018. Region-specific depletion of synaptic mitochondria in the brains of patients with Alzheimer's disease. Acta Neuropathol. 136 (5), 747–757.

Plascencia-Villa, G., Ponce, A., Collingwood, J.F., Arellano-Jiménez, M.J., Zhu, X., Rogers, J.T., et al., 2016. High-resolution analytical imaging and electron holography of magnetite particles in amyloid cores of Alzheimer's disease. Sci. Rep. 6, 24873.

Plassman, B.L., Langa, K.M., Fisher, G.G., Heeringa, S.G., Weir, D.R., Ofstedal, M.B., Burke, J.R., Hurd, M.D., Potter, G.G., Rodgers, W.L., Steffens, D.C., Willis, R.J., Wallace, R.B., 2007. Prevalence of dementia in the United States: the aging, demographics, and memory study. Neuroepidemiology 29, 125–132.

Polito, J.M., Mendeloff, A.I., Harris, M.L., Bayless, T.M., Childs, B., Rees, R.C., 1996. Preliminary evidence for genetic anticipation in Crohn's disease. Lancet 347 (9004), 798–800.

Polymeropoulos, M.H., Lavedan, C., Leroy, E., Ide, S.E., Dehejia, A., Dutra, A., Pike, B., Root, H., Rubenstein, J., Boyer, R., Stenroos, E.S., Chandrasekharappa, S., Athanassiadou, A., Papapetropoulos, T., Johnson, W.G., Lazzarini, A.M., Duvoisin, R.C., Di Iorio, G., Golbe, L.I., Nussbaum, R.L., 1997. Mutation in the alpha-synuclein gene . identified in families with Parkinson's disease. Science 276 (5321), 2045−2047.

Power, J.H.T., Barnes, O.L., Chegini, F., 2017. Lewy bodies and the mechanisms of neuronal cell death in Parkinson's disease and dementia with lewy bodies. Brain Pathol. 27 (1), 3−12.

Price, D.L., Tanzi, R.E., Borchelt, D.R., Sisodia, S.S., 1998. Alzheimer's disease: genetic studies and transgenic models. Annu. Rev. Genet. 32, 461−493.

Priller, C., Bauer, T., Mitteregger, G., Krebs, B., Kretzschmar, H.A., Herms, J., 2006. Synapse formation and function is modulated by the amyloid precursor protein. J. Neurosci. 26, 7212−7221.

Pringsheim, T., Wiltshire, K., Day, L., Dykeman, J., Steeves, T., Jette, N., 2012. The incidence and prevalence of Huntington's disease: a systematic review and meta-analysis. Mov. Disord. 27 (9), 1083−1091.

Queralt, R., Ezquerra, M., Lleó, A., Castellví, M., Gelpí, J., Ferrer, I., Acarín, N., Pasarín, L., Blesa, R., Oliva, R., 2002. A novel mutation (V89L) in the presenilin 1 gene in a family with early onset Alzheimer's disease and marked behavioural disturbances. J. Neurol. Neurosurg. Psychiatry 72 (2), 266−269.

Rademakers, R., Cruts, M., Van Broeckhoven, C., 2003. Genetics of early-onset Alzheimer dementia. Sci. World J. 3, 497−519.

Radi, R., Beckman, J.S., Bush, K.M., Freeman, B.A., 1991. Peroxynitrite oxidation of sulfhydryls. The cytotoxic potential of superoxide and nitric oxide. J. Biol. Chem. 266 (7), 4244−4250.

Rall Jr., S.C., Weisgraber, K.H., Mahley, R.W., 1982. Human apolipoprotein E. The complete amino acid sequence. J. Biol. Chem. 257 (8), 4171−4178.

Rebeck, G.W., Reiter, J.S., Strickland, D.K., Hyman, B.T., 1993. Apolipoprotein E in sporadic Alzheimer's disease: allelic variation and receptor interactions. Neuron 11 (4), 575−580.

Reddy, P.H., 2008. Mitochondrial medicine for aging and neurodegenerative diseases. Neuro-Molecular Med. 10, 291−315.

Reddy, P.H., Shirendeb, U.P., 2011. Mutant huntingtin, abnormal mitochondrial dynamics, defective axonal transport of mitochondria, and selective synaptic degeneration in Huntington's disease. Biochim. Biophys. Acta 1822 (2), 101−110.

Ries, V., Silva, R.M., Oo, T.F., Cheng, H.C., Rzhetskaya, M., Kholodilov, N., et al., 2008. JNK2 and JNK3 combined are essential for apoptosis in dopamine neurons of the substantia nigra, but are not required for axon degeneration. J. Neurochem. 107, 1578−1588.

Rizzu, P., Hinkle, D.A., Zhukareva, V., et al., 2004. DJ-1 colocalizes with tau inclusions: a link between parkinsonism and dementia. Ann. Neurol. 55 (1), 113−118.

Rocca, W.A., 2017. Time, sex, gender, history, and dementia. Alzheimer Dis. Assoc. Disord. 31, 76−79. 3.

Rocca, W.A., 2018. The future burden of Parkinson's disease. Mov. Disord. 33, 8−9.

Rong, J., McGuire, J.R., Fang, Z.H., Sheng, G., Shin, J.Y., Li, S.H., Li, X.J., 2006. J. Neurosci. 26, 6019.

Roos, P.M., Vesterberg, O., Nordberg, M., 2006. Metals in motor neuron diseases. Exp. Biol. Med. 231, 1481−1487.

Ross, R.A.C., 2010. Huntington's disease: a clinical review. J. Rare Dis. 5 (40), 1−8.66.

Russel, R., Eliezer, D., 2001. Residual structure and dynamics in Parkinson's disease-associated mutants of alpha-synuclein. J. Biol. Chem. 276 (49), 45996−46003.

Safaiyan, S., Kannaiyan, N., Snaidero, N., Brioschi, S., Biber, K., Yona, S., Edinger, A.L., Jung, S., Rossner, M.J., Simons, M., 2016. Age-related myelin degradation burdens the clearance function of microglia during aging. Nat. Neurosci. 19 (8), 995−998.

Safia, S.S., Ejazul, H., Mir, S.S., 2012. Neurodegenerative Diseases: multifactorial conformational diseases and their therapeutic interventions. J. Neurodegener. Dis. 2013, 563481. https://doi.org/10.1155/2013/563481, 8 pages.

Samura, E., Shoji, M., Kawarabayashi, T., Sasaki, A., Matsubara, E., Murakami, T., Wuhua, X., Tamura, S., Ikeda, M., Ishiguro, K., Saido, T.C., Westaway, D., St George Hyslop, P., Harigaya, Y., Abe, K., 2006. Enhanced accumulation of tau in doubly transgenic mice expressing mutant betaAPP and presenilin-1. Brain Res. 1094 (1), 192−199.

Sandal, M., Valle, F., Tessari, I., Mammi, S., Bergantino, E., Musiani, F., et al., 2008. Conformational equilibria in monomeric alpha-synuclein at the single-molecule level. PLoS Biol. 6 (1), e6.

Saxton, W.M., Hollenbeck, P.J., 2012. The axonal transport of mitochondria. J. Cell Sci. 125, 2095−2104.

Scazufca, M., Menezes, P.R., Vallada, H.P., Crepaldi, A.L., Pastor-Valero, M., Coutinho, L.M., Di Rienzo, V.D., Almeida, O.P., 2008. High prevalence of dementia among older adults from poor socioeconomic backgrounds in São Paulo, Brazil. Int. Psychogeriatr. 20 (2), 394−405.

Schellenberg, G.D., Bird, T.D., Wijsman, E.M., Orr, H.T., Anderson, L., Nemens, E., White, J.A., Bonnycastle, L., Weber, J.L., Alonso, M.E., Nov, 1992. Genetic linkage evidence for a familial Alzheimer's seasesease locus on chromosome 14. Science 258 (5082), 668−671.

Scherzinger, E., Sittler, A., Schweiger, K., Heiser, V., Lurz, R., Hasenbank, R., Bates, G.P., Lehrach, H., 1999. Wanker EE Self-assembly of polyglutamine-containing huntingtin fragments into amyloid-like fibrils: implications for Huntington's disease pathology. Proc. Natl. Acad. Sci. U. S. A. 96 (8), 4604−4609.

Schulz-Schaeffer, W.J., 2012. Neurodegeneration in Parkinson disease: moving Lewy bodies out of focus. Neurology 79, 2298−2299.

Schuman, E.M., Madison, D.V., 1991. A requirement for the intercellular messenger nitric oxide in long-term potentiation. Science 254 (5037), 1503−1506.

Selkoe, D.J., 1991. The molecular pathology of Alzheimer's disease. Neuron 6 (4), 487−498.

Semaka, A., Creighton, S., Warby, S., et al., 2006. Predictive testing for Huntington disease: interpretation and significance of intermediate alleles. Clin. Genet. 70, 283−294.

Serrano, F., Kolluri, N.S., Wientjes, F.B., Card, J.P., Klann, E., 2003. NADPH oxidase immunoreactivity in the mouse brain. Brain Res. 988 (1−2), 193−198.

Shen, J., 2004. Protein kinases linked to the pathogenesis of Parkinson's disease. Neuron 44 (4), 575−577.

Shendelman, S., Jonason, A., Martinat, C., Leete, T., Abeliovich, A., 2004. DJ-1 Is a redox-dependent molecular chaperone that inhibits α-synuclein aggregate formation. PLoS Biol. 2 (11) article e362.

Sheng, Z.H., Cai, Q., 2012. Mitochondrial transport in neurons: impact on synaptic homeostasis and neurodegeneration. Nat. Rev. Neurosci. 13, 77−93.

Sheng, D., Qu, D., Kwok, K.H.H., Ng, S.S., Lim, A.Y.M., Aw, S.S., Lee, C.W.H., Sung, W.K., Tan, E.K., Lufkin, T., et al., 2010. Deletion of the WD40 domain of LRRK2 in zebrafish causes parkinsonism-like loss of neurons and locomotive defect. PLoS Genet. 6.

Silberberg, D., Anand, N., Michels, K., et al., 2015. Brain and other nervous system disorders across the lifespan — global challenges and opportunities. Nature 527. S151−S154.

Silvestri, L., Camaschella, C., 2008. A potential pathogenetic role of iron in Alzheimer's disease. J. Cell Mol. Med. 12, 1548–1550.

Singh, P.P., Singh, M., Mastana, S.S., 2006. APOE distribution in world populations with new data from India and the UK. Ann. Hum. Biol. 33 (3), 279–308.

Smith, C.D., Carney, J.M., Starke-Reed, P.E., Oliver, C.N., Stadtman, E.R., Floyd, R.A., Markesbery, W.R., 1991. Excess brain protein oxidation and enzyme dysfunction in normal aging and in Alzheimer disease. Proc. Natl. Acad. Sci. U. S. A. 88 (23), 10540–10543.

Smith, M.A., Rottkamp, C.A., Nunomura, A., Raina, A.K., Perry, G., 2000. Oxidative stress in Alzheimer's disease. Biochim. Biophys. Acta 1502 (1), 139–144.

Snider, B.J., Norton, J., Coats, M.A., Chakraverty, S., Hou, C.E., Jervis, R., Lendon, C.L., Goate, A.M., McKeel Jr., D.W., Morris, J.C., 2005. Novel presenilin 1 mutation (S170F) causing Alzheimer disease with Lewy bodies in the third decade of life. Arch. Neurol. 62 (12), 1821–1830.

Soos, J., Engelhardt, J.I., Siklos, L., Havas, L., Majtenyi, K., 2004. The expression of PARP. NF-kappa B and parvalbumin is increased in Parkinson disease. Neuroreport 15, 1715–1718.

Spillantini, M.G., Schmidt, M.L., Lee, V.M., Trojanowski, J.Q., Jakes, R., Goedert, M., 1997. Alpha-synuclein in Lewy bodies. Nature 388 (6645), 839–840.

Spillantini, M.G., Crowther, R.A., Jakes, R., Hasegawa, M., Goedert, M., 1998. α-Synuclein in filamentous inclusions of Lewy bodies from Parkinson's disease and dementia with Lewy bodies. Proc. Natl. Acad. Sci. U. S. A. 95 (11), 6469–6473.

Stankiewicz, J.M., Brass, S.D., 2009. Role of iron in neurotoxicity: a cause for concern in the elderly? Curr. Opin. Clin. Nutr. Metab. Care 12, 22–29.

Steffan, J.S., Kazantsev, A., Spasic-Boskovic, O., Greenwald, M., Zhu, Y.Z., Gohler, H., Wanker, E.E., Bates, G.P., Housman, D.E., Thompson, L.M., 2000. The Huntington's disease protein interacts with p53 and CREB-binding protein and represses transcription. Proc. Natl. Acad. Sci. U. S. A. 97 (12), 6763–6768.

Sterky, F.H., Lee, S., Wibom, R., Olson, L., Larsson, N.G., 2011. Impaired mitochondrial transport and Parkin-independent degeneration of respiratory chain-deficient dopamine neurons in vivo. Proc. Natl. Acad. Sci. U. S. A. 108 (31), 12937–12942.

Stokin, G.B., Lillo, C., Falzone, T.L., Brusch, R.G., Rockenstein, E., Mount, S.L., Raman, R., Davies, P., Masliah, E., Williams, D.S., Goldstein, L.S., 2005. Axonopathy and transport deficits early in the pathogenesis of Alzheimer's disease. Science 307, 1282–1288.

Sumimoto, H., Miyano, K., Takeya, R., 2005. Molecular composition and regulation of the Nox family NAD(P)H oxidases. Biochem. Biophys. Res. Commun. 338 (1), 677–686.

Suzuki, N., Cheung, T.T., Cai, X.D., et al., 1994. An increased percentage of long amyloid β protein secreted by familial amyloid β protein precursor (βAPP717) mutants. Science 264 (5163), 1336–1340.

Tahirbegi, I.B., Pardo, W.A., Alvira, M., Mir, M., Samitier, J., 2016. Amyloid Abeta 42, a promoter of magnetite nanoparticle formation in Alzheimer's disease. Nanotechnology 27, 465102.

Takahashi, K., Taira, T., Niki, T., Seino, C., Iguchi-Ariga, S.M.M., Ariga, H., 2001. DJ-1 positively regulates the androgen receptor by impairing the binding of PIASx alpha to the receptor. J. Biol. Chem. 276 (40), 37556–37563.

Tamaoka, A., 1994. Apolipoprotein E4 and late-onset Alzheimer's disease. Nihon Rinsho 52 (12), 3257–3265.

Tanzi, R.E., 2012. The genetics of Alzheimer disease. Cold Spring Harb. Perspect. Med. 2 (10), a006296.

Tanzi, R.E., Bertram, L., 2005. Twenty years of the Alzheimer's disease amyloid hypothesis: a genetic perspective. Cell 120 (4), 545–555.

Telenius, H., Kremer, B., Goldberg, Y.P., Theilmann, J., Andrew, S.E., Zeisler, J., Adam, S., Greenberg, C., Ives, E.J., Clarke, L.A., 1994. Somatic and gonadal mosaicism of the Huntington disease gene CAG repeat in brain and sperm. Nat. Genet. 6 (4), 409–414.

Tharp, W.G., Sarkar, I.N., 2013. Origins of amyloid-β. BMC Genom. 14 (1), 290.

Thomas, D.B., 2008. Genetic aspects of Alzheimer disease. Genet. Med. 10 (4).

Thomas, D.B., 2018. Alzheimer disease overview. In: Adam, M.P., Ardinger, H.H., Pagon, R.A., et al. (Eds.), Gene Reviews. University of Washington, Seattle (WA), pp. 1993–2020.

Tiraboschi, P., Hansen, L.A., Masliah, E., Alford, M., Thal, L.J., Corey-Bloom, J., 2004. Impact of APOE genotype on neuropathologic and neurochemical markers of Alzheimer disease. Neurology 62 (11), 1977–1983.

Trzesniewska, K., Brzyska, M., Elbaum, D., 2004. Neurodegenerative aspects of protein aggregation. Acta Neurobiol. Exp. 64, 41–52.

Turner, P.R., O'Connor, K., Tate, W.P., Abraham, W.C., 2003. Roles of amyloid precursor protein and its fragments in regulating neural activity, plasticity and memory. Prog. Neurobiol. 70, 1–32.

Urbanska, M., Blazejczyk, M., Jaworski, J., 2008. Molecular basis of dendritic arborization. Acta Neurobiol. Exp. 68 (2), 264–288.

U.S. National Library of Medicine, 2020a. PSEN I Gene. Available at: https://medlineplus.gov/genetics/gene/psen1/#conditions.

U. S. National Library of Medicine, 2020b. Huntington Disease. Available at: https://ghr.nlm.nih.gov/condition/huntington-disease.

United States National Library of Medicine, 2020c. HTT Gene. Available at:http://medlineplus.gov/genetics/gene.

Van Giau, V., Pyun, J.-M., Suh, J., Bagyinszky, E., An, S.S.A., Kim, S.Y., 2019. A pathogenic PSEN1 Trp165Cys mutation associated with early-onset Alzheimer's disease. BMC Neurol. 19, 188.

Van Houten, B., Woshner, V., Santos, J.H., 2006. Role of mitochondrial DNA in toxic responses to oxidative stress. DNA Repair 5 (2), 145–152.

van Leeuwen, F.W., de Kleijn, D.P., van den Hurk, H.H., Neubauer, A., Sonnemans, M.A., Sluijs, J.A., Köycü, S., Ramdjielal, R.D., Salehi, A., Martens, G.J., Grosveld, F.G., Peter, J., Burbach, H., Hol, E.M., 1998. Frameshift mutants of beta amyloid precursor protein and ubiquitin-B in Alzheimer's and Down patients. Science 279 (5348), 242–247.

Velier, J., Kim, M., Schwarz, C., Kim, T.W., Sapp, E., Chase, K., Aronin, N., DiFiglia, M., 1998. Wild-type and mutant huntingtins function in vesicle trafficking in the secretory and endocytic pathways. Exp. Neurol. 152 (1), 34–40.

Venkatraman, P., Wetzel, R., Tanaka, M., Nukina, N., Goldberg, A.L., 2004. Eukaryotic proteasomes cannot digest polyglutamine sequences and release them during degradation of polyglutamine-containing proteins. Mol. Cell. 14, 95–104.

Virág, L., Szabó, C., 2002. The therapeutic potential of poly(ADP-ribose) polymerase inhibitors. Pharmacol. Rev. 54 (3), 375–429.

Vito, P., Wolozin, B., Ganjei, J.K., Iwasaki, K., Lacanà, E., D'Adamio, L., 1996. Requirement of the familial Alzheimer's disease gene PS2 for apoptosis. Opposing effect of ALG-3. J. Biol. Chem. 271 (49), 31025–31028.

Vlad, S.C., Miller, D.R., Kowall, N.W., Felson, D.T., 2008. Protective effects of NSAIDs on the development of Alzheimer disease. Neurology 70 (19), 1672−1677.

Voet, D., Voet, J.G., Pratt, C.W. (Eds.), 2008. Fundamentals of Biochemistry. Life at the Molecular Level. John Wiley and Sons, New York.

von Strauss, E., Viitanen, M., De Ronchi, D., Winblad, B., Fratiglioni, L., 1999. Aging and the occurrence of dementia: findings from a population-based cohort with a large sample of nonagenarians. Arch. Neurol. 56 (5), 587−592.

Wakabayashi, K., Rinsho, S., 2008. Lewy body formation in Parkinson's disease: neurodegeneration or neuroprotection? Rinsho Shinkeigaku 48 (11), 981−983.

Walker, F.O., 2007. Huntington's disease. Lancet 369 (9557), 218−228. https://doi.org/10.1016/S0140-6736(07)60111-1.

Wallace, S.S., 2013. DNA glycosylases search for and remove oxidized DNA bases. Environ. Mol. Mutagen. 54, 691−704.

Walsh, D.M., Hartley, D.M., Kusumoto, Y., Fezoui, Y., Condron, M.M., Lomakin, A., Benedek, G.B., Selkoe, D.J., Teplow, D.B., 1999. Amyloid beta-protein fibrillogenesis. Structure and biological activity of protofibrillar intermediates. J. Biol. Chem. 274 (36), 25945−25952.

Wang, H., Shimoji, M., Yu, S.W., Dawson, T.M., Dawson, V.L., 2003. Apoptosis inducing factor and PARP-mediated injury in the MPTP mouse model of Parkinson's disease. Ann. N. Y. Acad. Sci. 991, 132−139.

Wang, L.H., Besirli, C.G., Johnson Jr., E.M., 2004. Mixed lineage kinases: a target for the prevention of neurodegeneration. Annu. Rev. Pharmacol. Toxicol. 44, 451−474.

Wang, X., Su, B., Perry, G., Smith, M.A., Zhu, X., 2007. Insights into amyloid-β-induced mitochondrial dysfunction in Alzheimer disease. Free Radic. Biol. Med. 43 (12), 1569−1573.

Wang, X., Su, B., Fujioka, H., Zhu, X., 2008. Dynamin-like protein 1 reduction underlies mitochondrial morphology and distribution abnormalities in fibroblasts from sporadic Alzheimer's disease patients. Am. J. Pathol. 173 (2), 470−482.

Wang, X., Su, B., Lee, H.G., Li, X., Perry, G., Smith, M.A., et al., 2009. Impaired balance of mitochondrial fission and fusion in Alzheimer's disease. J. Neurosci. 29 (28), 9090−9103.

Ward, R.J., Zucca, F.A., Duyn, J.H., Crichton, R.R., Zecca, L., 2014. The role of iron in brain ageing and neurodegenerative disorders. Lancet Neurol. 13, 1045−1060.

West, A.B., Moore, D.J., Choi, C., Andrabi, S.A., Li, X., Dikeman, D., Biskup, S., Zhang, Z., Lim, K.-L., Dawson, V.L., et al., 2007. Parkinson's disease-associated mutations in LRRK2 link enhanced GTP-binding and kinase activities to neuronal toxicity. Hum. Mol. Genet. 16, 223−232.

Wolozin, B., Iwasaki, K., Vito, P., Ganjei, J.K., Lacanà, E., Sunderland, T., Zhao, B., Kusiak, J.W., Wasco, W., D'Adamio, L., 1996. Participation of presenilin 2 in apoptosis: enhanced basal activity conferred by an Alzheimer mutation. Science 274 (5293), 1710−1713.

Wong, E., Cuervo, A.M., 2010. Autophagy gone awry in neurodegenerative diseases. Nat. Neurosci. 13, 805−811.

Wyss-Coray, T., 2016. Ageing, neurodegeneration and brain rejuvenation. Nature 539 (7628), 180−186.

Yamatsuji, T., Matsui, T., Okamoto, T., et al., 1996. G protein-mediated neuronal DNA fragmentation induced by familial Alzheimer's disease-associated mutants of APP. Science 272 (5266), 1349−1352.

Yanagida, T., Tsushima, J., Kitamura, Y., et al., 2009. Oxidative stress induction of DJ-1 protein in reactive astrocytes scavenges free radicals and reduces cell injury. Oxid. Med. Cell. Longevity 2 (1), 36–42.

Yoon, Y., Krueger, E.W., Oswald, B.J., McNiven, M.A., 2003. The mitochondrial protein hFis1 regulates mitochondrial fission in mammalian cells through an interaction with the dynamin-like protein DLP1. Mol. Cell Biol. 23 (15), 5409–5420.

Zarranz, J.J., Alegre, J., Gómez-Esteban, J.C., Lezcano, E., Ros, R., Ampuero, I., Vidal, L., Hoenicka, J., Rodriguez, O., Atarés, B., Llorens, V., Gomez Tortosa, E., del Ser, T., Muñoz, D.G., de Yebenes, J.G., 2004. The new mutation, E46K, of alpha-synuclein causes Parkinson and Lewy body dementia. Ann. Neurol. 55 (2), 164–173.

Zecca, L., Youdim, M.B.H., Riederer, P., Connor, J.R., Crichton, R.R., 2004. Iron, brain ageing and neurodegenerative disorders. Nat. Rev. Neurosci. 5, 863.

Zhou, B., Westaway, S.K., Levinson, B., Johnson, M.A., Gitschier, J., Hayflick, S.J., 2001. A novel pantothenate kinase gene (PANK2) is defective in Hallervorden-Spatz syndrome. Nat. Genet. 28, 345–349.

Zhou, R.M., Huang, Y.X., Li, X.L., Chen, C., Shi, Q., Wang, G.R., Tian, C., Wang, Z.Y., Jing, Y.Y., Gao, C., Dong, X.P., 2010. Molecular interaction of α-synuclein with tubulin influences on the polymerization of microtubule in vitro and structure of microtubule in cells. Mol. Biol. Rep. 37 (7), 3183–3192.

Zoccarato, F., Cavallini, L., Bortolami, S., 2007. Alexandre A Succinate modulation of H_2O_2 release at NADH:ubiquinone oxidoreductase (Complex I) in brain mitochondria. Biochem. J. 406 (1), 125–129.

Zuccato, C., Ciammola, A., Rigamonti, D., Leavitt, B.R., Goffredo, D., Conti, L., MacDonald, M.E., Friedlander, R.M., Silani, V., Hayden, M.R., Timmusk, T., Sipione, S., Cattaneo, E., 2001. Loss of huntingtin-mediated BDNF gene transcription in Huntington's disease. Science 293 (5529), 493–498.

Zucconi, G.G., Cipriani, S., Scattoni, R., Balgkouranidou, I., Hawkins, D.P., Ragnarsdottir, K.V., 2007. Copper deficiency elicits glial and neuronal response typical of neurodegenerative disorders. Neuropathol. Appl. Neurobiol. 33, 212–225.

Zuo, L., van Dyck, C.H., Luo, X., Kranzler, H.R., Yang, B.Z., Gelernter, J., 2006. Variation at APOE and STH loci and Alzheimer's disease. Behav. Brain Funct. 2 (1), 13.

Further reading

Bäckman, L., Jones, S., Berger, A.K., Laukka, E.J., Small, B.J., 2004. Multiple cognitive deficits during the transition to Alzheimer's disease. J. Intern. Med. 256 (3), 195–204.

Carlesimo, G.A., Oscar-Berman, M., June 1992. Memory deficits in Alzheimer's patients: a comprehensive review. Neuropsychol. Rev. 3 (2), 119–169.

Desai, V., Kaler, S.G., 2008. Role of copper in human neurological disorders. Am. J. Clin. Nutr. 88 (3), 855S–858S.

Förstl, H., Kurz, A., 1999. Clinical features of Alzheimer's disease. Eur. Arch. Psychiatr. Clin. Neurosci. 249 (6), 288–290.

Frank, E.M., 1994. Effect of Alzheimer's disease on communication function. J. S. C. Med. Assoc. 90 (9), 417–423.

Genetic and Rare Diseases Information Center(GARD), 2020. Spinocerebellar Ataxia. Available at: https://rarediseases.info.nih.gov/diseases/10748/spinocerebellar-ataxia.

Grundman, M., Petersen, R.C., Ferris, S.H., Thomas, R.G., Aisen, P.S., Bennett, D.A., et al., 2004. Mild cognitive impairment can be distinguished from Alzheimer disease and normal aging for clinical trials. Arch. Neurol. 61 (1), 59–66.

Hoehn, M.M., Yahr, M.D., 1967. Parkinsonism: onset, progression and mortality. Neurology 17 (5), 427–442.

Ince, P.G., Clark, B., Holton, J., Revesz, T., Wharton, S.B., 2008. Chapter 13: diseases of movement and system degenerations. In: Greenfield, J.G., Love, S., Louis, D.N., Ellison, D.W. (Eds.), Greenfield's Neuropathology. 1, eighth ed. Hodder Arnold, London, p. 947.

Jankovic, J., 2008. Parkinson's disease: clinical features and diagnosis. J. Neurol. Neurosurg. Psychiatr. 79 (4), 368–376.

Murray, E.D., Buttner, N., Price, B.H., 2012. Depression and psychosis in neurological practice. In: Bradley, W.G., Daroff, R.B., Fenichel, G.M., Jankovic, J. (Eds.), Bradley's Neurology in Clinical Practice, sixth ed. Elsevier/Saunders, Philadelphia, PA.

Nehls, P., Seiler, F., Rehn, B., Greferath, R., Bruch, J., 1997. Formation and persistence of 8-oxoguanine in rat lung cells as an important determinant for tumor formation following particle exposure. Environ. Health Perspect. 105 (Suppl. 5), 1291–1296.

Opal, P., Zoghbi, H.Y., 2016. The Spinocerebellar Ataxias. http://www.uptodate.com/contents/the-spinocerebellar-ataxias.

Robakis, N.K., 2011. Mechanisms of AD neurodegeneration may be independent of Abeta and its derivatives. Neurobiol. Aging 32, 372–379.

Squitieri, F., Cannella, M., Simonelli, M., 2002. CAG mutation effect on rate of progression in Huntington's disease. Neurol. Sci. 23 (Suppl. 2), S107–S108.

Statland, J.M., Barohn, R.J., McVey, A.L., Katz, J.S., Dimachkie, M.M., 2015. Patterns of weakness, classification of motor neuron disease, and clinical diagnosis of sporadic amyotrophic lateral sclerosis. Neurol. Clin. 33 (4), 735–748.

Taler, V., Phillips, N.A., 2008. Language performance in Alzheimer's disease and mild cognitive impairment: a comparative review. J. Clin. Exp. Neuropsychol. 30 (5), 501–556.

Waldemar, G., Dubois, B., Emre, M., Georges, J., McKeith, I.G., Rossor, M., Scheltens, P., Tariska, P., Winblad, B., 2007. Recommendations for the diagnosis and management of Alzheimer's disease and other disorders associated with dementia: EFNS guideline. Eur. J. Neurol. 14 (1), e1–26.

Yao, S.C., Hart, A.D., Terzella, M.J., 2013. An evidence-based osteopathic approach to Parkinson disease. Osteopath. Fam. Physician 5 (3), 96–101.

Role of caspases, apoptosis and additional factors in pathology of Alzheimer's disease

Introduction

Alzheimer's disease is the leading global health hazard in the aged population worldwide Alzheimer's Association (2016). It is estimated that approximately 50 million populations are inflicted with Alzheimer's disease or AD-related dementia (ADI, 2018). The manifestations in Alzheimer's disease are the prominent disability conditions jeopardizing the quality of life in the aged population.

Alzheimer's disease is a progressive neurodegenerative disease. The principal signs and symptoms of Alzheimer's disease may include a progressive decline in memory, impaired routine activities; the early onset of Alzheimer's disease is manifested by cognitive dysfunction, altered behavior, and changes in language and speech (Tarawneh and Holtzman, 2012). Additionally, around 30% of patients with early-onset AD manifest psychiatric disorders like manic depression (Zubenko et al., 2003). The late-onset of Alzheimer's disease is associated with intense memory loss, disorientation, hallucinations, and apathy with respiratory syndrome leading to the death of the patient (Kalia, 2003).

The pathological hallmark of Alzheimer's is the accumulation of amyloid-β, senile plaques, neurofibrillary tangles, loss of synapses, and progressive neuronal loss leading to a gradual decline in the quality and expectant of life in the affected population.

The pathology of Alzheimer's is widely described but lacks consistency, universality, and unanimity in the predisposing factors of Alzheimer's disease in the minds of neurologists, scientists, academicians, and clinicians. An attempt has been made to explore the role of caspases, apoptosis, and collateral factors implicated in the pathology of Alzheimer's disease that would help to furnish extensive as well as intensive knowledge over the subject matter.

Amyloid cascade hypothesis in Alzheimer's disease

Alzheimer's disease is a neuronal progressive disorder manifested as continuous worsening of the cognitive ability and behavior of affected persons. The amyloid cascade hypothesis has been in existence past 3 decades explaining the partial role in the pathogenesis of Alzheimer's disease. Initially, The β-amyloid deposition is followed by the formation of senile plaques and neurofibrillary tangles leading to apoptosis of a specific group of neurons and manifestations of Alzheimer's disease (Reitz, 2012). According to the amyloid cascade hypothesis, during the pathogenesis of Alzheimer's, the deposition of β-amyloid is the pioneer event that triggers the subsequent pathological changes in the neuron structure and associated functions.

In 1992, the deposition of amyloid-β peptides in the brain parenchyma was described by Hardy and Higgins (1992) and its implication in the pathogenesis of Alzheimers disease was forwarded.

Strong evidence supporting the pathology of early-onset Alzheimer's disease (EOAD) are the presence of manifestations of the disease in patients suffering from Down syndrome and mutations in genes amyloid precursor protein (APP), PSEN1, and PSEN2 controlling the formation of the amyloid precursor protein, and activity of the γ-secretase complex.

Evidences in favor of amyloid cascade hypothesis in Alzheimer's disease

Amyloid-beta (Aβ) in brain tissues in Down syndrome and Alzheimer's disease

The first substantive evidence for substantiating the "amyloid cascade hypothesis" is provided in terms of identification of β-amyloid (Aβ) in the senile plaques in the brain tissues of the person affected with Down syndrome and Alzheimer's disease.

The presence of a common sequence of amyloid fibril protein both in Alzheimer's disease and Down syndrome is provided by the following studies.

A study by Glenner and Wong (1984) was conducted on the relationship between Alzheimer's disease and Down syndrome having a common sequence of amyloid fibril protein in an adult patient with Down syndrome.

Authors isolated cerebrovascular amyloid protein from an adult person with Down syndrome. The protein was purified and sequence analysis was performed. The authors reported that the sequence of the cerebrovascular amyloid protein was homologous with the β-amyloid protein in Alzheimer's disease.

The findings are claimed to be the first of its kind chemical evidence to prove homology between Alzheimer's disease and Down syndrome in terms of a sequence of proteins. Additionally, the genes implicated in Down syndrome and a few hereditary cases of Alzheimer's disease are located on chromosomes 21. Authors suggested that Down syndrome might be viewed as a prognostic model for Alzheimer's disease.

Additional evidence in substantiating the "amyloid cascade hypothesis" and its potential in the formation of senile plaques and neurofibrillary tangles in Alzheimer's can be derived from genetic mutations in APP gene and PSEN1, and PSEN2 genes.

Mutations in amyloid precursor protein

The APP gene is mapped to chromosome 21 (Goldgaber et al., 1987). It was initially identified during 1987 by sequence analysis method from purified β-amyloid (Kang et al., 1987). The gene encodes APP protein that is composed of a large extracellular domain and short cytosolic domain. The Two β-secretase catalyzed cleavage in the extracellular domain and γ-secretase catalyzed cleavage in the transmembrane domain result in the formation of β-amyloid from the amyloid precursor protein (Sandbrink et al., 1996).

The APP695 is the most prominent spliced protein in neurons (Sandbrink et al., 1996). Otherwise, alternate splicing of the APP gene can lead to the formation of several APP proteins.

Owing to the flexibility of the catalytic ability of γ-secretase on the selection of a site for splitting in the APP protein, variants of β-amyloids are generated showing variations in the C-terminals (Ling et al., 2003).

The formation of β-amyloid 40 is the most prominent activity of γ-secretase on APP protein during normal and healthy conditions.

However, the formation of β-amyloid 42 that is only 10% of the total formation of β-amyloid (Wiltfang et al., 2002) is attributed to several factors including genetic mutation in the APP gene. The Aβ42 is neurotoxic and is implicated in the formation of neuro-fibrils and enhances the aggregate formation (Wolfe and Guenette, 2007).

It can be concluded that duplication of *APP* gene and missense mutations in APP gene is directly implicated in the formation of β-amyloid peptide (β-amyloid 42) with neurotoxicity involved in the pathology of Alzheimer's.

Furthermore, the pathology in the brains of patients with Down syndrome and Alzheimer's bear close homology in terms of pathological deposition of β-amyloid leading to a decline in cognitive functions and altered behaviors. The APP gene is located on the HSA21 and occurs in three copies in patients with Down syndrome resulting in overexpression of the APP gene and surplus formation of β-amyloid (Julia and Goate, 2017).

Surprisingly, the pathology of Alzheimer's exhibits high prevalence rate in patients with Down syndrome. The progression rates of down syndrome and Alzheimer's are age-dependent (Oyama et al., 1994).

The neuropathological lesions in the brains in patients with Down syndrome owing to trisomy of HSA21 bear close resemblance with those in patients with Alzheimer's in the age group of 30—40 years (Burger and Vogel, 1973); additionally, the majority of patients with Down syndrome in the age group in 70—75 years develop dementia (Wisniewski et al., 1985).

Interestingly, patients with Alzheimer's and Down syndrome exhibit deposition of β-amyloid, progression to the formation of neurofibrillary tangles, expression of inflammation, hyperphosphorylation of the τ-protein (microtubule-associated protein) leading to a degeneration of cholinergic neurons in the basal forebrain (Hof et al., 1995; Sadowski et al., 1999).

Thus, excessive deposition of Aβ in the brains of patients with Alzheimer's and Down syndrome mediate neural cell death and neural dysfunction.

Duplication mutation of APP gene

The duplication mutation in HSA21 containing the *APP* gene has been implicated in Alzheimer's disease in several families (Rovelet-Lecrux et al., 2006).

The pathological investigation of the brains with Alzheimer's presented with high depositions of Aβ, formation of neurofibrillary tangles in the parenchyma. The brains of such family members showed a high predisposition to cerebral amyloid angiopathy (deposition of β-**amyloid** in the vessel walls in the brain) in blood vessels in the brain (Ellis et al., 1996; Pfeifer et al., 2002; Guyant-Marechal et al., 2007).

Another study (Sleegers et al., 2006) analyzed the potential of APP gene locus duplications in prematurity onset of Alzheimer's disease in a Dutch population.

Authors identified APP gene locus duplications in 1 family out of 10 multigenerational families affected with early-onset Alzheimer's disease (Sleegers et al., 2006).

The family was showed established features of cerebral amyloid angiopathy with the early-onset Alzheimer's disease. Authors reported the prevalence of <2% APP gene locus duplications that are familial, nonautosomal dominant Alzheimer's disease traits. It is responsible for de novo mutation (Sleegers et al., 2006).

Mutations in PSEN1 gene

Till date, according to an estimate, higher than 200 mutations in gene *PSEN1* with a wide disease spectrum across the globe, nearly 2/3rd of the total pathogenic mutations have been localized in the exons 5, 6, 7, and 8 of gene *PSEN1* (Dai et al., 2018).

The genetic mutation in PSEN1 is associated with the most frequent factor in the pathogenesis of autosomal dominant early-onset Alzheimer disease. Cruts et al. (2012) prepared databases for neurodegenerative brain diseases on the basis of locus-specific mutations. The prevalence rate of 70%–80% (Cruts et al., 2012) of autosomal dominant early-onset Alzheimer's disease has been attributed to mutations in the PSEN1 gene.

Comparative study delineated that mutations in PSEN1 gene are the causative factors in the manifestations of early onset of Alzheimer with around 8.4 years (Cruts et al., 2012) (manifestations occurrence at average 42.9 (Cruts et al., 2012) years than average 51.3 years) before the *APP* gene mutations induced manifestations early onset of Alzheimer's.

The missense mutations in gene *PSEN1* are implicated in the pathology of early onset of Alzheimer's. The missense mutations are associated with the substitution of one amino acid residue with another leading to the formation of a variant protein with altered function. Mutations of domains in gene PSEN1 controlling the formation of transmembrane domains 2 and 4 are responsible for manifestations of early onset of Alzheimer's (Lippa et al., 2000).

Mutations in the PSEN1 gene are insinuated in the formation of a higher amount of Aβ42 in comparison to Aβ40 leading to a rise in the ratio of Aβ42/Aβ40 (Sun et al., 2017).

The β-amyloid-42 is implicated closely in the formation of amyloid aggregates in brain tissues in comparison to Aβ40 (Jan et al., 2008).

Contrary findings were provided in the study involving the effects of mutations in the PSEN1 gene on the synthesis and activity of γ-secretase and formation of Aβ42 and Aβ40. Sun et al. (2017) studied the total 138 mutations reported human PSEN1 gene.

The anterior-pharynx-defective protein 1 was initially reported in the *C. elegans* in its Notch signaling pathway. The protein is involved in controlling the expression of nicastrin (Goutte et al., 2002). Furthermore, the homologs of anterior-pharynx-defective protein 1 are found in humans and represent elements of the γ-secretase complex (polymeric protease) possessing presenilin as catalytic subunit and PEN2 and nicastrin as regulatory subunits (Goutte et al., 2002).

Sun et al. (2017) reconstituted individual mutant PS1 proteins into anterior-pharynx-defective protein 1 and in vitro study revealed their influence on the formation of Aβ42 and Aβ40. Authors reported that nearly 90% of the total 138 mutations resulted in a decline in the formation of Aβ42 and Aβ40. Additionally, the remaining 10% of the studied mutations resulted in a reduction in the ratio of Aβ42/Aβ40.

It can be concluded that the effect of PSEN1 gene mutations on the γ-secretase is not always associated with a persistent increase in the ratio of Aβ42/Aβ40.

Furthermore, the study revealed the absence of substantive association between the role of *PSEN1* gene mutations and the increase in the ratio between Aβ42 and Aβ40.

Inconsistency prevails in several publications related to mutations in the PSEN1 gene causing an abnormal gain or partial loss of function of γ-secretase (Bentahir et al., 2006).

The cells with deficient PSEN1 gene were selected to analyze the effects of mutations on amyloid precursor protein, syndecan-3, Notch, and N-cadherin substrate processing, and formation of γ-secretase complex (Bentahir et al., 2006).

Bentahir et al. (2006) reported that presenilin mutants showed a loss in amyloid precursor protein and Notch substrate processing at epsilon and S3 cleavage sites in variants.

The PSEN1 gene mutations namely PS1-Delta9 and PS1-L166P led to a decline in the formation of Aβ40 while another mutant PS1-G384A led to an increase in the formation of Aβ42 (Bentahir et al., 2006). The authors reported several ways affecting the processing of amyloid precursor protein at the γ site in the study.

The study concluded that several mutations in the PSEN1 gene influence the structure and or function of γ-secretase via multiple ways.

Overall, it can be inferred that studies involved in the effect of mutations in gene *PSEN1* that in turn influence processing of amyloid precursor protein by γ-secretase have exhibited inconsistency in their findings that might be contributed by variation in the experimental systems involved in these studies.

Mutations in PSEN2 gene

Gene PSEN-2 is mapped to chromosome 1 (1q42.13). It exhibits around 60% homology to gene PSEN-1 (Rogaev et al., 1995). The PSEN 2 gene contains 12 exons controlling the synthesis of 448-amino-acid residues containing PSEN-2 protein. It contains nine transmembrane domains and a large cytosolic loop domain located in between the sixth and seventh transmembrane domains (Levy-Lahad et al., 1995).

PSEN2 protein provides the catalytic ability to the γ-secretase complex. The PSEN2 expresses limited distribution as opposed to the wide distribution of PSEN1 and the former protein is located in late endosomes and lysosomes (Sannerud et al., 2016).

A study involving the overexpression of gene PSEN2 in Neuro2a cells described the presence of higher predisposition to chromatin condensation and decline in the viability of Neuro2a cells. The gene PSEN2 additionally influences the expression of Bax protein linked with apoptotic cell death (Kumar et al., 2016).

Several studies indicated the neurotoxic potential of β-amyloid in mediating neuro-inflammation via **the enhanced release of proinflammatory cytokines.**

Qin et al. (2017) predicted that deficient gene PSEN2 promotes Aβ-mediated neuroinflammation. It was described in the PSEN2 knockout model of mice that showed a rise in cognitive abnormalities and a higher predisposition to cerebral injury. The knockout model of gene PSEN2 showed upregulated expression of P2X7 in in vitro as well as in vivo studies. The PSEN2 gene knockout showed an increase in expression of P2X7 in neurons and microglia.

The study delineated the potential of Aβ-mediated release of proinflammatory cytokines namely TNF-α, IL-1β, and IL-1α that was further increased in the PSEN2 knockout model of mice in microglia (Qin et al., 2017).

Thus, PSEN-2 has a protective role against the Aβ-mediated neuroinflammation and cerebral injury via **Down-regulation of P2X7 gene expression.**

However, unlike 138 mutations in gene PSEN1, only 40 mutations in gene *PSEN2* have been reported (Dai et al., 2018).

Patients with Familial AD due to mutant PSEN2 show delayed onset and with longer duration of Alzheimer's disease in comparison to the early onset of disease in patients with mutations in gene *PSEN1* (Jayadev et al., 2010). Few other disorders namely Parkinson's disease with dementia, frontotemporal dementia, and dementia with Lewy bodies have been reported with PSEN2 mutations (Cai et al., 2015).

The penetrance (in genetics, signifies the proportion of persons with the mutant allele expressing into disease manifestations, 90% penetrance means 90% persons with mutant allele develop disease manifestations) of Alzheimer's disease in patients inflicted with gene *PSEN2* mutations shows wide variability in the age of onset of disease varying from age of 40 to age of 80 in patients (Cai et al., 2015).

It is reported that out of identified total of 40 mutations in PSEN2 gene, only 10 mutants are nonpathogenic (Dai et al., 2018) with wide uncertainty among the researchers related to the potential of remaining mutations toward manifestations of Alzheimer disease in affected populations.

However, the study showed (Walker et al., 2005) **that most of the mutations in gene PSEN2 that result in decline in β-amyloid formation can be termed as mutations with partial loss of function.**

Additionally, probably mutant PSEN-2 M239I was implicated in the surplus formation of amyloid-β-42 coupled with impaired intracellular calcium homeostasis in the affected neurons.

Zatti et al. (2004) described the consequence of the PSEN2-familial AD-linked mutant M239I on the expression of cytosolic calcium homeostasis in terms of calcium storage potential of the endoplasmic reticulum and activation level of capacitative calcium entry (calcium influx pathway in plasma membrane owing to intracellular calcium storage).

Authors reported a comparative decline in the release of calcium ions from the endoplasmic reticulum in fibroblasts in patients with Alzheimer's disease in comparison to that in fibroblasts in healthy persons or patients with sporadic forms of Alzheimer's disease.

Authors conducted the study in HEK293 and HeLa cell lines transiently expressing the PSEN2-FAD linked mutant M239I and reported a decline in the calcium release from the endoplasmic reticulum in affected cells.

Similar findings in a study by Zatti et al. (2006) were described conveying PSEN-FAD linked mutations reduce the intracellular calcium storage in the cytoplasmic organelles, while the cytosolic lowering of calcium was most pronounced in the cells expressing mutant of PSEN2.

Additionally, the aforementioned effect was weakly expressed in cells with PSEN2 mutant, D366A (nonpathogenic mutant).

Thus, PSEN 2 can modulate the cytosolic calcium homeostasis that might be implicated in the pathogenesis of Alzheimer's disease.

The aforementioned studies involving mutations of (genes *APP*, *PSEN1*, and *PSEN2*) posited the association of mutants of genes APP, PSEN1, *and* PSEN2 with a higher predisposition to formation of β-amyloid-42 and impaired cytosolic calcium storage of endoplasmic reticulum and mitochondria toward the pathogenesis of the EOAD. These genetic studies further support the amyloid cascade hypothesis in the pathogenesis of Alzheimer's disease.

Furthermore, contrary evidences were forwarded against the amyloid cascade hypothesis in Alzheimer's disease. The sporadic forms of AD are much more prevalent than familial Alzheimer's disease. The sporadic forms of the disease show the concomitant occurrence of β-amyloid plaques and τ-neurofibrillary tangles. Thus, both forms of Alzheimer's disease share separate etiopathogenesis (Ricciarelli, and Fedele 2017).

Thus, advocates of the amyloid cascade hypothesis and its opponents hold their views on firm findings to conclude the absence of the exclusive role of amyloid-β in the pathogenesis of Alzheimer's disease.

Role of apoptosis in the neuronal death in Alzheimer's disease: a possibility

Alzheimer's disease is the progressive neurodegenerative disorder manifested in terms of neuronal loss in specific regions of the brain in the affected persons resulting in impaired cognitive function, behavior changes, and dementia. The potential of amyloid-β, amyloid oligomers, and neurofibrillary tangles possessing hyperphosphorylated τ (Golde et al., 2006) in the initiation and progression of Alzheimer's is crucial.

Neurodegenerative diseases including Alzheimer's, extensive neuronal loss, and programmed cell death (apoptosis) might be interlinked.

Apoptosis is the conserved and evolutionary programmed cell death that serves a prime role in the developmental biology of an organism. It serves to maintain dynamic equilibrium in rapidly proliferating tissues. the apoptosis regulates the normal development of the nervous system in vertebrates (Kuan et al., 2000).

Apoptosis is marked by nuclear alterations including condensation of chromatin, margination, pyknosis of nucleus, and fragmentation followed by cytoplasmic changes covering blebbing of the plasma membrane, cell shrinkage, and formation of apoptotic bodies (Kerr et al., 1972).

Additional studies further explained the nuclear changes in apoptosis. Initially, chromatin is split into fragments of 50−300 kb and designated as high molecular weight (Timmer and Salvesen, 2007) DNA fragmentation. Afterward, HMW DNA fragments are degraded into still smaller fragments of 180−200 bp (oligonucleosomal size) and are termed as low molecular weight DNA fragments. These are identified as **laddering pattern** in an agarose gel electrophoresis (DNA ladder) (Jacobson et al., 1997).

Moreover, dyshomeostasis in the apoptotic machinery might be implicated in the manifestations of disease including neurodegenerative diseases (Nicholson, 2000; Gavrieli et al., 1992).

Dysregulation of apoptosis can be attributed to multiple factors acting individually in synergism by several factors including amyloid-β, Bax, Bcl2, caspases, tumor necrosis factor-α, and reactive oxygen species (Obulesu and Lakshmi, 2014).

Gervais et al. (1999) **commented that "The loss of hippocampal neurons by apoptotic cell death is a prominent feature of Alzheimer's disease."**

Wellington and Hayden (2000) **remarked that "Unregulated apoptosis underlies many pathological conditions, including neurodegenerative diseases".**

Prominent facts related to the probable role of apoptosis in neuronal cell death in Alzheimer's disease are discussed in the forthcoming section.

In situ detection of DNA fragments in apoptotic cells

The DNA fragmentation into oligo-nucleosomal size is the specific feature of cells passing through the stage of apoptotic cell death (Hengartner, 2000).

Su et al. (1994) studied the surgical biopsy tissues from the brains of patients with Alzheimer's. Further, the brain tissues from the entorhinal cortex and hippocampi were obtained from the patients with Alzheimer's and age-matched healthy controls.

The brain tissues and surgical biopsies were examined with the help of the ApopTag peroxidase system (in situ apoptosis detection kit). The system helps to identify DNA fragmentation and morphology of nuclei associated with apoptosis.

Authors identified characteristics of apoptosis in the nuclei of neurons in both tangles containing neurons and nontangle-containing neurons in the brains in patients with Alzheimer's (Su et al., 1994). The neurons in the brains from controls were without the presence of neurofibrillary tangles and features of DNA fragmentation (Su et al., 1994).

Thus, the above in vitro study delineates that apoptosis is an important mechanism in neuronal cell death and is closely linked with the pathology of Alzheimer's.

Another study by Smale et al. (1995) furnished strong evidence in favor of the critical role of apoptosis in neuronal cell death in Alzheimer's disease. Authors utilized in situ labeling technique for DNA fragments that is named as "terminal transferase-mediated dUTP-biotin nick end labeling" abbreviated as "TUNEL" (Smale et al., 1995).

The technique employs the terminal transferase to add biotinylated nucleotides into the DNA fragments of apoptotic cells.

Smale et al. (1995) obtained the tissues specimens from the hippocampi of brains from patients with Alzheimer's disease and from brains of healthy, non-Alzheimer disease.

Authors identified with the help of the TUNEL technique that neurons and astrocytes as well as the majority of microglial cells were in the process of apoptosis (Smale et al., 1995). The intensity of apoptosis was much raised in the tissues specimens from hippocampi of brains of patients with Alzheimer's disease than non-AD and age-matched participants.

Furthermore, on the basis of findings in the study, apoptosis is certainly associated with neuronal loss and activation of microglial cells that are the essential elements in the pathology of Alzheimer's.

A study by Jellingera and Stadelmann (2001) was conducted on the tissues derived from the brains of patients ($n = 09$) with neurodegenerative diseases including Alzheimer's disease. All patients were selected on the basis of CERAD criteria of definite Alzheimer's disease (Mirra et al., 1991). The study also included tissues from the brains of ($n = 7$) age-matched healthy controls.

The brain tissues were preserved in the formalin solution (Jellingera and Stadelmann, 2001) and blocks of the tissues from different regions of the brains were prepared and were dipped in paraffin. Later on, Jellingera and Stadelmann (2001) utilized a 5 μm deparaffinized tissue section for histochemical studies.

Authors made use of the terminal deoxynucleotidyl transferase-mediated incorporation of digoxigenin-labeled nucleotides (TUNEL technique) to identify fragments of DNA in the tissues specimens from brains of patients and participants (Jellingera and Stadelmann, 2001).

Jellingera and Stadelmann (2001) reported the around 50 times higher presence of DNA fragments in the neurons and 25-fold higher DNA fragments in the microglial cells from the brain tissues from Alzheimer's disease than those from the healthy controls.

Furthermore, the authors identified a decline in neurons size, chromatin condensation, and the formation of apoptotic bodies in the hippocampus regions of brains in Alzheimer's patients.

Additionally, Jellingera and Stadelmann (2001) localized maximum TUNEL positive neurons in the medial temporal allocortex region of brains in Alzheimer's patients.

Conclusively, it can be posited that apoptosis constitutes the genetically regulated and evolutionary conserved programmed cell death in the body of an organism (Majno and Joris, 1995)**. However, dysregulated apoptosis is implicated in neuronal death affecting particular regions of the brains in specific neurodegenerative diseases. Several factors including** (Cotman et al., 2000) **chronic exposure to amyloid-β, fibrillar amyloid, and tangles (amyloidogenic breakdown products of amyloid-β-protein precursor** (Reed, 2000) **contribute to dysregulation of apoptosis mechanism in the brains of susceptible population groups.**

In Alzheimer's disease, the expression of proapoptotic factors namely c-Fax, c-Jun, Bax, p53, and APO-1/Fas-DC95 (Anderson et al., 1996; de la Monte et al., 1997) and antiapoptotic factors such as Bcl-2, and Bcl-X have been reported to be activated in the brain tissues (Anderson et al., 1996; de la Monte et al., 1997).

These factors of apoptosis and antiapoptosis contribute to ferrous ion-mediated dysfunction of mitochondria in the brains in Alzheimer's disease (Duan et al., 1999).

Several studies (Tortosa et al., 1998; Kitamura et al., 1998; MacGibbon et al., 1997) examined the biopsy tissues obtained from postmortem brains from Alzheimer's patients. These indicated the presence of fragments of DNA in the brain regions as a manifestation of apoptosis of neurons and microglial cells (Kitamura et al., 1998). All studies employed the terminal deoxynucleotidyl transferase dUTP and labeling (TUNEL) technique. The brain specimens from biopsies also revealed the presence of higher concentrations of proteins like c-Jun, Bax, and Bcl-2 in Alzheimer brain tissues with reduced levels of protein Bcl-2 in tangle-bearing neurons (Giannakopoulos et al., 1999).

Furthermore, several neuropathological publications in the field of Alzheimer's disease focus on the dyshomeostasis between factors of apoptosis and antiapoptotic factors in the brain tissues in patients. There is a higher inclination toward the proapoptotic environment in the brains of sufferers (Stadelmann et al., 1998; Adamec et al., 1999).

Contrarily, studies point to the nonspecificity of TUNEL positive neurons in determining perfect cause and effect relation in Alzheimer's disease. Possibly, the author posited that DNA fragments revealed by the TUNEL technique in the brains of Alzheimer's might be representative of DNA damage without any significant clinical contribution to the induction of neuronal apoptosis in Alzheimer's disease (subclinical, sublethal DNA injury) (Roth, 2001).

Another study by Stadelmann et al. (1998) provided contradictory evidence stating that in situ labeling technique is nonspecific for the apoptotic cascade in the nuclei of neurons in the brains of the Alzheimer population. The DNA fragments identified by the TUNEL technique in neurons and microglial cells in apoptosis might indicate the raised susceptibility of neurons' altered metabolism than actual apoptosis.

A study with findings contradictory to those in the aforementioned studies is conducted by Stadelmann et al. (1998). Authors posited that several studies consistently point to the higher levels of DNA fragments in the neurons in the brains of patients with Alzheimer's disease in comparison to brains of age-matched non-Alzheimer persons.

Moreover, in Alzheimer's disease, either necrosis or apoptosis is incriminated in the neuronal loss, is still needs explanation.

Stadelmann et al. (1998) analyzed the brain tissues from patients with Alzheimer's and compared the results with those obtained from the examination of brain tissues derived from patients suffering from pontosubicular neuron necrosis (a condition marked with apoptosis).

Authors identified apoptotic neurons in the pontosubicular neuron necrosis exhibiting condensation of chromatin, DNA fragmentation, and cytoplasmic condensation (Stadelmann et al., 1998).

Surprisingly Stadelmann et al. (1998), reported the presence of massive neurons displaying DNA fragments in the hippocampus of brains in Alzheimer's disease. But a few neurons in the hippocampus in Alzheimer's exhibited morphological features of apoptosis and absence of staining for apoptotic specific proteins in the neurons in Alzheimer's.

Thus, finds in the cited study delineate that neuronal loss coupled with the presence of DNA fragments in the hippocampus region of brains are characteristics of Alzheimer's disease. But the absence of apoptotic-related nuclear features and lack of detection of apoptotic specific proteins in the neurons impose uncertainty over the precise role of dysfunctional apoptosis in the neuronal death associated with Alzheimer's disease.

Furthermore, the biopsies of the postmortem brain tissues from the patients with Alzheimer's disease coupled with the identification of apoptotic neurons after excluding the nonapoptotic lesions in the postmortem brains might be the perfect specimens for the analysis of apoptosis in Alzheimer's disease. Till date, it is a difficult task to perform.

Conclusively, several studies in favor of in situ **identification of DNA fragments in the brain tissues and neurons in Alzheimer's disease and detection of higher levels of proapoptotic proteins in the brains in Alzheimer's could not be rebutted. Additional research is essential in determining the precise**

involvement of neuronal apoptosis in the pathogenesis of Alzheimer's disease and could pave the way for the therapeutic interventions.

Pattern of neuronal loss in brain in Alzheimer's disease

Neurodegenerative diseases including Alzheimer's disease are marked with progressive and massive neuronal loss in specific regions of the brains (Mattson, 2000).

Several factors are implicated in the neuronal loss in the brain in the Alzheimer. It can be a lack of essential synthesis and control by neurotrophic factors like neutrophils, glial cell-line derived neurotrophic factor family ligands, and neuropoietic cytokines in the brain.

It may also be overexpression of glutamate receptors and overproduction of reactive oxygen species in the mitochondria in neurons leading to oxidative damage of mitochondria and death of neurons (Mattson, 2000).

Prominently, necrosis and apoptosis are involved in the neuronal loss in the brains in neurodegenerative diseases (Gorman, 2008).

The role of necrosis in the neuronal death in Alzheimer's can be ruled out due to the apoptotic specific pattern of dead neurons. The apoptotic neurons exhibit cell shrinkage, chromosome condensation, and DNA fragmentation (Toné et al., 2007). The contents of neurons are not leaked out leading to the formation of apoptotic bodies in the brains.

Conversely, necrosis can be mainly differentiated from apoptosis by the leak of neuronal contents in the extracellular space (Chan et al., 2015).

Moreover, additional pathways involved in neuronal death have been described in several studies that indirectly rule out the exclusive possibility of apoptosis in the cause of neuronal death in Alzheimer's disease.

The authors mentioned an alternative form of programmed cell death that is different from apoptosis on the basis of cell morphology and biochemical tests. This specific type of programmed cell death might be involved in neurodegenerative diseases and the normal growth and development of organisms (Sperandio et al., 2000).

Larke (1990) described essentially the involvement of three pathways in the death of cells in the organisms. These can be either apoptotic, autophagic, or nonlysosomal vesiculate pathways. Therefore, in the neuronal death in Alzheimer's, exclusive involvement of apoptosis in the neuronal death cannot be supported; at the same time, the role of apoptosis in the neuronal death in Alzheimer's cannot be denied.

Still another study by Kegel et al. (2000) described the overexpression of the endosomal-lysosomal system leading to activation of autophagic machinery and neuronal loss either in Huntington's disease or similar endosomal-lysosomal-mediated autophagic neuronal death in another neurodegenerative disease.

Still another study by Adamec et al. (2000) posited that a large number of neurons in Alzheimer's disease are degenerated. This accounts for potent up-regulation of the endosomal-lysosomal system of autophagy of neurons. Adamec et al. (2000)

conducted in vitro study on rat hippocampal neurons in culture to assess the role of several types of experimental injury and activation of the endosomal-lysosomal system.

Authors tested the apoptotic, oncotic, and mixed (apoptotic and oncotic) pathways in the neuronal death in rat hippocampal neurons in culture (Adamec et al., 2000).

Adamec et al. (2000) reported elevated size and number of late endosomes and lysosomes in the slowly developing apoptotic or slowly developing mixed types of experimental injury in the rat hippocampal neurons in culture.

Thus, endosomal-lysosomal system activation is involved in the massive neuronal death in Alzheimer's disease and experimental neuronal injury.

Conclusively, massive neuronal death in the specific regions of brains in Alzheimer's disease and other neurodegenerative diseases is induced by multiple pathways including apoptosis. The role of exclusive neuronal apoptosis in the pathology of Alzheimer's disease cannot be established.

Upregulation of expression of proapoptotic factors in Alzheimer's disease

Apoptosis is the genetically regulated programmed cell death. It is controlled by the expression of proapoptotic factors and antiapoptotic factors in the cells.

The apoptosis is mediated by proapoptotic proteins namely, Bax (**Apoptosis regulator BAX**), Bak (**Bcl-2 homologous antagonist/killer**), ICH-1, and CPP32, while it is suppressed by Bcl-2 (B-cell lymphoma 2) and Bcl-x proteins (B-cell lymphoma-extra large). The Bcl-2 and Bcl-x protein serve as antiapoptotic proteins.

A study by Kitamura et al. (1998) analyzed the concentrations of proapoptotic as well as antiapoptotic factors in the temporal cortex in participants with Alzheimer and compared with those in the controls.

Kitamura et al. (1998) reported raised concentrations of Bak, Bcl-2, and Bcl-x in the brains in Alzheimer's disease.

Thus, upregulated expression of Bak, Bcl-2, and Bcl-x in the brains is involved in the apoptosis of neurons in Alzheimer's disease.

Another study (Suzuki et al., 2000) identified a novel gene implicated in the apoptosis of neurons in the brain in Alzheimer's disease.

The cDNA of the novel gene was sequenced and the gene designated as NCKAP1 (human Nap1) (Suzuki et al., 2000). Authors localized the gene to human chromosome 2q32 (Suzuki et al., 2000). The novel gene was reported to be orthologous with rat Nap1 (Suzuki et al., 2000). The authors identified downregulation of the NCKAP1 gene in brains in Alzheimer's. Thus, suppressed expression of the antiapoptotic novel gene human Nap1 might be involved in neuronal death via apoptosis (Mattson, 2000).

Thus, studies point to the altered expressions of genes controlling the synthesis of apoptotic factors and antiapoptotic factors leading to dysregulated apoptosis in the brain tissues in Alzheimer's disease.

Overall, some studies support the role of apoptosis as the major mechanism in cell death in the brain in Alzheimer's disease, while other studies denied apoptosis as the main pathway in inducing cell death in the brain in Alzheimer's disease.

But, all studies point to massive loss of neurons in the brain in Alzheimer's disease. Further research is needed to establish whether apoptosis is directly involved in inducing neuronal loss in the brain in Alzheimer's disease.

Thus, the role of apoptosis in the neurodegeneration in the brain linked to Alzheimer's disease is controversial.

Several recent studies now advocate the activation of apoptotic machinery along with the activation and involvement of caspases as major protein molecules implicated in the pathology of Alzheimer's disease.

The next section is focused on the implication of caspases in apoptosis of neurons in Alzheimer's.

Role of caspases in apoptotic neuronal-death in Alzheimer's disease

Caspases are cysteinyl aspartate-specific proteases that catalyze the splitting of the substrate after an aspartic acid residue.

Caspases are essentially involved in apoptotic cell death in neurodegeneration diseases including Alzheimer's disease. These protein molecules can function as transducers of apoptotic cell death as well as the ultimate executioners of neuronal death (Roth, 2001).

The association between caspases and apoptotic cell death was delineated initially by Yuan et al. (1993). The caspase-1 is the homolog to CED-3 protein that is encoded by ced-3 gene involved in the programmed cell death in the nematode, *Caenorhabditis elegans* (Yuan et al., 1993).

Additional evidence between caspase activation and apoptotic cell death comes from the increased caspase activity leading to apoptosis in cells and inhibition of caspase activity leads to suppression of apoptosis in the tissues as is shown by both in vitro as well as in vivo studies (Yuan et al., 1993; Kuida et al., 1995; Schwartz and Milligan, 1996; Alnemri, 1997).

Caspases are synthesized and exist in cells in inactive and zymogen forms (Roth, 2001).

The zymogen or inactive caspases possess a prodomain, a large subunit, and a small subunit (Nicholson, 1999). After exposure to a stimulus, inactive caspase undergoes proteolysis to exclude the N terminal prodomain and splitting between large subunit and small subunit leading to the formation of active caspase in the form of a tetramer of two large subunits and two small subunits (Hengartner, 2000).

Several immuno-histochemical substantiated the occurrence of activated caspases and caspase-processed substrates in the senile plaques and neurofibrillary tangles (Shimohama, 2000; Behl, 2000). **Both activated caspase-3 and activated caspase -6 are the centers of focus in the apoptotic cell death in Alzheimer's disease.**

Role of caspase-3 in apoptotic neuronal-death in Alzheimer's disease

The caspase-3 is a protein that is encoded by the *CASP3* gene (Alnemri et al., 1996). **The caspase-3 activates caspase-6 and caspase-7. It is the main caspase that is involved in the splitting of amyloid-β 4A precursor protein that is linked with the neuronal apoptosis in Alzheimer's disease** (Gervais et al., 1999).

The caspase-3 is the executioner caspase and it is activated by upstream initiator caspases as 8, 9, and 10 in response to the apoptotic signaling events. It is an executioner caspase and is activated by an initiator caspase by proteolytic cleavage in response to apoptotic signals (Walters et al., 2009). **In its activation, the extrinsic pathway and (death receptor) and intrinsic pathway (mitochondrial) are involved** (Ghavami et al., 2009).

Several studies investigated the involvement of caspases including caspase 3 in the pathogenesis of neurodegenerative diseases. The proteins namely amyloid precursor protein, huntingtin, and presenilin-1, 2 implicated in the neurodegenerative diseases including Alzheimer are additionally cleaved by caspase-3 (D'Amelio et al., 2010).

This activity of caspase-3 might serve as the basis for the formation of mutant protein that may promote neurodegeneration.

A study by Gastard et al. (2003) investigated the localization of activated caspase-3 (a biomarker of apoptosis) in the medial temporal lobes in the brains in old aged persons with minimally impaired cognitive function (a sign of early-onset Alzheimer's disease).

Authors identified activated caspase-3 in the parahippocampal gyrus in the brains of patients with Alzheimer's disease (Gastard et al., 2003).

Gastard et al. (2003) suggested that the apoptotic pathway might be activated in response to the activation of caspase-3 and the events occur in early-stage in the medial temporal lobe in brains with Alzheimer's disease.

Another study (Cribbs et al., 2004) identified the presence of procaspase-3 and activated caspase-3 in the postsynaptic densities in patients with Alzheimer's disease as well as in the age-matched controls.

But the levels of caspase-3 and synaptic procaspase-3 were significantly higher in patients with Alzheimer's (Cribbs et al., 2004).

The authors suggested the role of caspase 3 in synapse degeneration in the course of progression of the disease.

Caspase-3 and presenilin-1 and 2 in apoptosis

The caspase-3 can enhance the formation of pathological amyloid-β via cleavage of presenilin-1 and 2 proteins.

The presenilin-1 and 2 proteins are associated with early-onset familial Alzheimer's disease (D'Amelio et al., 2010). These proteins are cleaved by caspase-3 cells at the time of apoptosis. In vitro study demonstrated the catalytic role of caspase-3 over the purified presenilin-1 and 2 (Kim et al., 1997; Loetscher et al., 1997). Possibly, it is assumed that caspase-catalyzed cleavage of presenilin-1 and 2 enhance the formation of pathological amyloid-β.

Moreover, contradictory findings in the study (Brockhaus et al., 1998) showed that caspase-induced cleavage is essential for the amyloidogenesis potential of the presenilin-1 and 2. The change in the caspase-mediated cleavage site in presenilin-1 and 2 did not alter the formation of pathological amyloid-β.

Caspase-3 and β-site APP-cleaving enzyme in apoptosis

The ADP-ribosylation factor-binding protein, GGA3 **is an adaptor protein. This protein is involved in the trafficking of** β-site APP-cleaving enzyme (BACE) that has a role in the formation of amyloid-β implicated in Alzheimer's disease. The levels of BACE are raised in Alzheimer's disease.

A study by Tesco et al. (2007) posited showed that caspase-3 cleaves the adaptor protein, GGA3, thus leading to a decline in the concentration of GGA3 in the brain. The GGA3 is essential for the intracellular trafficking of BACE for its lysosomal degradation. In the presence of reduced GGA3 level, the BACE lysosomal degradation is suppressed resulting in stabilization of BACE.

Furthermore, β-secretase also named as BACE is the transmembrane aspartic protease. It is actively involved in the cleavage of amyloid precursor protein into amyloid-β peptide that is closely associated with the etiology of Alzheimer's disease (Vassar, 2005).

Caspase-3-mediated cleavage of GGA3 elevates the BACE amyloidogenic activity in brain regions that subsequently is manifested into raised production of amyloid-β peptides and exaggeration of neuronal loss and loss of synapse in the neo-cortex in the Alzheimer disease (Fukumoto et al., 2002).

Caspase-3 and TAR DNA-binding protein-43 in apoptosis

Localization of caspase-3 cleaved TDP-43 protein (TAR DNA-binding protein-43) in abundance in brain regions is the hallmark pathological finding in Alzheimer's disease.

The transactive response DNA binding protein (TAR DNA-binding protein 43) (TDP-43) **is made up of 414 amino acid residues. It contains four domains** (Afroz et al., 2017).

The TDP43 protein contains nuclear localization signal from amino acid residues 82 to 98 nuclear export signal from amino acid residues 239 to 250 and three caspase-3 cleavage sites at residues 219, 89, and 13 (Vega et al., 2019).

TDP-43 is a transcriptional repressor protein that binds with the transactivation response element (TAR) in DNA and suppresses the transcription of HIV-1 (Ou et al., 1995).

Additionally, TDP-43 regulates alternate splicing of the CFTR gene and the apoA-II gene (Buratti and Baralle 2001).

TDP43 protein.

Caspase-3 cleaved, hyperphosphorylated, and ubiquitinated TDP43 protein is the pathological and abnormal protein that is involved in α-synuclein negative frontotemporal dementia (FTLD-TDP) (Mackenzie et al., 2011) **and Alzheimer's disease** (Tremblay et al., 2011).

A study by Rohn (2008) involved the synthesis of site-directed caspase-cleavage antibody to TDP-43 on the basis of caspase-3 cleavage consensus site within TDP-43 at position D219 by authors.

Rohn (2008) applied the antibody in brain tissues in postmortems and reported the localization of caspase-3 cleaved TDP-43 protein in tangles, Hirano bodies, reactive astrocytes, and neuritic plaques in the brains in Alzheimer's disease. Caspase-3 cleaved TDP-43 was additionally found in ubiquitinated neurons in Alzheimer's disease.

Caspase-3 and caspase-activated DNase in apoptosis

The caspase-activated DNase (CAD) protein synthesis is controlled by the DFFB gene (Liu et al., 1997).

The CAD protein is the heterodimer possessing endonuclease activity (Yuste et al., 2005).

It induces breakage in the DNA molecule during the course of apoptosis.

Under normal conditions, the caspase-activated DNase exists as a monomer and remains in the inactive state in the cells. The caspase-activated DNase is associated with an inhibitor of CAD named ICAD to form a complex as ICAD-CAD (Sakahira et al., 1999).

The ICAD protein contains two caspase recognition sites. The first site is located at Asp117 and the second site is at Asp224.

The activated caspase-3 catalyzes the cleavage at Asp117 and Asp224 in the ICAD and brings about dissociation of the ICAD-CAD complex (McCarty et al., 1999).

The released CAD undergoes dimerization and exits as an active form of caspase-activated DNase (Jog et al., 2012).

The active endonuclease CAD brings about the degradation of DNA (Fernando and Megeney 2007) during apoptosis. The **CAD causes the initiation of the DNA strand breakage** (Lai et al., 2011).

A study by Enari et al. (1998) supplemented the role of activated caspase-3-mediated apoptosis via enhancing the CAD activity and suppressing the ICAD inhibitory role.

Authors identified CAD and its inhibitor (ICAD) in the cytoplasmic fraction of lymphoma cells in the mouse.

Possibly, the ICAD serves as a chaperone (Enari et al., 1998) **during the synthesis of CAD. Its complexity with the CAD suppresses the endonuclease activity of CAD. Further, apoptotic stimuli-induced activated caspase-3 results in the cleavage of ICAD following the nuclear translocation of CAD bringing about the degradation of chromosomal DNA during apoptosis in Alzheimer's disease.**

Caspase-3 and ρ-associated coiled-coil-containing protein kinase 1 (ROCK1) in apoptosis

Caspase-3-mediated cleavage of the ROCK1 protein might be additionally involved in the caspase-3 stimulated neuronal apoptosis in Alzheimer's disease.

Caspase-3 cleaves the ROCK1 protein leading to activation of its constitutive kinase activity. The ROCK1 protein is involved in membrane blebbing at the time of apoptosis. The terminal phase of apoptosis is marked with altered morphology of apoptotic cell including cell contraction and membrane blebbing.

The rho-associated coiled-coil-containing protein kinase 1 (ROCK1) is a serine/threonine kinase enzyme. It is the main effector protein of the Ras homolog family member A(RhoA), that is, a small GTPase protein that acts on the ROCK1 protein. The ROCK1 is the main regulator of the cytoskeleton actin-myosin contractility (Rath and Olson, 2012).

The ROCK1 is the homodimer comprising the N-terminus kinase domain with catalytic activity (amino acid residues 76 to 338) (Nakagawa et al., 1996), the coiled-coil region between amino acid residues 425 to 1100 possessing RhoA-binding domain, the pleckstrin homology domain (Nakagawa et al., 1996), and cysteine-rich domain.

The kinase activity of ROCK1 is suppressed after the pleckstrin-homology domain and ρ-binding domain in the C-terminus are linked separately with the N-terminus catalytic kinase domain) (Nakagawa et al., 1996). These events occur in the absence of substrate.

After the GTP-associated RhoA attaches with the ρ-binding domain located in the coiled-coil region of ROCK1 resulting in dissociation of the pleckstrin-homology domain and the ρ-binding domain in the C-terminus from the N-terminus catalytic kinase domain.

Furthermore, caspase-3-mediated cleavage of the C-terminus inhibitory domain during the execution phase of apoptosis activates the kinase domain of ROCK1 (Jacobs et al., 2006).

A study by Sebbagh et al. (2001) posited that phosphorylation of the myosin light chain is crucial in the membrane blebbing that is the prominent feature of alteration in the cell morphology during apoptosis.

Authors reported the significance of ROCK I protein as an effector molecule of the small GTPase RhoA and as the substrate for the catalytic activity of caspase-3 during apoptosis (Sebbagh et al., 2001).

The ROCK1 is split by caspase-3 at a conserved DETD1113/G motif (Sebbagh et al., 2001) resulting in the removal of the C-terminal inhibitory domain and activating the constitutive kinase potential of the ROCK1.

The activated ROCK1 subsequently controls the phosphorylation of the myosin light chain. The levels of phosphorylated myosin light chain have been reported to enhance in the cells undergoing apoptosis (Sebbagh et al., 2001).

Furthermore, phosphorylation of myosin light chain and membrane blebbing was reported to be retracted after the suppression of caspase-3 cleavage activity.

It is suggested that the expression of caspase-3-mediated truncation and activation of ROCK1 is associated with membrane blebbing during apoptosis.

Thus, caspase-3 is suggested to have a role in the altered cell morphology as membrane blebbing during the execution of apoptosis.

Furthermore, caspase-3 activated ROCK1 in turn is implicated in the production of amyloid-β peptides and affects the progression of Alzheimer's disease.

The upregulated expression of ROCK1 has been reported to be associated with mild cognitive function deficiency in Alzheimer's disease than the controls (Henderson et al., 2016).

The mutant Aβ42 oligomers (Henderson et al., 2016) exhibit two types of actions. These oligomers induced a mild rise in the levels of ROCK1 and ROCK2 proteins in the neurons in Alzheimer's brains, as well as led to enhanced phosphorylation of Lim kinase 1 (actin-binding kinases).

These findings are further supplemented by a study involving ROCK1 heterozygous knock-out mice that showed a reduction in the formation of Aβ40 in the brain of ROCK1 knockout mice (Henderson et al., 2016) than those in the brain in wild-type littermate mice.

The study reported (Henderson et al., 2016) that knockdown of gene ROCK1 in mice showed a reduction in the amyloid precursor protein, and further application of macrolide group of antibiotic (bafilomycin) led to the accumulation of amyloid precursor protein in neurons with knockout ROCK1 gene.

Henderson et al. (2016) **suggested that decline in the ROCK1 protein, in turn, reduces the production of amyloid-β** via **promoting degradation of the amyloid precursor protein.**

Conclusively, it can be posited that proteins ROCK1 and ROCK2 can serve as the target of pharmacological intervention to suppress the formation of amyloid-β peptide in Alzheimer's disease.

Further analysis of the potential of caspase-3 in the apoptotic pathway in Alzheimer's disease revealed that caspase-3 has a role in the activation of mammalian sterile 20-like kinase 1 (MST1).

Caspase-3 and mammalian sterile 20-like kinase 1 (MST1) in apoptosis

The MST-1 protein represents the serine/threonine kinase. It is the main component in the Hippo signaling pathway (contains Hippo (Hpo) protein) Maejima et al., (2013)

The MST1 is closely involved in the biological events in response to reactive oxygen species and cell apoptosis (Qu et al., 2018).

The ROS-induced oxidative stress activates the MST1 protein that in turn induces neuronal apoptosis and death of microglial cells (Yun et al., 2011).

A study by Wang et al. (2017) showed that deleted MST1 led to the diminished loss of neurons in a mice model of spinal cord injury and offered neuroprotection in the experimental animal.

It is putative that ROS-induced oxidative stress is associated with several age-related disorders including Alzheimer and other neurodegeneration diseases (Gonfloni et al., 2012)

A study by Khan et al. (2019) involved a high-fat diet fed experimental mice and stress-induced hippocampal HT22 cells to assess the MST1-mediated neuronal apoptosis via p-JNK/Casp-3 dependent pathway.

Khan et al. (2019) identified that high-fat diet and stress in the hippocampal HT22 cells resulted in activation of MST1/JNK/caspase-3 apoptotic signaling pathway leading to neuronal apoptosis and exaggerated expression of BACE1 (β-amyloid-cleaving enzyme) (Khan et al., 2019) and associated impairment in cognitive function.

Furthermore, the interaction between caspase-3 and activation of MST1 is additionally explained in the study by Ura et al. (2001).

The MST1 protein possesses two nuclear export signals located toward the C-terminal domain of the protein. The C-terminal domain inhibits the kinase activity of the MST1 protein (Creasy et al., 1996). The N-terminal domain of MST1 exhibits consistent homology in terms of its kinase domain with the kinase domain of Ste20 protein and p21-activated kinase (Creasy et al., 1996).

However, no homology was identified by Creasy et al. (1996) outside the N-terminal kinase domain of MST1 with other kinases (Creasy et al., 1996).

It is further asserted that an inhibitory element is present inside the 63-amino acid residues-rich central region of the protein (Creasy et al., 1996) that inhibits the kinase activity of MST1.

The MST1 protein in its untruncated form is localized in the cytosol in an inactive state. Upon response to the apoptotic signal, caspase-3 catalyzes the cleavage of the **C-terminal domain resulting in activation of MST1 and subsequent its nuclear translocation** (Graves et al., 2001).

Thus, caspase-3-induced cleavage and constitutive nuclear translocation of truncated MST1 leads to chromatin condensation that is a marker of nuclear apoptosis.

Still another study by Qi et al. (2020) posited that both MST1 protein and MST2 protein have a significant role in neuronal cell death. The activated MST1 is involved in the apoptosis of microglial cells, and astrocytes (Qi et al., 2020).

Either MST1 protein or MST2 protein is implicated in the pathogenesis or promotes the progression of neurodegenerative disease including Alzheimer's disease.

Role of caspase-6 in apoptotic neuronal-death in Alzheimer's disease

Caspase-6 belongs to the cysteine aspartyl-specific peptidase and is involved in neuronal apoptosis and inflammation in neurodegenerative disease including Alzheimer's disease. The caspase-6 is encoded by the CASP6 gene (Tiso et al., 1996). The gene is mapped to human chromosome 4q25 (NCBI, 2020).

The inactive procaspase-6 has a molecular weight of nearly 34 kDa. After its splitting, it is demarcated into p20 fragment with mw of nearly 20 kDa while p10 fragment with mw around 10 kDa (Fernandes-Alnemri et al., 1995) these fragments assemble into a tetramer.

The procaspase-6 is cleavage into activated caspase-6 with splitting sites located at Asp-23 in prodomain and on either side of intersubunit linker at residues Asp-179 and Asp-193) (Dagbay et al., 2017).

The intersubunit linker binds the large subunit with the small subunit of procaspase-6. The large subunit of caspase-6 contains active site Cys-163 (Dagbay et al., 2017).

The active site caspase-6 is made up of four loops designated as L1, L2, L3, and L4) (Dagbay et al., 2017). These loops have the flexibility and pass-through conformational rearrangement that helps to bind with the substrate. The enzyme contains four substrate-binding sites designated as S1, S2, S3, and S4) (Dagbay et al., 2017).

Procaspase-6 is activated by caspase-3 (Simon et al., 2012), and it can undergo self-activation that has been proved in both in vitro and in vivo studies (Klaiman et al., 2009; Wang et al., 2010).

The caspase-3 cleavages caspase-6 at residue Asp-179 leading to activation of caspase-6 during the course of apoptosis (Dagbay et al., 2017).

Recent studies showed that caspase-6 undergoes a self-activation pathway (Dagbay et al., 2017). It is mediated by self-cleavage at the residue Asp-193 located in the intersubunit linker (Dagbay et al., 2017). The self-activation pathway has been reported by Dagbay et al. (2017) to be involved in the neurodegeneration in the brain.

Caspase-6 has a critical role in the pathogenesis of Alzheimer's disease (LeBlanc, 2013).

Proteins in the neurons namely τ (Horowitz et al., 2004), amyloid precursor protein (Albrecht et al., 2009), and presenilin I and II (Albrecht et al., 2009) are the important substrates of caspase-6 enzymatic activity. The truncated forms of these neuronal proteins are involved in Alzheimer's disease.

Caspase-6 functions as an executioner caspase as well as an inflammatory caspase (Guo et al., 2006).

Activated caspase-6 in amyloidogenesis and apoptosis in Alzheimer's disease

Alzheimer's disease is characterized by progressive neuronal loss proportional to the impaired cognitive function in the affected population (LeBlanc et al., 1999).

The entorhinal cortex region of the brain in Alzheimer's disease is marked with nearly 50% loss of neurons, as reported by LeBlanc et al. (1999) in most of the mild forms of cognitive impairment in Alzheimer's disease. As a result, loss of link between the hippocampus and the neocortex characterizes as impaired memory and learning potential of the affected population (LeBlanc et al., 1999; Gomez-Isla et al., 1997).

Furthermore, progressive loss of neurons in Alzheimer's is associated with deposition of amyloid-β in specific regions of the brain and formation of neurofibrillary tangles. Additionally, the formation of both the forms of amyloid-β as Ab40 and Ab42 (LeBlanc et al., 1999) is enhanced in the course of the disease and is the characteristic feature in the etiology of all familial forms of Alzheimer's disease.

The sporadic form constitutes the most prevalent form (nearly 90%) (LeBlanc et al., 1999) of Alzheimer's disease in the affected population; however, its precise etiology unlike the familial form is still not substantiated.

Furthermore, familial as well as sporadic forms of Alzheimer's disease exhibit homology in terms of increased production of amyloid-β peptide (LeBlanc et al., 1999).

In vitro study (LeBlanc, 1995) of human primary neurons programmed to cell death generate a larger amount of amyloid-β that can be around four times higher in comparison to the normal neurons suggesting the role of neuronal apoptosis in the increase in the production of amyloid-β peptide.

The amyloid-β has neurotoxic potential (Yankner et al., 1990); possibly inflicting irreversible injury on the healthy neurons resulting in further neuronal loss and hence, neuronal apoptosis-induced rise in the formation of amyloid-β might trigger a vicious cycle of adverse events in the brain in Alzheimer's disease.

The amyloid-β peptide and neurofibrillary tangles are the prominent biomarkers in the pathology of Alzheimer's disease.

Caspase-6-mediated rise in the formation of β-amyloid peptide occurs in specific regions of the brain in Alzheimer's, and neuronal apoptosis is incriminated in the enhanced formation of the β-amyloid peptide. These events are closely interconnected in the pathology of Alzheimer's advocating the potential of neuronal apoptosis in the activation of caspase-6 followed by involvement of caspase in the cleavage of the amyloid precursor protein.

Furthermore, LeBlanc et al. (1999) reported the localization of procaspase-6 in the brains in adults and localization of active caspase-6 fragment (p10) in the brain tissues in patients with Alzheimer's disease.

Caspase-6-mediated cleavage of amyloid precursor protein: possibility of either direct or indirect action

The APP exists in multiple isoforms including the APP695 form. The APP 695 contains caspase-6 sites (LeBlanc et al., 1999).

The authors analyzed the direct or indirect role of caspase-6 on the APP695 protein as substrate. Active recombinant caspases 3, -6, 7, and -8 were procured and incubated with protein extracts containing a high quantity of APP695 derived from neuron cultures (LeBlanc et al., 1999).

Authors reported that caspase-7 and caspase-8 catalyzed cleavage of full-length APP and caspases-mediated APP cleavage were suppressed with the addition of caspase 7 and 8 inhibitors in the cultures (LeBlanc et al., 1999).

Moreover, inactivity of caspase-3 and caspase-6 in the cleavage of full-length APP695 was reported, despite the localization of caspase-3 and caspase-6 sites in the APP695 protein (LeBlanc et al., 1999).

Authors suggested that possibly caspase-3 and caspase-6 sites in the APP695 could not be targeted by caspase 3 and 6 leading to failure in direct action of caspases on the APP (LeBlanc et al., 1999).

Additionally, the author posited the localization of endogenous inhibitors of caspase-3 and 6 in the neuronal extracts that suppressed recombinant caspase-3 and 6-mediated cleavage of APP695 (LeBlanc et al., 1999).

The additional possibility was posited in the involvement of the indirect effect of caspase-6 on the cleavage of APP695 protein.

An alternate cleavage pathway is suggested for amyloidogenic processing of amyloid precursor protein neurons in the brain regions that are implicated in the neuronal apoptosis in Alzheimer's disease.

Caspase-6-mediated formation of amyloidogenic fragment (Capp6.5)

A study by LeBlanc et al. (1999) utilized immuno-precipitated (precipitating an antigen from solution by interaction with antibody) neuronal APP695 (LeBlanc et al., 1999) to assess caspase-3 and caspase-6 cleavage products in the absence of other neuronal proteins.

Authors reported that recombinant caspase-3 and caspase-6 cleaved the immuno-precipitated APP695 protein suggesting the presence of endogenous inhibitors of caspases in the neuronal extracts (LeBlanc et al., 1999).

The study confirmed the formation of Capp6.5 and Capp3 fragments in low quantity in response to cleavage by recombinant caspase-6 enzyme suggesting the potential of caspase-6 in the formation of Capp6.5, amyloidogenic fragment; however, caspase-6 could not produce 4-kDa amyloid-β (LeBlanc et al., 1999).

LeBlanc et al. (1999) identified that caspase-6 induces cleavage of amyloid precursor protein at the carboxy-terminal producing C-terminus fragment with 3 kDa molecular weight designated as "Capp3" and Aβ-fragment with 6.5-kDa (LeBlanc et al., 1999) molecular weight designated as "Capp6.5." The amount of Capp6.5 fragment is raised in the neurons with serum deprivation.

The caspase-6 generated Capp6.5 fragment contains the sequence of amyloid-β peptide (LeBlanc et al., 1999) but caspase-6 is unable to directly produce 4 kDa amyloid β. Furthermore, Capp6.5 fragment can in turn induce the formation of 4 kDa amyloid-β fragment in the neurons (LeBlanc et al., 1999).

Additionally, it is asserted that caspase-6 can cleave amyloid precursor protein similar to the cleavage induced by β-secretase and γ-secretase enzymes.

A study by Gervais et al. (1999) involving the Swedish mutant APP with substitution of sequence VKMD653 into sequence VNLD653 at the enzyme β-secretase site, the caspase-6 induces much higher cleavage potential on the substrate VNLD-AMC (nearly 6 times higher) than the VKMD-AMC as revealed in the in vitro study.

The caspase-6, additionally, can cleave the amyloid precursor protein at sequence VEVD720/A, the γ-secretase site of APP (Gervais et al., 1999).

Caspase-6, therefore is involved in the proteolytic cleavage of amyloid precursor protein leading to a higher propensity to the formation of amyloid-β peptide and inducing apoptotic neuronal death incriminated in the pathology of Alzheimer's disease.

Study (Zhang et al., 2000) provided valuable findings related to the role of caspase-6 in neuronal apoptosis in Alzheimer's disease. Authors directly microinjected the active recombinant caspase-6 in the human primary neurons.

Authors reported nearly 20% neuronal apoptosis within 16 days after the microinjection of caspase-6 in dose <0.25 pg/cell (Zhang et al., 2000). Moreover, microinjected neurons displayed signs of oxidative stress before the execution stage of apoptosis suggesting the susceptibility of neurons to oxidative stress in response to activation of caspases (Zhang et al., 2000).

Contrarily, microinjection of caspase-6 could not show any lethal effect on the astrocytes suggesting cell-specific effects of activation of caspases including caspase-6 on the cells in the central nervous system (Zhang et al., 2000).

Thus, microinjection of recombinant active caspase-6-induced extended duration of neuronal apoptosis in cell-specific, dose, and time-dependent pattern in the brain.

Thus, activation of caspases in neurons in the brain leads to an extended course of neuronal apoptosis coupled with increased production of amyloid-β peptide and additional neurotoxic fragments implicated in the protracted, dose, and time-dependent apoptosis of neurons in Alzheimer's disease.

Caspase-6-mediated cleavage of τ in Alzheimer's disease

The τ-proteins constitute a class of six proteins existing in the native folded state with high solubility (Goedert et al., 1988). In humans, the gene MAPT controls the synthesis of τ-proteins and the gene is mapped to human chromosome 17q21. The gene MAPT contains 16 exons (Neve et al., 1986).

The six isoforms of τ-proteins are produced by alternative splicing (in genetics, single gene encodes multiple proteins) of gene MAPT (microtubule-associated protein). τ-Proteins function as a regulator of the stability of microtubules in axons. τ-Proteins are prominently located in the neurons in the brain and spinal cord.

The alternative RNA splicing of exon 2, exon 3, and exon 10 in the MAPT gene encodes the six isoforms of τ-protein (Sergeant et al., 2005).

τ contains several (79) serine and threonine-rich phosphorylation sites and nearly 30 sites are phosphorylated in the normal functioning of τ-proteins (Billingsley and Kincaid, 1997).

Kinases prominently protein kinase C control the phosphorylation of τ-proteins (Taniguchi et al., 2001; Mawal-Dewan et al., 1994).

Under normal conditions, τ-protein binds with microtubules and serves to stabilize the microtubule organization. In a hyperphosphorylated state, τ-proteins cannot bind with microtubules resulting in their disintegration. The unbound and hyperphosphorylated τ-proteins aggregate into an insoluble and misfolded structure called as neurofibrillary tangle (Calafate et al., 2015; Roman et al., 2019).

The pathological hyperphosphorylation (saturation of multiple phosphorylation sites in the compound) of τ-proteins (neurofibrillary tangles) is implicated in the pathology of neurodegenerative disease including Alzheimer's disease (Lei et al., 2010). The neurofibrillary tangles are also called as paired helical filaments.

The neurofibrillary tangles (Alonso et al., 1997) spread from one neuron to another with uncertain mechanisms (Frost et al., 2009).

Furthermore, τ hypothesis also supplements the aforementioned events in the formation of neurofibrillary tangles or paired helical filament (Mohandas et al., 2009; Alonso et al., 2001).

Caspases activation including caspase-6 and caspase-3 precede the formation of neurofibrillary tangles in the brains in Alzheimer's disease.

A study by de Calignon et al. (2010) involved examination of postmortem brain tissues in Alzheimer's disease. The regions of the brain with neurofibrillary tangles displayed the occurrence of massive neuronal loss, biomarkers of caspases activation, and apoptosis in the study.

But, the postmortem analysis of brain tissues could not predict whether caspase activation preceded the neurofibrillary tangles or occurred following the hyperphosphorylation and misfolding of τ-proteins in the progression of Alzheimer's.

In vivo study by authors de Calignon et al. (2010) on the Tg4510 strain (living τ-transgenic mice) with the help of multiphoton microscopy (visualization of living and intact biological tissues) visualized the activity of neurofibrillary tangles and activation of executioner caspases.

The activation of caspases preceded the formation of neurofibrillary tangles. The tangles formation occurred after a period of a few hours or a day following (de Calignon et al., 2010). caspase activation and that new tangle formation led to suppression of caspase activity in the intact and alive neuron.

Insertion of mutant **4R-**Tau isoform (de Calignon et al., 2010) in the wild-type animals resulted in executioner caspases activation, enzymatic cleavage of τ-protein, and formation of neurofibrillary tangles in the neurons.

Thus, a study by the authors suggested a new model involving executioner caspase-6 activation and cleavage of τ-protein into truncated τ that recruits nontruncated τ leading to the formation of tangles in the neurons inflicting neurotoxicity and inducing apoptosis in Alzheimer's disease.

Colocalization of active caspase-6 and its cleaved product (TauDeltaCsp6) has been reported in the neuritic plaques, and neurofibrillary tangles during the end-stage Alzheimer's disease (Guo et al., 2004).

Furthermore, neoepitope antibodies (detect the newly formed C-terminus or N-terminus after protein truncation) detected nearly threefolds increase (Guo et al., 2004) in the active caspase-6 in the temporal and frontal cortical regions of brains in Alzheimer's coupled with the spotting of neurofibrillary tangles, neuropil threads, and the neuritic plaques.

The caspase-6 accumulates in the tangles and neuritis in Alzheimer's leading to the impaired cytoskeleton of neurons, dysfunction, and neuronal apoptosis. Thus, active caspase-6 and its τ cleavage forms are involved in neuronal apoptosis in Alzheimer's disease.

Additionally, activated caspase-6 can cleave the τ-protein at the D421 (Delta Tau). An additional study by Roberts (1988) showed that delta-Tau serves as a nucleation center for the aggregation of τ-proteins into neurofibrillary tangles.

The glycogen synthase kinase-3β can induce (Roberts, 1988) phosphorylation of delta-Tau and is implicated in the formation of NFTs and its level of phosphorylation is proportionate to the severity of the cognitive decline in Alzheimer's disease.

Roberts (1988) identified the association of delta-tau (truncated τ) with biomarkers of neurofibrillary tangles in early and late onset of cognitive impairment in transgenic mice and brains in patients with Alzheimer's disease. **Interesting findings in the study revealed the correlated occurrence of delta-tau with the amyloid β42** (Roberts, 1988).

Suggested that caspase-6 and other executioner caspase are activated by ambulation of amyloid-β in the neurons resulting in caspase-induced cleavage of τ-proteins and concomitant hyperphosphorylation of τ-proteins are synchronously implicated in the pathology of Alzheimer's disease.

The colocalization and functional correlation between amyloid-β peptide and neurofibrillary tangles in the neurons are posited by several studies.

The executioner caspase including caspase-6 has a role in the cleavage of τ-protein. In vitro study by Chris Gamblin et al. (2003) delineated caspase-mediated cleave at Asp^{421} (conserved) aspartate residue located in the C-terminus of τ-protein. Also, similar activity was reported by authors in neurons treated with amyloid-β peptide (Aβ42) (Chris Gamblin et al., 2003). The caspase-mediated cleavage at Asp^{421} is followed by neuronal apoptosis.

Further, C-terminal 20 amino acid residues deficient and truncated τ-protein exhibited much higher potential than wild-type τ-protein to form insoluble

aggregates and exist as neurofibrillary tangles suggesting the harmful synergistic relation between amyloid-β42 and truncated τ implicated in the pathology of Alzheimer's disease.

Caspase-cleaved τ in mitochondrial dysfunction in Alzheimer's disease

Mitochondria are essential organelles for energy supply to brain cells, normal brain functions, and offer protection of brain tissues against neurodegeneration.

Mitochondrial dysfunction in the neurons is an additional causative factor involved in the initiation of disease and is largely implicated in the progression of Alzheimer's disease. Mitochondrial dysfunction can be manifested in terms of loss of membrane integrity and function, impaired bioenergetics, and impaired mitochondrial transport.

Possibly, neurons with altered mitochondrial function have a higher propensity to injury in response to oxidative stress that in turn exaggerate the function of mitochondria via interplay in a vicious cycle in the course of Alzheimer's disease.

Several studies posited interrelation between the caspase-induced truncated τ (Asp421) and mitochondrial dysfunction in the neurons in Alzheimer's.

A study by Quintanilla et al. (2009) identified the higher toxic potential of truncated τ (caspase cleavage at Asp-421) in the immortalized cortical neurons in comparison to the effects of wild-type τ-protein (full-length τ) and their effects on the mitochondrial dysfunction in terms of membrane integrity and function. The caspase cleaved τ (Asp421) exhibits inducible expression in the neurons and cells and is involved in the neurotoxic injury on the organelles like mitochondria and endoplasmic reticulum in neurons (Matthews-Roberson et al., 2008).

Quintanilla et al. (2009) utilized immortalized cortical neurons that, in response to induction, expressed either the full-length τ designated as "T4" or caspase-truncated τ at Asp421 designated as "T4C3" to delineate the toxic effects of truncated τ on the mitochondrial function.

Furthermore, authors identified raised levels of oxidative stress, higher mitochondrial fragmentation, and impaired membrane integrity in response to T4C3 τ in comparison to cells with T4 expression (Quintanilla et al., 2009).

The thapsigargin (**sesquiterpene lactone, noncompetitive inhibitor of the sarcoplasmic Ca^{2+} ATPase, induces a rise in cytosolic calcium level**) was applied to T4 cells and T4C3 cells (Quintanilla et al., 2009).

Quintanilla et al. (2009) reported a greater decline in the function of mitochondrial membranes in T4C3 cells than T4 cells.

In another attempt by Quintanilla et al. (2009), the T4C3 cells were pretreated with cyclosporine A that prevented the toxic effects of thapsigargin in the pretreated cells.

Thus, findings suggested that caspase cleaved Asp421 τ might be involved in mitochondrial dysfunction in terms of higher mitochondrial fragmentation,

impaired membrane integrity, and sensitizing neurons to oxidative stress that enhance the neuronal apoptosis in the progression of Alzheimer's disease.

Role of synaptic loss in Alzheimer's disease

Alzheimer's disease is associated with the reduction in density of cortical neurons. Latest studies were conducted utilizing [^{18}F]-fluorodeoxyglucose positron emission tomography (^{18}F-FDG) (a radiopaque compound used in medical imaging) to have images of the brain regions. The study indicated a reduction in the use of 2-deoxy-2-[fluorine-18]fluoro-D-glucose by the neurons as revealed in the positron emission tomography. It is suggestive of loss of neuronal synapses in brain regions of patients (Mosconi et al., 2010).

Authors claimed that loss of synapses occurs much before (nearly 20–30 years) the clinical manifestations of cognitive dysfunction and Alzheimer's disease (Jack and Holtzman, 2013; Jack et al., 2013).

Several studies claim the potential of synaptic loss in the impairment of cognitive function associated with Alzheimer's disease.

The study was conducted involving 18 patients (Terry et al., 1981) with Alzheimer's senile dementia. With the help of microscopy, counts of glial and neuronal perikarya were estimated in the midfrontal region and superior temporal gyrus of the participants with senile dementia, and findings were compared with 12 age-matched normal participants (Terry et al., 1981).

Terry et al. (1981) reported the presence of an 8% reduction in brain weight in participants with senile dementia.

The reduction of 46% neurons in the temporal region and 40% neuronal reduction in the frontal cortex were reported (Terry et al., 1981).

Furthermore, the thickness of senile plaques was inconsistent with the reduction in brain weight and neuronal count in different regions of the brain in patients with senile dementia.

Additionally, apoptosis and neuronal death are incriminated in the pathology of Alzheimer's disease by several researchers (Araki et al., 2000; Abe et al., 2003; Benaki et al., 2005).

But, apoptosis and neuronal death are not primarily associated with cognitive impairment in Alzheimer's disease.

The neocortical synaptic loss is closely implicated in the pathology of cognitive function in Alzheimer whose occurrence has been reported much earlier than the manifestation of the reduced count of cortical neurons, apoptosis, and neuronal death in Alzheimer's (Terry, 2000).

It is assumed that the loss of hundreds of axonal terminals in the cortical neurons results in synaptic loss, consequently leading to clinical manifestations of Alzheimer's. Despite the loss of axonal terminals, the cell bodies still survive. But the pronounced loss of synapses in the cortical neurons is followed by the impaired release of neurotransmitters, disturbed impulse conduction, deficiency of trophic factors leading to apoptosis, and death of cell bodies (Terry, 2000).

A study by Kidd (1963) and Gonatas et al. (1967) reported the presence of abnormal neurofibrils (helices) in the dendritic processes and postsynaptic processes in neurons in the brains of patients with Alzheimer. Another study by Luse and Smith (1964) reported enlarged axon terminals in the senile plaques containing large and dense vesicles and fibrils.

Furthermore, a study by Gonatas et al. (1967) showed alterations in the axons, presynaptic nerve endings, postsynaptic nerve endings, and dendritic processes in the senile plaques.

The aforementioned studies suggest the implication of altered synapses in the pathogenesis of impaired cognitive function in Alzheimer's.

Another study by Terry et al. (1991) provided the physical factor in the impaired cognitive function in Alzheimer's disease and proved that loss of synapses was significantly associated with the cognitive dysfunction in Alzheimer's. In the study by Terry et al. (1991), 15 patients with Alzheimer's disease and nine healthy participants were selected.

The authors reported a poor correlation among the presence of tangles and senile plaques with the psychometric indices. Moreover, a strong correlation was reported among the density of neocortical synapses with tangles and senile plaques with the psychometric indices.

Thus, the study revealed that a reduction in the density of synapses in the cortical region is associated with higher chances of cognitive dysfunction in Alzheimer's disease.

Still another study by Jacobsen et al. (2006) predicted early-onset of synaptic deficits and altered behavioral patterns in a mouse model of Alzheimer's disease.

Jacobsen et al. (2006) studied the advancement of neuronal dysfunction in terms of morphological, functional, and behavioral expressions in the Tg2576 mouse model of Alzheimer's disease.

Authors reported a reduction in density of the dendritic spine, disturbed long-term potentiation, and behavioral alterations that were observed to occur much earlier than the formation of senile plaques (Jacobsen et al., 2006).

The authors further elaborated the time duration for the expression of several defects in the brain regions in the mouse model of Alzheimer's (Jacobsen et al., 2006). The reduction in dendritic spine density was identified at age of 4 months, while the decrease in long-term potentiation was reported at 5 months old in the Tg2576 mouse model of Alzheimer's disease (Jacobsen et al., 2006).

Surprisingly, the rise in the ratio between $A\beta42$ and $A\beta40$ was reported during 4−5 months old (Jacobsen et al., 2006). Furthermore, the formation of plaques containing β-amyloids was reported at the age of 14 months in the Tg2576 mouse model of Alzheimer's disease.

Thus, synaptic loss and dysfunction are closely implicated in the pathology of cognitive dysfunction associated with Alzheimer's disease (Bastrikova et al., 2008).

The dendritic spine represents the small protrusion from the dendrite of the neuron. It receives a signal from the axon at the synapse. It functions to transmit a signal to the cell body of neurons (Alvarez and Sabatini, 2007). The major cytoskeleton named as filamentous actin (F-actin) maintains the shape and function of dendritic spines.

Furthermore, loss of the dendritic spine is the early characteristic of Alzheimer's disease.

A study was conducted by Kommaddi et al. (2018) in APPswe/PS1ΔE9 male mice (a mouse model of Alzheimer's disease) and determined the neurotoxic effect of β-amyloid on the F-actin disassembly in dendritic spines and its potential in the pathology of cognitive dysfunction in Alzheimer's disease.

Kommaddi et al. (2018) reported depolymerization of synaptosomal filamentous-actin and a rise in the concentration of globular-actin (G-actin). These events were observed at age of 1 month in the AD model of a male mouse suggesting the role of equilibrium between the levels of F-actin and G-actin optimal behavior.

Furthermore, Kommaddi et al. (2018) identified that depolymerization process involving F-actin in dendritic spines results in the disruption of the normal organization of outwardly oriented F-actin rods in cortical neurons in the mouse model, APPswe/PS1ΔE9.

Thus, it can be inferred that cytoskeletal protein, F-actin, is helpful in supporting the normal structure and function of synapses in the brain. The altered state of F-actin is responsible for the loss of dendritic spines and synaptic loss in specific regions of the brain in Alzheimer's.

Thus, synaptic loss and dysfunction constitute the basis of cognitive impairment associated with Alzheimer's disease.

Role of microglia in the pathogenesis of Alzheimer's disease

Microglia constitutes the fixed macrophages in the brain and spinal cord.

These represent around 15% of the total cells in the brain. Microglia serves as the first defense line immune cells in the central nervous system.

Microglia cells sense the microenvironment of the central nervous system and are the prominent immune cells in CNS (Xie et al., 2017). In response to external stimuli, the microglial cells are activated to release proinflammatory cytokines and lead to modulation in the secretion of neurotransmitters in CNS (Nimmerjahn et al., 2005).

Contrary to the above-mentioned role of microglia, the senile brain contains activated microglia cells coupled with the presence of raised levels of cytokines namely interleukin-1 β, interleukin-6, and tumor necrosis factor-α (Sierra et al., 2010).

The study revealed different expression patterns of microglia cells in the brains of aged mice in comparison to young mice with a wide difference in the response to lipopolysaccharides (Holtman et al., 2015). The senile brains in mice models showed upregulated expression of the major histocompatibility complex II molecules and complement receptor 3 suggesting their roles in the activation and aggregation of microglia (Barrientos et al., 2006).

Interestingly, similar changes in the expression of microglia in terms of activated microglia associated with proinflammatory cytokines have been reported in brains in patients with Alzheimer's disease and other neurodegenerative diseases (Xie et al., 2017).

The activated microglia can be classified into M1 phenotype with proinflammatory effect and M2 phenotype with immunosuppressant effect. The microglia cells with M2 phenotype possess phagocytic potential toward β-amyloid, and it was shown in a study involving APP/PS1 transgenic mice. The study further described the transdifferentiation potential of M2 phenotype microglia at the age of 6 months. These cells passed through transdifferentiation and acquire the M1 phenotypes at the age of 18 months. The transdifferentiation capability of the M2 phenotype is dependent on the levels of soluble β-amyloids oligomers.

A study by Colton et al. (2006) involved analysis of expression patterns of microglia alternative activation genes in Alzheimer's and in mouse models Alzheimer.

Colton et al. (2006) posited that microglia are linked with a higher tendency for neuritic plaques in brain tissues in Alzheimer's. Microglia are the main immune cells in the brain. The amyloid-β peptide fragments are assembled to form the neuritic plaques. Authors further commented that microglia on exposure to amyloid-β peptides induce expression of proinflammatory cytokines (Colton et al., 2006). Thus, microglia after exposure to amyloid-β induce T-cell-mediated (subset Th-1) immune response with release of proinflammatory cytokines.

Colton et al. (2006) studied expression patterns of genes namely arginase I and mannose receptor-1, found in inflammatory zone 1, and chitinase 3-like 3 linked to alternative activation in microglia in the brain in Alzheimer's.

The expression patterns of genes were studied in a mouse model of Alzheimer's, APPsw (Tg-2576) (Colton et al., 2006) for amyloid deposition and a mouse model of Alzheimer's, Tg-SwDI for cerebral amyloid angiopathy (Colton et al., 2006).

Authors reported a rise in levels of Arginase-1 mRNA, mannose receptor-1-mRNA, and chitinase 3-like 3 mRNA in the brain tissues in the Tg-2576 mouse model.

Furthermore, TNFα-mRNA levels were increased with NOS2 mRNA levels remained unchanged in the brain tissues in the Tg-2576 mouse model. The levels of TNFα mRNA were raised in the brains of the Tg-SwDI mouse model.

Conclusively, the mRNA levels for TNF-α, MRC1, AGI, and chitinase-3 like 3 were significantly elevated, whereas mRNA levels of inducible NOS2 and IL-1β remained unchanged.

It can be inferred that microglia cells in brains with Alzheimer's disease exist in a composite activation state including features of both classical activation as well as alternative activation.

Microglial TREM2 receptor in Alzheimer's disease

The triggering receptor expressed on myeloid cell 2 (TREM-2) is the membrane receptor. The TREM2 gene controls the synthesis of TREM2 protein in humans (Bouchon et al., 2000). The TREM2 receptor is prominently expressed on the surface of microglial cells and macrophages (Schmid et al., 2002).

Ligand binding with the extracellular domain of TREM2 activates its intracellular domain and transduces the signals affecting the phosphorylation of downstream effectors including Phosphoinositide 3-kinase, phospholipase C γ, and **Guanine nucleotide exchange factor** (Vav2/3) (Xing et al., 2015; Peng et al., 2010).

However, variants of TREM2 are linked to a higher propensity to the development of Alzheimer's disease suggesting the potential of microglial cells in the pathology of AD (Gratuze et al., 2018).

The most frequent variant of TREM2 protein is the rs75932628 that mediates the arginine to histidine substitution at position 47 (R47H) Korvatska et al., (2015); Lill et al., (2015) and exhibited a much higher predisposition to the development of AD.

The microglial cell surface receptor, TREM2 binds with the amyloid-β peptide and mediates the receptor-mediated phagocytosis of amyloid-β and hastens the proteasomal-dependent degradation of amyloid-β.

Furthermore, the TREM2 receptor interacts with transmembrane DAP12 protein in response to amyloid-β and consequently activates the downstream effectors and GSK3β pathway. But, the study revealed that the deletion of TREM2 led to the impaired phagocytic potential of microglial cells resulting in the reduced clearance of amyloid-β (Zhao et al., 2018; Zhong et al., 2018).

Other studies involved deleted the TREM2 5XFAD model of mice (Wang et al., 2015; Lee et al., 2018). The 5XFAD mice express human APP and PSEN1 transgenes coupled with five AD-associated mutations; human APP transgene with K670N/M671L mutation (Swedish mutation) (Mullan et al., 1992), I716V (Florida mutation (Eckman et al., 1997), and V717I (London mutation (Goate et al., 1991), and human PS1 gene with M146L and L286V mutations (Citron et al., 1998).

Wang et al. (2015) and Lee et al. (2018) reported that deleted the TREM2 5XFAD model of mice resulted in raised amyloid plaques and a rise in the number of dystrophic neuritis.

The overexpression of TREM2 as linked with the reduction in amyloid-β lesion and improved cognitive deficiency in mice with Alzheimer's (Lee et al., 2018).

Thus, findings suggested that the TREM2 receptor is essential in microglial-mediated degradation and clearance of amyloid-β peptide in brain regions and further mutants of TREM2 might be implicated in the pathology of Alzheimer's disease via **hampering microglia-mediated degradation and accumulation of Aβ in brain regions.**

Microglial LRP1 receptor in Alzheimer's disease

The low-density lipoprotein receptor-related protein 1 (LRP1) is also called as α-2-macroglobulin receptor (A2MR) or cluster of differentiation 91 (CD91), or apolipoprotein E receptor (APOER).

The gene LRP1 controls the synthesis of LRP1 protein in humans (Herz et al., 1988).

The LRP1 receptor represents the type I transmembrane protein (attached to phospholipid bilayer in the plasma membrane with stop-transfer anchor sequence, with N-terminal domains oriented to the ER lumen during synthesis) and it mediates phagocytosis and degradation of ligands like amyloid-β and APOE (Kowal et al., 1989).

The LRP1 receptor is abundantly expressed in microglia Yang et al., (2016), neurons (May et al., 2004), and astrocytes (Wyss-Coray et al., 2003).

Several studies provide information related to the involvement of LDLR-related protein (LRP)1 in the pathology of late-onset Alzheimer's disease.

These studies describe that the LRP1 receptor controls the formation and clearance of amyloid-β peptides in brain regions and maintains the homeostasis between the rate of formation of amyloid-β and its clearance via phagocytosis and degradation of Aβs in the brain. Thus, impaired activity of LRP1 contributes to the pathology of Alzheimer's in Aβ-independent and Aβ-dependent pathways (Shinohara et al., 2017).

Furthermore, several preclinical have yielded findings showing the pathological role of LRP1, APOE, and gene APOE in the pathology of Alzheimer's.

Astrocytes are the nonneuronal cells representing subtypes of glial cells in the brain involved in the clearance of amyloid-β.

A study by Chuang et al. (2016) described that expression of LRP1 in the microglial cells offers a protective function against the proinflammatory cytokines. It is further supplemented in the LRP1 gene deletion model that showed suppressed expression of LRP1 in microglial cells, rise in the level of proinflammatory cytokines, and LPS-mediated inflammatory injury in microglial cells.

Other study by Liu et al. (2017) used LRP1 knockout mice models and cell cultures assessing the role of low-density lipoprotein receptor-related protein 1 (LRP1) in amyloid-β phagocytosis in astrocytes.

Liu et al. (2017) reported the critical function of LRP1 in astrocytes in Aβ clearance and described the role played by several enzymes under the control of LRP1 for degradation and clearance of Aβ in the brain and suggested that LRP1 could be the novel drug target in AD.

In the presence of contrasting findings, the role of the LRP1 receptor in microglial cell-mediated amyloid degradation and its role in APOE metabolism remains uncertain with an uncertain contribution to the pathology of Alzheimer's.

Microglial advanced glycation end-product receptor in Alzheimer's disease

The receptor for advanced glycation end products (RAGE) binds with multiple ligands (multiligand receptor). The RAGE belongs to the superfamily of immunoglobulins. It serves as a cell surface receptor for binding with advanced glycation end products (AGEs) of proteins, lipids, and nucleic acids (Schmidt et al., 1994).

Furthermore, the RAGE also functions as a receptor for the amyloid-β-peptide in the brain region. The study revealed upregulated expression of RAGE in microglial cells in patients with Alzheimer's disease leading to neuronal dysfunction (Li et al., 2012). Additionally, the study delineated the role of the RAGE receptor in mediating the harmful effects of accumulated AGEs in brain regions on the hyperphosphorylation of τ-protein that is closely implicated in the pathology of Alzheimer's (Yan et al., 1996).

Several studies (Lue et al., 2001; Sasaki et al., 2001; Miller et al., 2008) analyzed the brain tissues in patients with Alzheimer's disease and controls without dementia for the distribution pattern and extent of expression of RAGE. Collectively, studies pointed to the distribution of receptors for AGEs prominently in the neurons and microglial cells in cortex and hippocampus regions of the brain in both Alzheimer patients and controls without dementia. However, studies demonstrated the overexpression of RAGE in the brain tissues in patients with Alzheimer's in comparison to the controls.

Studies involved the use of a transgenic mice model of AD with targeted overexpression of mutant APP in neurons. The transgenic mice model showed overexpression of RAGE that further intensified the harmful effects of already overexpressed mutant APP protein in the neurons in terms of impaired cognitive function, synaptic plasticity, and enhanced expression of biomarkers for the pathology of Alzheimer's disease (Arancio et al., 2004; Cho et al., 2009).

Later on, treated transgenic mice showed reduced expression of RAGE in neurons with overexpressed mutant APP and resulted in improved cognitive impairment and decline in biomarker expression of neuropathological changes in neurons (Arancio et al., 2004; Cho et al., 2009).

Thus, it can be concluded that the receptor for AGEs expresses prominently in the neurons, and microglial cells in the cortex and hippocampus in the brain and is closely involved in the internalization, degradation, and clearance of amyloid-β and irregularity in its function is associated with the pathology of Alzheimer's disease. Furthermore, RAGE might be the novel drug target in the management of Alzheimer's.

Microglial Fc γ receptors in Alzheimer's disease

The Fc fragment of IgG receptor IIb is encoded by the FCGR2B gene in humans. The Fc γ receptor (FcγRIIb) is the receptor with low affinity for the Fc domain of IgG. It is involved in the immune cell-mediated phagocytosis and formation of antibodies by B-lymphocytes (Adolfsson et al., 2012).

Fc γ receptors are predominantly expressed on the surface of microglial cells in the brain.

The FcγRIIb receptor has been reported to possess inhibitory properties. The FcγRIIb receptor can bind with the IgG-antigen immune complex and has been identified on the surface of B-lymphocytes, neutrophils, and macrophages (Katz, 2002). The FcγRIIb receptor inhibits B cell receptor-mediated immunity and suppresses the autoimmunity (Pritchard and Smith, 2003).

In a study involving mice with *Fcgr2b* gene knockout, exaggerated B-lymphocyte-mediated humoral response was reported that predisposed the B6 strain of mice (laboratory inbred mice model) to autoimmune disease (Bolland and Ravetch, 2000).

Point mutation leading to substitution of Ile232 with Thr (I232T) in the transmembrane domain of FcγRIIb leads to the impaired inhibitory activity of the receptor that manifests as autoimmune disorder (Pritchard and Smith, 2003).

The expression of FcγRIIb receptors was reported to be enhanced in hippocampal regions of the brains in patients with Alzheimer's disease (Kam et al., 2013).

Further, the monomeric and oligomeric forms of amyloid-β inside the neurons in the hippocampus region in brains in patients with AD were identified close to the immuno-reactivity of the FcγRIIb receptor (Kam et al., 2013). In the study, western blotting revealed the overexpression of FcγRIIb in the brain in the mouse model of AD, while the *Fcgr2b* Knockout mice model showed suppressed expression of FcγRIIb in the brain in the mice model (Kam et al., 2013). The DNA sequence analysis in the study revealed that western blot analysis showed that $A\beta_{1-42}$ in the SH-SY5Y neuroblastoma cells and primary cultured neurons induced the overexpression of gene *Fcgr2b* leading to a rise in the levels of FcγRIIb protein (Kam et al., 2013). Authors suggested $A\beta_{1-42}$ stimulated phosphorylation and activation of c-Jun N-terminal protein kinase leading to activation of downstream c-Jun (transcription factor) in cortical neurons in the wild-type mouse model and the aforementioned events were not reported in the neurons in the *Fcgr2b* Knockout mice model suggesting the involvement of FcγRIIb protein (Kam et al., 2013).

Furthermore, inhibition of c-Jun N-terminal protein kinase by inhibitors led to suppression of $A\beta_{1-42}$-mediated neurotoxicity in primary cortical neurons (Kam et al., 2013).

A study by Roberson et al. (2007) delineated the wide distribution of the FcγRIIb receptor in the microglial cells in the hippocampus, cortex, and cerebellum of the brain. Further, the expression of FcγRIIb was elevated in the cortex of the 6 and 17-month old of the mice model of AD that expresses the mutant amyloid precursor protein (hAPP) for the familial Alzheimer's disease (Roberson et al., 2007).

It can be concluded that the amyloid-β42 can control the expression of the *FCGR2B* gene via the regulation of transcription through the activity of JNK-c-Jun kinase. Thus, the FcγRIIb receptor is implicated in the Aβ42-induced neurotoxicity and might be implicated in the pathology of Alzheimer's disease.

Microglial CD36 transmembrane protein in Alzheimer's disease

The cluster of differentiation 36 (CD36 protein) is also called as platelet glycoprotein4, fatty acid translocase, and scavenger receptor class B member 3.

The class B scavenger receptor (CD36) is predominantly expressed in the vascular endothelial cells and microglial cells in the brains of normal persons and patients with Alzheimer's disease patients.

The CD36 protein binds fibrillar amyloid-β and is reported to be involved in neurotoxicity (Coraci et al., 2002). The class A scavenger receptor (CD36) can initiate the binding of β-amyloid fibrils to the surface of microglial cells and macrophages in the brain regions (Coraci et al., 2002).

The CD36 receptor mediates the formation of H_2O_2 by activated microglial cells after the binding of fibrillar amyloid-β.

Possibly, the dysregulation in the function of CD36 might be implicated in the surplus production of H_2O_2 by microglial cells in response to fibrillar amyloid-β that may contribute to oxidative stress and may serve as a factor in the pathology of Alzheimer's disease.

A study by Coraci et al. (2002) further showed that the application of antibodies to CD36 protein resulted in the suppression of fibrillar amyloid-β-induced production of H_2O_2 by microglial cells. These facts supplemented the role of CD36 in mediating H_2O_2 production by macrophages and microglial cells after binding with fibrillar amyloid-β and oxidative stress and inflammatory response in brains.

Thus, fibrillar β-amyloid can mediate inflammatory response via CD36 signaling pathway in microglial cells and macrophages in the brain.

Another evidence linking CD36 with Alzheimer's disease (Akiyama, 1994; Blass, 2002) was provided in the studies.

Neuroinflammation in response to accumulated amyloid-β is the basic feature in the pathology of Alzheimer's disease. During the early phase of Alzheimer's, amyloid-β binding with CD36 receptor activates microglial cells leading to enhanced phagocytosis, degradation, and clearance of amyloid-β from the brain regions. Sustained amyloid-β impairs the protective function of the body leading to persistent activation of microglial cells, formation of H_2O_2, oxidative stress, and inflammatory state of neurons implicated in the progression of Alzheimer's disease.

A study by Moore et al. (2002) demonstrated that fibrillar β-amyloid mediates binding of CD36 receptor with Lyn (belongs to Src family of protein tyrosine kinases) and subsequently phosphorylates and activates p44/42 mitogen-activated protein kinase and Fyn kinase.

Additionally, targeted deletion of Src kinases leads to suppression of macrophage-induced inflammatory cascade of events after binding with β-amyloid and reduced activation of microglial cells (Moore et al., 2002).

Thus, the findings suggest the possible potential of CD36 in mediating chronic activation of macrophages and microglial cells after binding with amyloid-β and formation of surplus H_2O_2 coupled with initiation of inflammatory response in the neurons that are implicated in the progression of Alzheimer's disease.

Astrocytic glial α 7 subtype of nAChR (α7nAChRs) in Alzheimer's disease

The nicotinic acetylcholine receptor (nAChRs) is widely expressed in the brain tissues including microglial cells, astrocytes, and oligodendrocytes. It is involved in several physiological functions including cognitive function (Velez-Fort et al., 2009).

The activation of the α7nAChRs receptor mediates the influx of calcium ions leading to the release of neurotransmitters. It has a role in the excitability of neurons (Gray et al., 1996).

Colocalization of Ab42 and the α7 nicotinic acetylcholine receptor (neuronal pentameric ligand-gated cation channel) was reported in the neuritic plaques in the brain regions in patients with sporadic Alzheimer's disease (Wang et al., 2000).

A study by Wang et al. (2000) involved overexpressing α7nAChR human neuroblastoma cells. These cells were destroyed in response to Ab42 neurotoxicity; additionally, nicotine and epibatidine (α7nAChR agonists) provided protection against the Ab42neurotoxicity.

Authors suggested that Ab42 suppresses α7nAChR-mediated influx of calcium ions and the associated release of acetylcholine from the neurons and astrocytes. The release of neurotransmitters determines the cognitive function and derailed neurotransmitter release is linked with cognitive deficits.

The colocalization of α7nAChR with neuritic plaques in brain regions, the high affinity of α7nAChR to Ab42, and the potential of α7nAChR in regulating neurotransmitter release are interrelated and might be involved in the pathology of Alzheimer's disease.

The Ab42 exhibits neurotoxicity via **α7 nicotinic acetylcholine receptor-mediated-mitogen-activated protein kinase pathway in the astrocytes and neurons.**

A study by Dineley et al. (2001) revealed that sustained deposition of Ab42 in neurons, as was reported in the AD mice model, resulted in an upregulated expression of α7 nAChR coupled with simultaneous (Dineley et al., 2001) downregulation of the 42 kDa isoform of the extracellular signal-regulated kinase (ERK2) the hippocampus regions in brains in aged animals.

Authors suggested that impaired signal transduction in the hippocampus region of the brain in Alzheimer's disease might be due to elevated deposition of Ab42 and prolonged activation of the ERK2 pathway via α7 nAChR-receptor **resulting in downregulation of ERK2 and reduced phosphorylation of CREB protein.**

Another probable mechanism in the Aβ42 oligomer-induced toxicity via α7nAChRs is mediated through the release of glutamate. The Aβ42 oligomer can directly bind with the α7nAChR in the astrocyte. The interaction induces the release of glutamate from the astrocyte. The glutamate in turn binds and activates the extrasynaptic glutamate receptor named as N-methyl-D-aspartate receptor in the astrocyte and neurons. This leads to efflux of calcium ions impairing the mitochondrial function, leading to activation of caspase 3, hyperphosphorylation of τ, and surplus accumulation of reactive oxygen species. The cascade of molecular events damages the dendritic spines and impairs impulse transmission and disturbs the neuron to astrocyte communication (Pirttimaki et al., 2013).

Overall, it can be concluded that Ab42 can mediate neurotoxicity via **the involvement of α7nAChR in the astrocytes and an associated cascade of events that might have a role in the pathology of Alzheimer's disease**.

Astrocytic glial calcium-sensing receptor in Alzheimer's disease

The calcium-sensing receptor (CaSR) belongs to the family of C G-protein coupled receptors.

The CaSR regulates the homeostasis of free calcium (Hofer and Brown, 2003). It is expressed in neurons, astrocytes, and microglial cells in most of the regions of the brain (Yano et al., 2004).

In normal physiology, it is essential in the normal development of axons and dendrites. It controls the migration of neurons and glial cells in the brain and modulates synaptic plasticity (Riccardi and Kemp, 2012).

The Aβ42 oligomers can bind with the CaSR in astrocytes and neurons leading to activation of downstream effectors that inhibit the degradation of Aβ42 oligomers. Subsequently, accumulated Ab42 exhibits its neurotoxicity (Chiarini et al., 2016).

Furthermore, the surplus deposition of Ab42 and its binding with CaSR can enhance the expression of nitric oxide synthase-2 and raise the production of nitric oxide in neurons and astrocytes.

Also, binding of Ab42 with the CaSR operates via activation of MEK/ERKdependent pathway and increases the production of vascular endothelial growth factor-A. All the molecular events contribute to neuroinflammation in the brain (Armato et al., 2013).

Furthermore, amyloid-β42 oligomers can bind with the calcium-sensing receptor (CaSR) on the surface of astrocytes and neurons leading to the formation of Ab42-CaSR complex that alters the amyloid precursor protein cleavage pathway from non-amyloidogenic to amyloidogenic pathway.

Thus, α-secretase-mediated cleavage of amyloid precursor protein is shifted to enhanced cleavage activity of β-secretase in response to Ab42- CaSR complex leading to the formation of Aβ42/Aβ42-os peptides that are closely linked to the pathology of Alzheimer's disease (Chiarini et al., 2017).

Interestingly, the study revealed that the addition of calcilytics (selective allosteric CaSR antagonist) to the Aβ42/Aβ42-os-exposed human neuron cultures) mediated (Chiarini et al., 2017) the α-secretase-induced cleavage of APP to form neuro-protective sAPPα. The CaSR antagonist, additionally, suppressed the neurotoxic effects of Aβ42-CaSR complex-mediated signaling cascade irrespective of the presence of proinflammatory cytokines secreted by activated microglial cells (Chiarini et al., 2017).

Furthermore, the study delineated that Aβ42-CaSR complex-mediated signaling human neuronal cell cultures (Chiarini et al., 2017) induced the release of hyperphosphorylated τ-proteins and nitric oxide.

A study (Chiarini et al., 2017) also revealed that prolonged exposure to Ab42 resulted in overexpression of vascular endothelial growth factor A in the vessels exposed to Aβ42.

A study (Chiarini et al., 2017) revealed neo-angiogenesis in the hippocampus in the brain in patients with Alzheimer's and amnestic minor cognitive impairment.

Thus, Aβ42-CaSR complex-mediated release of hyperphosphorylated τ-proteins, nitric oxide, vascular endothelial growth factor A collectively impairs the microenvironment brain favoring the disintegration of synapses, neuroinflammation, and neuronal apoptosis contributing to slowly spreading Alzheimer's disease.

Microglia-τ interaction in Alzheimer's disease

The hyperphosphorylated τ-proteins induce the activation of microglial cells. The activated microglial cells mediate the intracellular signaling pathway leading to a rise in the secretion of proinflammatory cytokines prominently IL-6, IL-1β, and TNF-α in the extracellular space, thus promoting neuroinflammation (Wang et al., 2013).

Possibly, hyperphosphorylated τ-proteins induce the activation and nuclear translocation of NF-κB leading to transcription of proinflammatory cytokines and formation of NLRP3-ASC complex and oligomerization of inflammasome and activation of caspase1 in the activated microglial cells (Stancu et al., 2019). Conversely, activated microglial cells in turn influence the formation of τ-proteins and their phosphorylation. Activated microglial cells can directly mediate receptor-mediated internalization of pathogenic τ-proteins, their proteasomal degradation, and clearance from the brain regions (Luo et al., 2015).

Several publications described the role of CX3CR1, chemokine receptor in τ clearance. The CX3CR1 binds with τ-protein and hyperphosphorylated τ and induces the microglial cell-mediated phagocytosis, degradation, and clearance (Bolos et al., 2017).

The target deletion of CX3CR1 impairs microglia-mediated phagocytosis of pathogenic τ-protein leading to surplus accumulation of τ-proteins in brain regions (Evans et al., 2018).

The microglial cells mediate the propagation of hyperphosphorylated τ via the formation of exosomes (extracellular vesicles).

Additionally, activated microglial cells can modulate the pathogenesis of τ-protein indirectly via a proinflammatory cascade of molecular events. The activated microglia secrete the proinflammatory cytokines into the extracellular space that further activates p38 and CDK5 (τ-kinases) and exaggerates the pathogenesis of pathogenic τ-proteins (Kitazawa et al., 2005).

Conclusively, pathogenic τ and activated microglial cells are interrelated via **series of molecular events that collectively contribute to the progression of Alzheimer's pathology**.

Microglial PU.1 transcription factor in Alzheimer's disease

The transcription factor PU.1 (SPI1) regulates the normal development of microglial cells and their function in the brain (Smith et al., 2013).

The PU.1 (SPI1) is expressed in the microglial cells in the CNS. The levels of PU.1 in microglia either deletion or overexpression in the BV2 cell line was reported to be linked with the expression of genes involved in regulating neuroinflammation (Huang et al., 2017).

A recent study involving the genome-wide survival analysis Huang et al. (2017) reported the presence of rs1057233g in the CELF1 AD risk locus. This haplotype showed downregulated gene expression of PU.1 in macrophages and microglial cells related to delayed onset of Alzheimer's disease (Huang et al., 2017).

The level of PU.1 gene expression is associated with the predisposition to the development of Alzheimer's disease. The downregulated PU.1 expression is protective against Alzheimer's disease while the overexpression of the PU.1 gene predisposes to a higher risk for Alzheimer's (Lavin et al., 2014).

Myocyte-specific enhancer factor 2C (MEF2C) determines cardiac morphogenesis, myogenesis, angiogenesis, and neurogenesis. It is involved in the normal growth of cortical neurons (Bi et al., 1999).

The SPI1/PU.1 and MEF2C interaction has been reported to be the prime regulator of microglial enhancer sequences, thus controlling the activation state and inflammatory response by microglial cells mediated by Ab42 and hyperphosphory-lated τ-proteins (Lavin et al., 2014).

Since the levels of the PU.1 gene are the important determinant of risk factors for Alzheimer's disease, its expression is regulated by transcription factors (Mak et al., 2011).

The PU.1 remains bound to myeloid-specific enhancers in microglia. Several neurotoxic ligands as Ab42, pathogenic τ mediate activation of stimulus-dependent transcription factors (Holtman et al., 2017; Pimenova et al., 2021) in the microglial cells. The activated transcription factors in turn associated with the enhancer sequences of PU.1 leading to higher expression of PU.1, raised activation of microglial cells, increased levels of proinflammatory cytokines, and higher neuro-inflammation in the brain regions.

Thus, the putative aforementioned cascade of molecular events and hierarchical contiguity in the gene expression is implicated in the activation of the microglial cell, phagocytosis, the release of proinflammatory cytokine, production of ROS, mitochondrial function, and lysosomal function.

PU.1 serves as the prime regulator of the development, function, and homeostasis of microglial cells. It also controls the expression of several genes including TREM2, CD36, APOE, and MS4A4 that have a role in Alzheimer's disease (Satoh et al., 2014; Pimenova et al., 2021).

The latest study revealed multiple roles of PU.1 in the regulation of activated microglial cells to control neuroinflammation in the brain associated with the manifestation of Alzheimer's disease (Pimenova et al., 2021).

A study by Pimenova et al. (2021) described that targeted deletion of the PU.1 gene was related to a reduced microglial-induced inflammatory response in the brain regions.

In the study (Pimenova et al., 2021) involving the use of a BV2 cell line, the Ab42 mediates the formation of reactive oxygen species via activated microglial cells and PU.1 expression is linked with the regulation of ROS production in micro-glial cells. Furthermore (Pimenova et al., 2021), PU.1 expression was further linked with LPS-induced production of nitric oxide and release of proinflammatory cyto-kine by activated microglial cells.

Thus, PU.1 expression regulates multiple factors involved in the mediation of microglial-induced inflammatory response in the brain regions and thus levels of PU.1 expression modulate the neuroinflammation that is a clinical sign in Alzheimer's disease.

TLR4 (toll-like receptor)-mediated signaling pathway in Alzheimer's disease

The Toll-like receptors are the single-pass transmembrane proteins that traverse across the lipid bilayer a single time and also referred to as Bitopic proteins or single-spanning transmembrane proteins (Takeda and Akira, 2005).

Toll-like receptors are expressed on the surface of macrophages, dendritic cells including microglial cells (Mahla et al., 2013). Probably, TLRs acquire a name from the Toll gene found in Drosophila that encodes a protein homologous to Toll-like receptors. The Tool gene was discovered by Christiane Nusslein-Volhard and Eric Wieschaus in 1985 (Medzhitov et al., 1997).

There are several Toll-like receptors from TLR1 to TLR13 but the Toll-like receptor-4 is actively involved in the signaling pathway in the brain tissues in Alzheimer's disease.

Toll-like receptor-4 expression is reported in the microglia. A study by Lehnardt et al. (2003) identified activation of Toll-like receptor 4-stimulated pathway in the central nervous system that was involved in the triggering innate response in neurodegeneration.

A study (Fellner et al., 2013) posited the role of TLR4 in mediating activation of microglial cells, phagocytosis, and release of proinflammatory cytokine coupled with the formation of reactive oxygen species in brain tissues. Prolonged exposure to amyloid-β leads to dysregulation of Toll-like receptor-4 coupled with microgliosis and astrogliosis are involved in the surplus accumulation of α-synuclein protein in the specific regions of the brain leading to α-synucleinopathies.

Precise role of TLR4 in Alzheimer's disease can further be supplemented by several studies including the study of genetic polymorphisms in the TLR4 gene.

A study by Minoretti et al. (2006) was conducted on the role of genetic polymorphisms of the TLR4 gene (either protective or neurotoxic) toward implication in Alzheimer's and age-related pathologies in the brain.

The variant protein Asp299 Gly is the result of the genetic polymorphism of the TLR4 gene. The study involved a cohort of total ($n = 277$) late-onset Alzheimer's disease (LOAD) and other ($n = 300$) healthy participants as controls. Genotypes of all participants were determined for the presence of variant Asp299 Gly TLR4 polymorphism.

The authors reported a very low frequency of variant Asp299 Gly polymorphism of the TLR4 in patients with LOAD and suggested its role in the attenuation in TLR4-mediated signaling and flattened proinflammatory response.

Thus, it can be inferred that variant TLR4 Asp299 Gly might be neuroprotective in the development of late-onset Alzheimer's disease.

Contrary findings were provided by a study involving genetic polymorphisms of the TLR4 gene in the Han Chinese population.

A study by Wang et al. (2011) determined the potential of variant *TLR*4/11367 polymorphism toward the development of late-onset Alzheimer's disease.

The case-control research design was adopted in which the presence of variant *TLR*4/11367 (Wang et al., 2011) polymorphism was studied in the Han Chinese population (Wang et al., 2011) that comprised a total of 137 LOAD patients and other ($n = 137$) healthy participants as controls.

Wang et al. (2011) **reported the high association between variant *TLR*4/11367 polymorphism and the development of LOAD in a Han Chinese population.**

The aforementioned studies predict the involvement of TLR4 with Alzheimer's disease in separate populations displaying varied effects on the progression of Alzheimer's disease suggesting the presence of additional causative factors in the pathology of Alzheimer's.

The TLRs are the pattern recognition receptors expressed on the surface of macrophages including microglial cells. The TLRs undergo dimerization that can be either homodimerization or TLR2 can dimerize with TLR1 or TLR6 (heterodimerization) (Yamamoto et al., 2003).

The TLRs can recognize both the exogenous ligands that may include lipopolysaccharides, bacterial lipopeptides, double-stranded DNA, flagellin, double-stranded RNA or single-stranded RNA (ssRNA) and endogenous ligands including amyloid-β peptides, and damage-associated molecular patterns from injured cells.

The fibrillary Aβ can bind with TLR4 and in turn activates the receptor on the surface of the microglial cell.

The TLR-mediated signaling pathways play a pivotal role in the microglial activation, the release of proinflammatory cytokines, and induction of neuroinflammation involved in the pathogenesis of neurodegenerative diseases including Alzheimer's disease (Hayward and Lee, 2014).

The TLR-4 activation induces initiation of the MyD88-dependent and TRIF-dependent pathways in the microglial cell.

TLR-4 activated MyD88-dependent pathway in microglia

The binding of fibrillar amyloid-β with the TLR4 leads to dimerization of the receptor and in turn conformational change in the cytosolic domain of the TLR4 (Bonnert et al., 1997).

The receptor recruits the MyD88 (adaptor protein) belonging to the TIR family of proteins (Lee et al., 2012).

The adaptor protein, MyD88 further recruits and activate the IRAK kinases (Kawai and Akira, 2010), subsequently activated IRAK kinase activates TRAF6 protein that further catalyzes the polyubiquitination (attaching several ubiquitin residues to target protein) the TAK1 protein and favor its binding with IKK-β (Lee et al., 2012).

The activated TAK1 further activates the NF-κB signaling pathway. This protein complex controls the transcription of genes and synthesis of proinflammatory cytokines (Albensi and Mattson, 2000).

Under normal and healthy conditions, the NF-κB dimers are localized in the cytosol in "inactive form" due to the IκB protein called an inhibitor of κB.

The IκB kinase contains ankyrin repeats that mask the nuclear localization signals in NF-κB and maintain it in the sequestered and inactive form in the cytosol (Jacobs and Harrison, 1998).

The IkB kinase (IKK) is made up of two subunits IKKα and IKKβ that have a catalytic function, while the third subunit is called IKKγ (NEMO) that serves as the master regulatory subunit.

The activated TAK1 in turn activates the IKK complex via binding of IKKγ to kinase subunits IKK-α and IKK-β.

The activated IkB kinase further phosphorylates serine residues (S32 and S36) located inside the amino-terminal region of IκBα (inhibitor of NF-κB). Subsequently, activated IκBα is ubiquitinated leading to its degradation by the proteasome (Ghosh et al., 1998).

Thereafter, NF-κB protein undergoes nuclear translocation where it binds with the promoter sequence of the specific genes that are involved in the formation of proinflammatory cytokines. The NF-κB protein serves as a transcription factor that enhances gene expression.

These events are well characterized in the brain regions in patients with Alzheimer's disease (Arancibia et al., 2007).

Thus, TLR4-induced and MyD88-dependent signaling pathway in the microglia has its effect in inducing the inflammatory response in the brains in Alzheimer's.

TLR-4 activated TRIF-dependent pathway in microglia

The binding of fibrillar amyloid-β with TLR4 in microglial cells activates the TRIF-dependent signaling pathway in microglial cells (Louis et al., 2018). The TRIF (TIR-domain-containing adapter-inducing interferon-β) is the adapter protein in the cytosol that is activated and recruited in response to activation of TLRs by ligands (Karin, 1999).

After binding with ligand, the TLR4 undergoes a conformational change and gets activated leading to subsequent recruitment and activation of TRIF protein (Helgason et al., 2013).

The TRIF adaptor in turn activates the **TANK-binding kinase 1** (TBK1) thus forming the TRIF/TBK1 complex that further phosphorylates Interferon regulatory factor 3 (IRF3) that serves as a transcription factor. The phosphorylation of IRF3 allows its nuclear translocation from the cytosol leading to activated transcription of genes involved in the formation of proinflammatory cytokines (Hiscott et al., 1999).

The interaction microglial activation and neurodegeneration including the pathogenesis of Alzheimer is intricate and involves multiple causative factors and is still under investigation by researchers across the world.

Although microglia activation occurs in response to injury to the brain and spinal cord, additionally, activation of microglia in a specific environment may contribute to the release of proinflammatory cytokines resulting in neurodegeneration. Thus, activated microglial cells are implicated in the pathology of immune-mediated neurodegenerative diseases. A study was conducted on purified murine neonatal microglia that were cocultured with neuronal cells obtained from fetal brain. The study showed reduced survival of neuronal cells (Chao et al., 1992).

It is suggested that microglia have the potential to induce the formation of free radicals that exhibit neurotoxic effects leading to reduced neuronal survival rate and higher neuronal death and further supplemented that these effects showed variations with microglial density (Chao et al., 1992).

Another study (Wu et al., 2002) showed that minocycline application suppressed the microglial activation in the MPTP mouse model of Parkinson's disease and prevented the MPTP-induced microglial activation and neuronal death.

This study additionally supplemented the role of microglial activation in neurodegeneration as described by several above-mentioned studies.

Furthermore, in vitro study (Meda et al., 1995) utilizing cocultures showed the synergistic harmful potential of the IFN-γ and Aβ. These can lead to microglial activation and production of highly reactive nitrogen species and tumor necrosis factor α. These inflammatory mediators are critically implicated in the onset of senile changes in brain tissues and the initiation and progression of Alzheimer's disease.

Still another study (González-Scarano and Baltuch, 1999) provided laboratory proof through detection of high concentrations of tumor necrosis factor α and IL-1β in the serum as well as in the cerebrospinal fluid of patients affected with neurodegenerative disease including Parkinson's disease, Alzheimer's disease, multiple sclerosis, and amyotrophic lateral sclerosis. This study further substantiated the activation of microglial cells and the release of inflammatory cytokines causing irreversible injury to the neurons in specific regions of the brains.

Thus, fibrillar amyloid-β is suggested to be the ligand for the Toll-like receptor-4 expressed on the surface of microglial cells in the brains. The TLR-4 after its activation leads to the mediation of two prominent intracellular signaling pathways in the microglial cells leading to overexpression of proinflammatory cytokines and their implication in the pathogenesis of Alzheimer's disease.

Furthermore, inflammasomes are the next target in the cascade of events initiated by accumulated amyloid-β in brains regions and certainly have a role in the pathogenesis of Alzheimer's disease.

Inflammasome-mediated signaling pathways in Alzheimer's disease

Inflammasome represents the multimeric protein complexes located in the cytosol of innate immune cells including microglia in the brain and spinal cord (Mariathasan et al., 2004). The inflammasomes after activation help in the maturation and secretion of proinflammatory cytokines predominantly interleukin 1β and interleukin 18 implicated in the proinflammatory cell death called as pyroptosis.

The count of microglia is elevated near β-amyloid plaques in Alzheimer's disease. Possibly, it can be assumed as an adaptation to remove the amyloid deposits by phagocytosis by microglia (El Khoury et al., 2007). The phagocytic potential of microglia is supported by studies (El Khoury et al., 2007; Bamberger et al., 2003) that showed that suppression of phagocytic action of microglia led to an increase in the amyloid deposits in brain regions. The phagocytic potential of microglia was inhibited through the suppression of chemokine receptor-2 (**Ccr2**) expression on the surface of microglia that is helpful in the proliferation of macrophages and microglia at the site and enhances their scavenging action.

Already, it is substantiated by the aforementioned studies that chronic exposure of microglial cells to β-amyloid deposits results in persistent activation of microglial cells in Alzheimer's disease (Prinz et al., 2011). Furthermore, the progression of Alzheimer's disease is associated with an increase in amyloid deposits, and these two events are subsequently related to the dysfunction of microglial cells (Heneka et al., 2013; Hickman et al., 2008).

The NOD-like receptors (nucleotide-binding oligomerization domain-like receptor) (NLRs) constitute a group of cytosolic proteins. These function as pattern recognition receptors (Martinon and Tschopp, 2005).

The NLRs are made up of three structural domains. The leucine-rich repeat domain (LRR) toward the carboxy terminal, the effector domain located at the amino terminal, and NACHT (NOD or nucleotide-binding domain) domain in the center that is common in all NLRs.

The central NACHT domain is a highly conserved evolutionary protein motif. This domain is the structural motif in proteins namely (Venegas and Heneka, 2017) NLP family apoptosis inhibitor protein, MHC class II transcription factor (C2TA), incompatibility locus protein from *Podospora anserine* (HET-E), and telomerase-associated protein (TEP1) (Koonin and Aravind, 2000).

NACHT domain exhibits ATPase activity. It can attach to ribonucleotides. It is essential in the ATP-dependent self-oligomerization of NLR and its subsequent activation (Duncan et al., 2007).

The C-terminal LRR domain is leucine-rich repeat domain made up of 20–30 leucine amino acid tandem repeats (Shaw et al., 2008). These repeats fold together and form a **leucine-rich repeat domain (solenoid protein domain). The LRR domain is located at the C-terminus. It serves in** recognition of pathogen-associated molecular patterns, amyloid-β, and danger-associated molecular patterns (Stutz et al., 2009).

N-terminal effector domain can possess caspase recruitment domain (CARD), pyrin domain (PYD), and baculovirus inhibitor repeats or acidic transactivating domain (Franchi et al., 2009). **It serves to function as** protein–protein interaction (homotypic) domain.

Variable N-terminal domains help to subcategorize the NLRs family. Till date, nearly 20 NLRs members have been reported among humans.

Several NLRs proteins can induce the formation of multimeric protein complexes termed as inflammasomes. These represent a cluster of several protein molecules around a central scaffold protein (Venegas and Heneka, 2017). The activation of inflammasomes is a crucial step after exposure to fibrillar amyloid-β and amyloid-β oligomers.

A study (Venegas and Heneka) focused on the potential of danger-associated molecular patterns (DAMPs), including amyloid-β, chromogranin A, high-mobility group box 1, and nucleic acids in activating the innate-immune response involved in the pathogenesis of Alzheimer's disease.

In healthy conditions in nonstimulated cells, the expression of NLRP3 is unable to activate the assembly of the inflammasome in the cytoplasm of microglial cells.

The fibrillar amyloid-β and amyloid-β oligomers serve as priming signals for the overexpression of NLRP3 and oligomerization of NLRP3 in microglial cells (Bauernfeind et al., 2009).

Pathway of activation of NLRP3 inflammasome in microglial cell

Recognition and binding of fibrillar amyloid-β and amyloid-β oligomers with pattern recognition receptor (TLR4) on the surface of microglial cell precedes the activation of NLRP3 inflammasome (Heneka et al., 2013).

Nuclear factor-κB nuclear translocation-mediated activation of NLRP3 inflammasome

The activation of pattern recognition receptor (TLR4) subsequently mediates activation and nuclear translocation of a **transcription factor named as nuclear factor-κB (NF-κB). Inside the nucleus, nuclear factor-κB induces the upregulation of the expression of the NLR family pyrin domain-containing 3 protein (NLRP3) and pro-IL-1β** (Heneka et al., 2018).

These events are manifested into oligomerization and activation of NLRP3 inflammasome in the microglial cell.

Deubiquitylation of NLRP3 (splitting of ubiquitin from NLRP3) serves as an additional priming signal for the activation of NLRP3 inflammasome. In a study, the BRCC3 as a deubiquitinating enzyme was identified as the essential regulator of NLRP3 inflammasome activation. Contrarily, therapeutics inducing NLRP3 ubiquitination can be the potential drugs in the management of neurodegeneration diseases.

The activation of NLRP3-mediated assembly of the inflammasome in the microglial cell can result in the action of caspase 1 and subsequent proteolysis of proin-terleukin-1β and its maturation into the interleukin-1β (Mangan et al., 2018).

Cathepsin B, mitochondrial oxidative stress, and activation of NLRP3 inflammasome

The leucine-rich repeats in NLRs can recognize the ligands namely fibrillar amyloid-β, amyloid-β oligomers, and α-synuclein. These products are internalized via receptor-mediated phagocytosis by the microglial cells (Martinon and Tschopp, 2007).

The fibrillar amyloid-β, amyloid-β oligomers, and α-synuclein are sequestered in the phagosomes inside the microglial cells. These phagosomes fuse with the cathepsin B containing primary lysosomes and lead to the formation of enlarged lysosomes called as phagolysosomes in the microglial cell (Sun et al., 2012).

Cathepsin B is the cysteine protease involved in the proteolysis of foreign peptides inside the lysosomes. In humans, the CTSB gene encodes the cathepsin B (Chan et al., 1986).

Impaired lysosomal degradation and cathepsin B are closely implicated in the neuroinflammation and Alzheimer's disease. Further, secretion of cathepsin B is activated in the microglia during aging (Nakanishi, 2003).

A study (Nakanishi, 2003) revealed that the imbalance in the lysosomal enzymes including overexpression of cathepsin B in the microglial cells in the aging brains is involved in the neurodegenerative disease including Alzheimer's.

Imbalance in the production of cathepsin in response to accumulated fibrillar amyloid-β, amyloid-β oligomers and α-synuclein and leakage of cathepsin B due to age-related fragility of lysosomal membranes initiates the **cathepsin B leakage-mediated pathway of neurodegeneration in Alzheimer's.**

Cathepsin B leakage from the lysosomes can occur in aging brains and is also the marked feature in the pathology of Alzheimer's and other neurodegenerative diseases. The cathepsin B follows redistribution from the lysosomes to the cytosol in the microglial cells leading to neuroinflammation, apoptosis, and progression of Alzheimer's disease (Hook et al., 2020).

Leakage of cathepsin B from phagolysosomes in the microglial cells acts through multiple mechanisms to induce neuroinflammation and neurodegeneration.

The cathepsin B can induce NLRP3 activation (Chevriaux et al., 2020) and controls microglial lysosome- and mitochondria-generated reactive oxygen species in the neurons in the brain (Nakanishi and Wu, 2009).

The cathepsin B interplays between activation of NLRP3 inflammasome and mitochondrial oxidative stress in neurons and microglial cells.

A study by Bai et al. (2018) posited that cathepsin B, reactive oxygen species-mediated oxidative stress in mitochondria, and activation of NLRP3 inflammasome in microglia are closely interrelated and implicated in the pathology of Alzheimer's disease and other neurodegenerative diseases.

Bai et al. (2018) demonstrated that reactive oxygen species including hydrogen peroxide enhance the expression of cathepsin B in microglial cells leading to increased activity of cathepsin B leading to activation and oligomerization of NLRP3 inflammasome that subsequently is involved in proteolytic processing of procaspase-1 into caspase-1 and secretion of IL-1 β.

An aforementioned cascade of molecular evens was inhibited through genetic deletion of cathepsin B gene that was manifested as inability to process procaspase-1 and suppression of IL-1 β secretion in response to the production of H_2O_2 (Bai et al., 2018).

Bai et al. (2018) reported promoted activity of cathepsin B and raised levels of IL-1β and malondialdehyde (highly reactive enol and biomarker of oxidative stress in neurons) in plasma in patients with Alzheimer's disease. These findings were compared with healthy controls and reported to be significantly within the normal range. Additionally, plasma levels of glutathione (antioxidant) were raised in patients with Alzheimer's than controls.

Thus, the inflammasome-induced release of proinflammatory cytokine and pyroptosis is the hallmark of Alzheimer's disease in response to chronic exposure of microglial cells to fibrillar amyloid-β and amyloid-β oligomers.

K$^+$ efflux-dependent activation of NLRP3 inflammasome

The K+ efflux-dependent pathway represents a universal model for the activation of the NLRP3 inflammasome in the microglial cells. It manifests as a fall in cytosolic K+ ions owing to potassium efflux across the membrane of the microglial cell. It leads to the activation of caspase-1 and its sequel.

A study by Pétrilli et al. (2007) described that decline in cytosolic low K+ can activate NALP3 that in turn recruits the apoptosis-associated speck-like protein containing a CARD (inflammasome adaptor protein) and caspase-1 within the assembly of the inflammasome.

A study involving a neuronal cell culture experiment provided evidence for the activation of caspase-1 and the rise in the release of cytokine, IL-1β by neuronal cells in culture after treatment with valinomycin.

It was suggested that valinomycin promoted the K+ efflux in the neuronal cells (de Rivero et al., 2008).

A study by Fernandes-Alnemri et al. (2007) showed that K$^+$ efflux-dependent activation of NALP3 induces dimerization of **apoptosis-associated speck-like protein containing a CARD** (ASC adapter protein). The dimer further undergoes oligomerization and finally characterizes into pyroptosomes (protein supra-molecules). The pyroptosomes further activate the caspase-1 enzyme leading to the splitting of procytokines named as pro-IL-18 and pro-IL-1 β into active cytokines for the induction of pyroptosis.

Several studies (Furukawa et al., 1996; Yu et al., 1998) emphasized that amyloid-β oligomers and fibrillar amyloid-β have the potential to impair the function of K$^+$ channels leading to potassium efflux and decline in cytosolic potassium ion concentration in microglial cells.

Another study (Yu et al., 2006) showed that raised activity of potassium channels was involved in triggering apoptotic cell death after exposure to Aβ (1−40)-treated rat cortical neurons.

Furthermore, the amyloid-β has been unanimously considered as the critical DAMP in the brains in Alzheimer's disease (Venegas and Heneka, 2017). Regarding its oligomerization, initially, the amyloid-β fragment is derived from the APP via its proteolytic splitting by γ-secretase (Salminen et al., 2008). Persistent deposition of amyloid-β fragment may characterize into a rise in Aβ1−40 to Aβ1−42 ratio in the brain regions. The Aβ aggregates and forms oligomers that accumulate as fibrils in the senile plaques. The Aβ oligomers act as potential DAMPS and activate the NLRP3 inflammasome (De Rivero et al., 2008).

Thus, **Aβ oligomers are neurotoxic and can induce the potassium ions efflux leading to low cytosolic potassium level and persistent activation of the NLRP3 inflammasome, manifesting into a constant rise in proinflammatory cytokines in the brain and sustaining an inflammatory state in Alzheimer's disease.**

Therefore, pharmacotherapeutics that can suppress the activation of NLRP3 inflammasome can serve as novel drug molecules to manage manifestations in Alzheimer's disease.

Interleukin-18 is produced in response to activation of NLRP3 inflammasome. A study (Sutinen et al., 2012) showed the role of interleukin-18 in enhancing the formation of amyloid-β in human neuron-like cells representing a model of Alzheimer's disease.

Further, elevated plasma Interleukin-18 levels were reported in the cerebrospinal fluid in Alzheimer's patients (Ojala et al., 2009).

The deposition of amyloid oligomers and neurofibrillary tangles represent the hallmark of Alzheimer's disease. The amyloid oligomers and neurofibrillary tangles are recognized by cell surface and intracellular pattern recognition receptors and initiate varied signaling cascades comprising microglial activation, oligomerization of NLRP3 inflammasomes resulting in the release of cytokine as interleukin-1β that further potentiate the neuroinflammation in brain tissues in Alzheimer's. Both NLRP3 and NLRP1 inflammasomes are implicated in the pathology of Alzheimer's in animal models. Thus, inflammasomes can be the drug targets in mitigating manifestations and intercepting the progression of Alzheimer's disease.

Role of P2X purinoreceptors in Alzheimer's disease

The P2X receptors are the purinoreceptors and also termed as ATP-gated P2X receptor cation channel family constitute the proteins that are ligand-gated cation channels. The binding of ATP with the P2X receptors activates the cation transporting potential of the purinoreceptor (Chen et al., 2011).

The P2X receptors serve in several biological activities including activation of macrophages (Wewers and Sarkar, 2009, North, 2002), neuronal apoptosis (Kawano et al., 2012), normal interaction between neurons, and glia in the brain (Burnstock, 2013).

The P2X receptors exhibit wide distribution including presynaptic neurons and postsynaptic neurons and microglial cells in the brain, peripheral nervous system, and autonomic nervous system. The P2X receptors can regulate impulse transmission (Burnstock, 2013).

The activation of P2X receptors induces an influx of Na^+ and Ca^{2+} ions and efflux of K^+ ions from the membrane of cells (Nicke et al., 1998; Francistiová et al., 2020).

Structure of P2X7 receptor

It belongs to the family of P2X purinoreceptors. The P2X7 is encoded by the P2RX7 gene mapped to the human chromosome 12 (12q24.31) spanning 53,733 bases (NCBI, 2021).

The human P2X7R protein contains 595 amino acid residues. The P2X7 protein is a trimer consisting of an assembly of three identical subunits (Habermacher et al., 2016; McCarthy et al., 2019). Each subunit is made up of N-terminal domain, C-terminal domain, two transmembrane domains, and a large extracellular loop that connects two transmembrane domains together (Nicke et al., 1998). The intersection of two transmembrane domains represents the agonist-binding region of the P2X7 receptor (Habermacher et al., 2016; McCarthy et al., 2019).

The millimolar concentration of ATP as ligand activates the P2X7 receptor thus confirming it as an ATP-gated, nonselective cation channel (Na+, K+, and Ca2+) permitting the influx of these ions leading to depolarization of the resting membrane potential.

Moreover, an anomaly in the aforementioned nonselective cation transport function was cited in the P2X7 purinoreceptor (Illes, 2020; Virginio et al., 1999; Khakh et al., 1999; Di Virgilio et al., 2018). It was reported that P2X7 purinoreceptor like other P2X receptors can change its ion selectivity in time-dependent pattern. Possibly, prolonged agonist binding to its agonist-binding pouch could be implicated in the dilation of the ion channel into a pore that subsequently permits the passage of those ions that could not pass through the P2X7 channel earlier (Virginio et al., 1999; Khakh et al., 1999; Di Virgilio et al., 2018).

Further studies (Illes, 2020) confirmed that the P2X7 receptor can transmit large organic cations (Harkat et al., 2017) and is endowed with the ability to organize into a large conductance pore (Virginio et al., 1999).

Activation of P2X7 receptor

The P2X7 purinoreceptor is activated via several cytosolic mediators including nuclear factor kappa-light-chain enhancer (NFκB) (Ferrari et al., 1997), reactive oxygen species production (Hewinson and MacKenzie, 2007), calcium calmodulin kinases II (Diaz-Hernandez et al., 2008), glycogen synthase kinase-3 (GSK3) (Diaz-Hernandez et al., 2008), and inflammasome "NACHT, LRR, and PYD domains-containing protein 3" (NLRP3) (Franceschini et al., 2015; Francistiová et al., 2020).

The P2X7 purinoreceptor is primarily implicated in the activation of inflammasomes and induction of neuroinflammation in Alzheimer's disease.

The P2X7 purinoreceptor is largely distributed in the cells of innate as well as adaptive immune systems in humans. The microglial cells constitute the predominant immune cells of CNS and are the preferred site of expression of P2X7 purinoreceptors in the brain (Sluyter 2017; Di Virgilio et al., 2017).

The microglial cells possess the pattern recognition receptors, prominently the toll-like receptor-4 (TLR4) on their cell surfaces. These receptors can recognize pathogen-associated molecular patterns (PAMPs) and DAMPs like ATP (Illes et al., 2020; Martin et al., 2019).

In response to the binding of PAMPs or DAMPs to the toll-like receptor (TLR-4) (Hemmi et al., 2000), it undergoes phosphorylation in the microglial cell. The downstream effector like inactive NF-κB is (Hoebe et al., 2003) activated and is translocated from cytosol to nucleus in the microglial cell. Activated NF-κB is the transcription factor (Smith et al., 2006) that promotes the transcription of genes involved in the formation of NLRP3 inflammasome and pro-IL-1β.

PAMPs (e.g., bacterial lipopolysaccharide (LPS)) act on toll-like receptor-4 (TLR4) (Hoebe et al., 2003) and cause its phosphorylation. In consequence, in the cell nucleus, NF-κB is activated, which promotes the synthesis of the NLRP3 inflammasome and pro-IL-1β in the cytoplasm of the cell.

P2X7 purinoreceptor: activation of TLR-4 is associated with simultaneous binding of ATP with the P2X7 purinoreceptor leading to its activation and induction of inward transport of calcium ions/Na^+ ions and efflux of K^+ ions from the microglial cell. These events result in a fall in the cytosolic K^+ level in the microglial cell (Nicke, 2008).

Two-pore domain potassium channels: furthermore, the membrane of microglial cell also harbors **two-pore domain potassium channels** (also termed as **tandem pore domain potassium channels,** a family of 15 channels constitutes the **Leak channels**) (Goldstein et al., 2005). The binding of ATP with the P2X7 receptor leads to efflux of potassium ions via the **two-pore domain potassium channels; additionally, it deteriorates the cytosolic K^+ levels in the microglial cell.**

Chloride intracellular channels: moreover, the membrane of microglial cells also contains the chloride intracellular channels (Valenzuela et al., 2000). The 2KP channels additionally induce the outward transport of potassium ions from the microglial cell in response to activation of the P2X7 receptor (Valenzuela et al., 2000; Burnstock et al., 2011).

Thus, **P2X7 purinoreceptor, two-pore domain potassium channels and chloride intracellular channels are contained in the membrane of the microglial cell and operate synergistically in response to ligands like PAMPs or DAMPs and ATP** (Pétrilli et al., 2007; He et al., 2016; Goldstein et al., 2005; Nicke, 2008; North, 2002).

NEC7 serine/threonine kinase in the microglial cell is the sensor for the cytosolic K^+ ions concentration and its decline activates the NEC7 kinase (mitotic kinase) and subsequently (Abbracchio et al., 2006), NEC7 kinase binds with the inactive NLRP3 to form NEC7- NLRP3 complex (He et al., 2016). The NEC7 kinase is an essential element in the activation of NLRP3 inflammasome (Shi et al., 2016). There is the recruitment of **apoptosis-associated speck-like protein (ASC,** adapter protein) through the PYD−PYD interactions (Pétrilli et al., 2007) **and procaspase-1 within the assembly leading to oligomerization and activation of inflammasome in the microglial cell.**

Thus, the decline in the cytosolic K^+ ions in the microglial cell constitutes the K^+-efflux dependent activation of inflammasome in the microglial cell after exposure of TLR-4 with PAMPs, or DAMPs and P2X7 receptor with ATP.

Consequently, inactive proCasp-1 is activated via proteolytic cleavage in the inflammasome assembly that in turn cleaves the pro-IL-1β to active IL-1βinside the cytoplasm of the microglial cell. Thereafter, active IL-1β is secreted by activated microglial cells into the extracellular space. The IL-1β is a proinflammatory cytokine responsible for inducing neuroinflammation.

Furthermore, chronic activation of the P2X7 receptor is induced by sustained elevated ATP levels and is implicated in the pathology of neuronal apoptosis or necrosis (Virginio et al., 1999) **implicated in the progression of Alzheimer's disease.**

Role of P2X7 receptor in activation of inflammasome in Alzheimer's disease

Prolonged activation of microglial cells in the brain is manifested in the release of proinflammatory cytokines in the tissues promoting neuroinflammation and neuronal loss in brain regions (Wang et al., 2015).

In normal health conditions, microglial cells function as immune innate cells and undertake phagocytosis of misfolded proteins in the brain and promote degradation and clearance of amyloid-β, α-synuclein to maintain the health of neurons and display neuroprotective potential (Wang et al., 2015; Francistiová et al., 2020).

In vitro and in vivo **studies postulated the involvement of P2X7R receptor in inducing inflammasome activation and subsequent mediation in the course of Alzheimer's disease.**

In vitro study by Kim et al. (2007) focused on the chronic activation of microglial cells isolated from rat brains.

Authors identified that fibrillar Aβ42 peptide-induced release of ATP from chronically exposed microglial cells and the molecular event followed the P2X7R-dependent pathway (Kim et al., 2007).

Furthermore, in vitro and in vivo studies revealed that activation of the P2X7 receptor is essential in the chronic activation of microglial cells in response to amyloid-β42 peptide and further secretion of proinflammatory IL-1β (Chiozzi et al., 2019).

The precise mechanism of P2X7 purinoreceptor-mediated activation of the inflammasome, and release of inflammatory cytokine IL-1β by activated

microglial cells into the extracellular space has been already elucidating in the previous topic.

Neuroinflammation and neuronal apoptosis are implications in the pathology of Alzheimer's disease.

Several studies (Sanz et al., 2009; Kim et al., 2007; Chiozzi et al., 2019; Gustin et al., 2015; Boche et al., 2013; Takeda and Akira, 2004; Venigalla et al., 2016) focused on the contribution of P2X7R/NLRP3/Caspase1 Axis in the intracellular signaling cascade that can induce activation of inflammasome after the priming of the microglial cell (activation).

Role of P2X7 purinoreceptor in oxidative stress and Alzheimer's disease

Oxidative stress constitutes the clinical condition characterized as dyshomeostasis between the prooxidants and antioxidants resulting in harmful manifestations in the tissues (Zuo et al., 2015).

The exaggerated production of reactive oxygen species in the cells, especially brain tissues is involved in the structural and functional abnormalities in the macromolecules (Lin MT, Beal 2006), and organelles that ultimately manifest into the body tissues.

A study by Albers and Beal (2000) and Sebastian-Serrano et al. (2019) identified raised levels of reactive oxygen species in the brain tissues in patients with neurodegenerative disorders including Alzheimer's disease (Dias et al., 2013). The ROS are highly reactive molecules that initiate oxidative injury in the mitochondria and endoplasmic reticulum in several tissues including brain tissues.

Mitochondria in neurons are highly susceptible to oxidative injury by ROS resulting in impaired membrane integrity, migration of mitochondria, and bioenergetics of ATP production collectively manifested as mitochondrial dysfunction that subsequently elevates the production of additional ROS (Martin, 2010).

The purinoreceptor, P2X7 might be the essential contributor to the production of hydrogen peroxide by the activated microglial cells (Nuttle and Dubyak, 1994).

A study by Parvathenani et al. (2003) demonstrated that activated microglial cells (isolated from rat brain) released superoxide via NADPH oxidase activation-dependent pathway. This activity required stimulation of P2X7 purinoreceptor in the microglial cells by ATP or BzATP suggested that priming of microglial cells is the prerequisite in the P2X7-mediated production of reactive oxygen species in the activated microglial cells.

Another study by Kim et al. (2007) identified that fibrillar amyloid-β42-induced upregulation of the expression of P2X7 purinoreceptor in the microglial cells. This activity was followed by P2X7-mediated excessive production of ROS in microglial cells and contributed to the synaptic toxicity that is reported in the early phase of Alzheimer's disease (Lee et al., 2010).

A recent study by Chiozzi et al. (2019) utilized the N13 microglia cell line and identified that amyloid-β-induced neurotoxicity resulted in mitochondrial dysfunction in the involved microglial cell line. The amyloid-β interaction with the P2X7 purinoreceptors resulted in the activation of microglial cells and excessive production of ROS.

Furthermore, the study described that the P2X7 receptor is essential in amyloid β generated activation of microglia and secretion of IL-1β in the extracellular space. Additionally, amyloid-β-mediated activation of P2X7 purinoreceptors and microglial cells was inhibited by the application of dihydropyridine nimodipine at the site distal to the P2X7 receptor.

It is concluded from the findings that nimodipine constitutes the inhibitor of calcium channels in the microglial cells and acts as the blocker of neuronal and microglial injury induced by fibrillar Aβ42 and spares the mitochondria from oxidative damage and point to the potential of P2X7 receptor in mediating mitochondrial oxidative stress in response to the monomeric and oligomeric forms of amyloid-β.

Role of upregulated expression of P2X7 receptor in Alzheimer's disease

Upregulation of P2X7 receptor expression is associated with increased activation of microglial cells surrounding the amyloid-β42 in the brain regions in Alzheimer's disease.

In vitro study by McLarnon et al. (2006) reported increased expression of the P2X7 receptor concomitant with elevated levels of ATP in adult microglia isolated from the brains regions in patients with Alzheimer's disease and from fetal human microglia that were exposed to amyloid-β42 peptide.

Further in vivo study by McLarnon et al. (2006) involving tissues in rat hippocampus injected with amyloid-β42 also showed elevated expression of the P2X7 receptor.

Still another study by Parvathenani et al. (2003) demonstrated in primary rat microglial cells activated by interaction with ligand either ATP or BzATP (2'- and 3'-O-(4-benzoylbenzoyl)-ATP, the elevated production of superoxide ions via activated P2X7 purinergic receptor via NADPH oxidase-dependent pathway.

Parvathenani et al. (2003) reported p42/44 MAP kinase and p38 MAP kinase were activated in the microglial cells in response to ATP or BzATP. Further, the application of an inhibitor of phosphatidylinositol 3-kinase led to a decline in the production of superoxide ions by microglial cells.

It was concluded that the expression of P2X7 purinoreceptors was upregulated in microglial cells surrounding the amyloid-β42 plaques in the Tg2576 mouse model (Parvathenani et al., 2003) of Alzheimer's disease.

Another study by Skaper et al. (2006) linked the upregulated expression of P2X7 receptor concomitant with elevated levels of ATP in activated microglial cells surrounding the amyloid-β42 with the higher susceptibility of injury to the cortical neurons.

These findings correlate the amyloid-β42 neurotoxic effect on the cortical neurons via **upregulation of P2X7 receptors and the elevation of ATP coupled with the hyperactivity of microglial cells contributing to injury to cortical neurons that are closely related to the initiation of Alzheimer's disease.**

Role of P2X7 purinoreceptor and amyloidogenic processing of APP in Alzheimer's disease

The proteolytic processing of amyloid precursor protein either by α-secretase or β-secretase determines the formation of cleavage products in brain regions and their implication in the pathology of Alzheimer's disease.

The α-secretase-induced cleavage of amyloid precursor protein results in the formation of sAPP and the P3 peptides (extracellular peptides) and C-terminal fragment, named as C88 (intracellular fragment) (Selkoe, 2001; Hardy and Selkoe, 2002), and the pathway involved is termed as a **nonamyloidogenic** pathway.

The β-secretase-mediated cleavage of amyloid precursor protein leads to the formation of sAPPβ and Aβ-peptides (extracellular peptide) and the C-terminal fragment C99s (intracellular peptide), and the mechanism of cleavage is termed as **amyloidogenic pathway** (Hardy and Selkoe, 2002).

In healthy brains, the APP is preferably processed via a nonamyloidogenic pathway (Tyler et al., 2002). However, causative factors induced dysregulation in the processing of APP favor cleavage via the amyloidogenic pathway resulting in the formation and accumulation of amyloid-β (Stockley and O'Neill, 2008).

A study by Tyler et al. (2002) described a small quantity of amyloid-β formation in normal and healthy brains. Authors identified a decline in the cleavage activity of α-secretase by 81% of the normal) (Tyler et al., 2002) and surprisingly enormous rise in the activity of β-secretase activity by 185% of the normal (Tyler et al., 2002) in the temporal cortex in patients with sporadic Alzheimer's disease.

Further reported by Tyler et al. (2002) that a rise in β-secretase activity or a decrease in α-secretase activity or both contribute to the development of Alzheimer's disease in the majority of the population.

A study by Delarasse et al. (2011) involving N2a cell line (mouse neuroblastoma cells) expressing human amyloid precursor protein identified the BzATP-induced activation of P2X7 receptor and subsequent release of sAPPα by the mouse neuroblastoma cells. Further, specific small interfering RNA (siRNA)-mediated selective P2X7R knock-down model showed the suppression in the formation of sAPPα (Delarasse et al., 2011).

Delarasse et al. (2011) identified the role of mitogen-activated protein kinases, Erk1/2 and JNK in the P2X7R-mediated α-secretase cleavage activity and release of sAPPα.

Hence, concluded that P2X7 receptors are highly expressed in neurons in the hippocampus and microglial cells and serve as the drug target in the management of Alzheimer's disease.

Contradictory findings were provided in a study by Leon-Otegui et al. (2011) and demonstrated that activation of the P2X7 receptor was associated with a decline in the cleavage potential of α-secretase in the Neuro-2a cells and reduction in the formation of sAPPα.

Still another study by Diaz-Hernandez et al. (2012) provided contrary findings related to P2X7R activity and α-secretase-mediated cleavage of APP.

In vivo study by Diaz-Hernandez et al. (2012) demonstrated that inhibition of the P2X7 receptor in J20 mice (transgenic mice expressing mutant human APP) led to a decline in the number of amyloid plaques in the hippocampus region of the brain.

The studies by Diaz-Hernandez et al. (2012) and Leon-Otegui et al. (2011) furnished that overexpression of P2X7 receptor was involved in a reduction in α-secretase-mediated cleavage of APP and reduced formation of activity of sAPPα with neuroprotective function, while contradictory findings from studies (Delarasse et al., 2011; Tyler et al., 2002) showed that overexpression of P2X7 receptor was involved in the increase in α-secretase-mediated cleavage of APP and elevated level of sAPPα in the brain.

These contrary findings suggested the dual function of the P2X7 receptor in the microglial cells.

Another study by Martin et al. (2019) posited that amyloid β peptide aggregates serve as a stimulus for the activation of microglial cells and release of ATP that in turn acts as a ligand for the activation of P2X7 receptors.

The authors studied the properties of P2X7 receptors in transgenic mice with AD genetic deletion of P2X7 receptors (Martin et al., 2019).

Martin et al. (2019) reported that genetic ablation of P2X7 receptor was associated with a reduction in amyloid-β lesion, betterment in the cognitive deficits, and enhanced synaptic plasticity in transgenic AD mice. Moreover (Martin et al., 2019), release of IL-1β and formation of sAPPα (nonamyloidogenic fragment) were remained unaltered in the mice model with deleted P2X7 receptors.

Authors suggested on the basis of findings that the P2X7 receptor crucially regulates amyloid-β-mediated activation of microglial cells and proinflammatory cytokine as CCL3 that in turn is related to the recruitment of pathogenic CD8[+] T cell (Martin et al., 2019).

Thus, findings confer a novel and detrimental role played by the P2X7 receptor in inducing the release of proinflammatory cytokine in the extracellular space by microglial cells and could be the novel target for the action of pharmacotherapeutics in patients with Alzheimer's disease.

Role of chronic psychological stress and Alzheimer's disease

Every stress leaves an indelible scar, and the organism pays for its survival after a stressful situation by becoming a little older

Hans Selye (1950).

Persistent external stress can modulate the body's response to the stressors and eventually lead to the expression of psychological stress (Jones et al., 2018).

Stress disturbs the hypothalamic-pituitary-adrenocortical (HPA) axis and its chronicity results in hyperactivity of the hypothalamic–pituitary–adrenocortical axis and increased secretion of glucocorticoids (Smith and Vale, 2006).

The glucocorticoid can cross the blood-brain barrier and stimulate the glucocorticoid receptor in the human brain. Persistently elevated levels of glucocorticoid in the brain may be related to loss of neurons in the hippocampus and deletion of glucocorticoid receptors in the brain. These features are involved in additional injury to the surviving neurons and are implicated as the aggravating factors of Alzheimer's disease.

Truly, the stress, HPA axis, hormones, and neurodegenerative diseases are trapped in the vicious cycle in which each element drives the other toward the progression of disease (Bjorntorp, 1997; Wahrborg, 1998; Girod and Brotman, 2004; Reiche et al., 2004; DiMicco et al., 2006).

The study involved injection of sufficient (Sapolsky et al., 1985) corticosterone to rats to induce a rise in corticosterone concentrations in the bloodstream within the high physiological range.

Rats injected with sufficient corticosterone for a period of 3 months (Sapolsky et al., 1985) appeared like aged rats with manifestations as massive loss of corticosterone receptors in the hippocampus region of the brain. Additionally, the rats with corticosterone injection showed the loss of neurons in hippocampi that displayed (Sapolsky et al., 1985) similar pattern as was noted in the loss of neurons in hippocampi in the brains of aged rats.

Authors concluded that consequences of injection of corticosterone for 3 months to rats as in the study by Sapolsky et al. (1985) might simulate the harmful effects developed in the brains in response to chronic and cumulative exposure to elevated glucocorticosteroids in bloodstream generated due to impaired HPA axis after exposure to prolonged psychological and environmental stress to the human population.

Chronic exposure to stress resulted in a rise in hyperphosphorylated τ-protein levels in the hippocampus region in the mouse brain (Korneyev, 1998; Okawa et al., 2003).

Alzheimer's disease-related mutations in τ resulted in a rise in hyperphosphorylated τ and the formation of neurofibrillary tangle in the mouse brain and enhanced neurodegeneration (Carroll et al., 2011; Rosa et al., 2005).

Chronic psychological stress and subsequent loss of neurons in the hippocampus region might be implicated in the pathology of Alzheimer's disease.

Thus, chronic stress operates through multiple mechanisms in the pathology of Alzheimer's. Stress resulted in hyperactivity of HPA axis, hypersecretion of glucocorticoids and associated loss of neurons in brain regions, **overexpression of APP gene leading to the surplus formation and accumulation of Ab42 peptides, hyperphosphorylation of τ-proteins, characterizing into the formation of neurofibrillary tangles involved in the pathology of Alzheimer's disease.**

Role of microRNAs (miRNAs) in Alzheimer's disease

The micro-RNAs (miRNAs) comprise 21—25 nucleotides and constitute the small noncoding RNAs. These are mainly involved in the posttranscriptional control of gene expression (Bartel 2004). The miRNAs bind with the coding domain and 3'- and 5'-untranslated region in the messenger RNA and exert their effect on the gene expression.

The micro-RNAs have been reported to be derived from human brains and are essential in regulating normal synapse formation, release of neurotransmitter, and impulse conduction (Lee et al., 2002).

However, miRNAs are additionally implicated in the pathogenesis of neurodegenerative diseases, cardiovascular diseases, diabetes, obesity, cancer, and schizophrenia (Reddy et al., 2017).

miRNAs in amyloidogenic processing of APP in Alzheimer's

In the pathology of Alzheimer's disease, The microRNA-346 has been reported in inducing upregulated expression of the APP gene through targeting the APP mRNA 5'-untranslated region. The resultant increase in Ab42 formation is linked with the pathology of Alzheimer's disease (Long et al., 2019).

Long et al. (2019) posited that microRNAs are the active factors in the pathology of neurodegenerative diseases including Alzheimer's. Further, the microRNA-128 (miR-128) is posited to be dysregulated in Alzheimer's disease. Study (Long et al., 2019) employed clinical samples from patients with Alzheimer's disease, amyloid-β-treated primary mouse cortical neurons, and Neuro2a cell line to determine the expression levels of miR-128 and PPAR-γ messenger RNA using the RT-qPCR assay.

Long et al. (2019) reported upregulated expression of MiR-128 and suppressed expression of PPAR-γ in patients with AD, in amyloid-β (Aβ)-treated primary mouse cortical neurons and Neuro2a cells.

An additional study (Liu et al., 2019) reported that upregulated expression of miR-128 was associated with downregulated expression of PPARγ in cerebral cortex of the mice model of AD. Thus, PPARγ represented the direct target of miR-128.

The further study involved the use of a miR-128 knockout model of mice and showed that PPARγ expression was upregulated with suppressed amyloidogenic processing of APP, restricted formation of Ab42, formation of senile plaques leading to reduced AB42 neurotoxicity manifested in terms of reduced secretion of proinflammatory cytokine, and decline in neuroinflammation in the brains in mice model of mice.

Thus, it can be concluded that miR-128 inhibition-induced suppression of Aβ42-mediated neurotoxicity via upregulating PPAR-γ expression and inactivation of transcription factor NF-κB might be the novel drug target in the treatment of AD.

miRNA-200a-3p in neuronal apoptosis in Alzheimer's disease

Zhang et al. (2017) assessed the role of upregulated expression of miRNA-200a-3p in the Aβ42-mediated neuronal apoptosis.

SIRT1 (silent information regulator transcript-1, NAD-**dependent deacetylase sirtuin-1) can regulate the deacetylation of nonhistones and histones proteins** involved in the regulation of oxidative stress, gene transcription, and energy balance (Frye, 1999).

The upregulated expression of miRNA-200a-3p was reported in the hippocampus region of the brain in APP/PSEN 1 mice and in pheochromocytoma PC12 cells exposed to $Aβ_{25-35}$.

The upregulated expression of miRNA-200a-3p was associated with a decline in the levels of silent information regulator transcript-1 (SIRT1) (Marwarha et al., 2014). The decline in levels of SIRT1 in brain regions is linked with a higher propensity to the development of neurodegenerative disorders (Donmez G., Outeiro 2013) including Alzheimer's disease.

miRNA-137 in regulation of calcium voltage-gated channel subunit α-1 C (CACNA1C) in Alzheimer's

The calcium voltage-gated channel subunit α-1 C (*CACNA1C*) gene controls the synthesis of α 1C subunit of the voltage-gated calcium channel belonging to type L (CaV1.2) (Jakobsson et al., 2016). The channel is involved in the regulation of cytosolic levels of calcium ions in neurons in the brain.

It is reported that a rise in Ca^{2+} influx via active CaV1.2 channel is attributed to neuronal dysfunction and progression in AD (Davare and Hell, 2003).

Davare and Hell (2003) reported the association of decline in expression of miRNA-137 and increase in the levels of Aβ42 peptide and CACNA1C protein in the cortex and hippocampus regions of brains in the mice model of AD.

Thus, downregulation of miRNA-137 leads to an increase in Ca^{2+} levels and higher neuronal dysfunction.

miRNA in cognitive dysfunction in Alzheimer's

Jian et al. (2017) described the potential of miRNA expression in the Aβ42-mediated cognitive dysfunction in the brains.

The study involved mice with APP/PSEN1 and miRNA-34a knockout APP/PSEN1 mice model. Authors reported the raised level of miRNA-34a in APP/PSEN1 mice coupled with increased production of Ab42 and exaggerated cognitive function, while miRNA-34a KO/APP/PSEN1 mice showed a decline in the deposition of amyloid plaques and reduced cognitive dysfunction in the mice model.

The findings of the study suggested that upregulated expression of miRNA-34a in brain regions is contributory to the pathology of Alzheimer's through several mechanisms.

Conclusion

The pathology of Alzheimer's disease is highly intricate involving the interaction of multiple causative factors. The increased production of Aβ42 or its reduced degradation and clearance coupled with hyperphosphorylated τ-proteins are the consequence of impaired expression of several genes leading to the synthesis of proinflammatory cytokines, activation of caspases, development of proinflammatory microenvironment in the vulnerable regions of brains enhancing the levels of oxidative stress, neuroinflammation, loss of synapses, activation of microglial cells, and neuronal apoptosis.

The chapter related to the pathology of Alzheimer's disease decodes probable interrelated pathways implicated in the initiation and progression of Alzheimer's disease that might be helpful in developing novel drug targets to mitigate the debilitating consequences of Alzheimer's disease.

References

Abbracchio, M.P., Burnstock, G., Boeynaems, J.-M., Barnard, E.A., Boyer, J.L., Kennedy, C., Knight, G.E., Fumagalli, M., Gachet, C., Jacobson, K.A., et al., 2006. International union of pharmacology LVIII: update on the P2Y G protein-coupled nucleotide receptors: from molecular mechanisms and pathophysiology to therapy. Pharmacol. Rev. 58, 281−341.

Abe, Y., Kouyama, K., Tomita, T., Tomita, Y., Ban, N., Nawa, M., Matsuoka, M., Niikura, T., Aiso, S., Kita, Y., Iwatsubo, T., Nishimoto, I., 2003. Analysis of neurons created from wild-type and Alzheimer's mutation knock-in embryonic stem cells by a highly efficient differentiation protocol. J. Neurosci. 23, 8513−8525.

Adamec, E., Vonsattel, J.P., Nixon, R.A., 1999. DNA strand breaks in Alzheimer's disease. Brain Res. 849, 67−77.

Adamec, E., Mohan, P.S., Cataldo, A.M., Vonsattel, J.P., Nixon, R.A., 2000. Up-regulation of the lysosomal system in experimental models of neuronal injury: implications for Alzheimer's disease. Neuroscience 100, 663−675.

Adolfsson, O., Pihlgren, M., Toni, N., Varisco, Y., Buccarello, A.L., Antoniello, K., et al., 2012. An effector-reduced anti-β-amyloid (Aβ) antibody with unique aβ binding properties promotes neuroprotection and glial engulfment of Aβ. J. Neurosci. 32, 9677−9689.

Afroz, T., Hock, E.M., Ernst, P., Foglieni, C., Jambeau, M., Gilhespy, L., Laferriere, F., Maniecka, Z., Plückthun, A., Mittl, P., Paganetti, P., Allain, F.H., Polymenidou, M., 2017. Functional and dynamic polymerization of the ALS-linked protein TDP-43 antagonizes its pathologic aggregation. Nat. Commun. 8 (1), 45.

Akiyama, H., 1994. Inflammatory response in Alzheimer's disease. Tohoku J. Exp. Med. 174, 295–303.

Albensi, B.C., Mattson, M.P., 2000. Evidence for the involvement of TNF and NF-κB in hippocampal synaptic plasticity. Synapse 35 (2), 151–159.

Albers, D.S., Beal, M.F., 2000. Mitochondrial dysfunction and oxidative stress in aging and neurodegenerative disease. J. Neural. Transm. Suppl. 59, 133–154.

Albrecht, S., Bogdanovic, N., Ghetti, B., Winblad, B., LeBlanc, A.C., 2009. Caspase-6 activation in familial Alzheimer disease brains carrying amyloid precursor protein or presenilin I or presenilin II mutations. J. Neuropathol. Exp. Neurol. 68, 1282–1293.

Alnemri, E.S., 1997. Mammalian cell death proteases: a family of highly conserved aspartate specific cysteine proteases. J. Cell. Biochem. 64, 33–42.

Alnemri, E.S., Livingston, D.J., Nicholson, D.W., Salvesen, G., Thornberry, N.A., Wong, W.W., Yuan, J., 1996. Human ICE/CED-3 protease nomenclature. Cell 87 (2), 171.

Alonso, A.D., Grundke-Iqbal, I., Barra, H.S., Iqbal, K., 1997. Abnormal phosphorylation of tau and the mechanism of Alzheimer neurofibrillary degeneration: sequestration of microtubule-associated proteins 1 and 2 and the disassembly of microtubules by the abnormal tau. Proc. Natl. Acad. Sci. U. S. A. 94 (1), 298–303.

Alonso, A., Zaidi, T., Novak, M., Grundke-Iqbal, I., Iqbal, K., 2001. Hyperphosphorylation induces self-assembly of tau into tangles of paired helical filaments/straight filaments. Proc. Natl. Acad. Sci. U. S. A. 98 (12), 6923–6928.

Alvarez, V.A., Sabatini, B.L., 2007. Anatomical and physiological plasticity of dendritic spines. Ann. Rev. Neurosci. (30), 79–97. https://doi.org/10.1146/annurev.neuro.30.051606.094222.

Alzheimer's Association, 2016. 2016 Alzheimer's disease facts and figures. Alzheimers Dement 12 (4), 459–509.

Alzheimer's Disease International (ADI), World Alzheimer Report 2018, 2018. The State of the Art of Dementia Research: New Frontiers. London.

Anderson, A.J., Su, J.H., Cotman, C.W., 1996. DNA damage and apoptosis in Alzheimer's disease: colocalization with c-Jun immunoreactivity, relationship to brain area and effect of post mortem delay. J. Neurosci. 16, 1710–1719.

Araki, W., Yuasa, K., Takeda, S., Shirotani, K., Takahashi, K., Tabira, T., 2000. Overexpression of presenilin-2 enhances apoptotic death of cultured cortical neurons. Ann. N. Y. Acad. Sci. 920, 241–244.

Arancibia, S.A., Beltrán, C.J., Aguirre, I.M., Silva, P., Peralta, A.L., Malinarich, F., Hermoso, M.A., 2007. Toll-like receptors are key participants in innate immune responses. Biol. Res. 40 (2), 97–112.

Arancio, O., Zhang, H.P., Chen, X., Lin, C., Trinchese, F., Puzzo, D., et al., 2004. RAGE potentiates Abeta-induced perturbation of neuronal function in transgenic mice. EMBO J. 23, 4096–4105.

Armato, U., et al., 2013. Calcium-sensing receptor antagonist (calcilytic) NPS 2143 specifically blocks the increased secretion of endogenous Abeta42 prompted by exogenous fibrillary or soluble Abeta25-35 in human cortical astrocytes and neurons-therapeutic relevance to Alzheimer's disease. Biochim. Biophys. Acta 1832 (10), 1634–1652.

Bai, H., Yang, B., Yu, W., Xiao, Y., Yu, D., Zhang, Q., 2018. Cathepsin B links oxidative stress to the activation of NLRP3 inflammasome. Exp. Cell Res. 362 (1), 180–187.

Bamberger, M.E., Harris, M.E., McDonald, D.R., Husemann, J., Landreth, G.E., 2003. A cell surface receptor complex for fibrillar beta-amyloid mediates microglial activation. J. Neurosci. 23 (7), 2665–2674.

Barrientos, R.M., Higgins, E.A., Biedenkapp, J.C., Sprunger, D.B., Wright-Hardesty, K.J., Watkins, L.R., Rudy, J.W., Maier, S.F., 2006. Peripheral infection and aging interact to impair hippocampal memory consolidation. Neurobiol. Aging 27, 723–732.

Bartel, D.P., 2004. MicroRNAs: genomics, biogenesis, mechanism, and function. Cell 116, 281–297.

Bastrikova, N., Gardner, G.A., Reece, J.M., Jeromin, A., Dudek, S.M., 2008. Synapse elimination accompanies functional plasticity in hippocampal neurons. Proc. Natl. Acad. Sci. U.S.A. 105, 3123–3127.

Bauernfeind, F.G., Horvath, G., Stutz, A., Alnemri, E.S., MacDonald, K., Speert, D., Fernandes-Alnemri, T., Wu, J., Monks, B.G., Fitzgerald, K.A., Hornung, V., Latz, E., 2009. Cutting edge: NF-kappaB activating pattern recognition and cytokine receptors license NLRP3 inflammasome activation by regulating NLRP3 expression. J. Immunol. 183, 787–791.

Behl, C., 2000. Apoptosis and Alzheimer's disease. J. Neural. Transm. 107 (11), 1325–1344.

Benaki, D., Zikos, C., Evangelou, A., Livaniou, E., Vlassi, M., Mikros, E., Pelecanou, M., 2005. Solution structure of Humanin, a peptide against Alzheimer's disease-related neurotoxicity. Biochem. Biophys. Res. Commun. 329, 152–160.

Bentahir, M., Nyabi, O., Verhamme, J., Tolia, A., Horré, K., Wiltfang, J., Esselmann, H., De Strooper, B., 2006. Presenilin clinical mutations can affect gamma-secretase activity by different mechanisms. J. Neurochem. 96 (3), 732–742.

Bi, W., Drake, C.J., Schwarz, J.J., 1999. The transcription factor MEF2C-null mouse exhibits complex vascular malformations and reduced cardiac expression of angiopoietin 1 and VEGF. Dev. Biol. 211 (2), 255–267.

Billingsley, M.L., Kincaid, R.L., 1997. Regulated phosphorylation and dephosphorylation of tau protein: effects on microtubule interaction, intracellular trafficking and neurodegeneration. Biochem. J. 323 (Pt 3), 577–591.

Bjorntorp, P., 1997. Stress and cardiovascular disease. Acta Physiol. Scand. Suppl. 640, 144–148.

Blass, J.P., 2002. Alzheimer's disease and Alzheimer's dementia: distinct but overlapping entities. Neurobiol. Aging 23, 1077–1084.

Boche, D., Perry, V.H., Nicoll, J.A., 2013. Review: activation patterns of microglia and their identification in the human brain. Neuropathol. Appl. Neurobiol. 39, 3–18.

Bolland, S., Ravetch, J.V., 2000. Spontaneous autoimmune disease in Fc(gamma)RIIB-deficient mice results from strain-specific epistasis. Immunity 13 (2), 277–285.

Bolos, M., et al., 2017. Absence of CX3CR1 impairs the internalization of tau by microglia. Mol. Neurodegener. 12 (1), 59.

Bonnert, T.P., Garka, K.E., Parnet, P., Sonoda, G., Testa, J.R., Sims, J.E., 1997. The cloning and characterization of human MyD88: a member of an IL-1 receptor related family. FEBS Lett. 402 (1), 81–84.

Bouchon, A., Dietrich, J., Colonna, M., 2000. Cutting edge: inflammatory responses can be triggered by TREM-1, a novel receptor expressed on neutrophils and monocytes. J. Immunol. 164 (10), 4991–4995.

Brockhaus, M., Grünberg, J., Röhrig, S., Loetscher, H., Wittenburg, N., Baumeister, R., et al., 1998. Caspase-mediated cleavage is not required for the activity of presenilins in amyloidogenesis and NOTCH signaling. Neuroreport 9, 1481−1486.

Buratti, E., Baralle, F.E., 2001. Characterization and functional implications of the RNA binding properties of nuclear factor TDP-43, a novel splicing regulator of CFTR exon 9. J. Biol. Chem. 276 (39), 36337−36343.

Burger, P.C., Vogel, F.S., 1973. He development of the pathologic changes of Alzheimer's disease and senile dementia in patients with Down's syndrome. Am. J. Pathol. 73 (2), 457−476.

Burnstock, G., 2013. Introduction to purinergic signalling in the brain. Adv. Exp. Med. Biol. 986, 1−12.

Burnstock, G., Krügel, U., Abbracchio, M.P., Illes, P., 2011. Purinergic signalling: from normal behaviour to pathological brain function. Prog. Neurobiol. 95, 229−274.

Cai, Y., An, S.S., Kim, S., 2015. Mutations in presenilin 2 and its implications in Alzheimer's disease and other dementia-associated disorders. Clin. Interv. Aging 10, 1163−1172.

Calafate, S., Buist, A., Miskiewicz, K., Vijayan, V., Daneels, G., de Strooper, B., et al., 2015. Synaptic contacts enhance cell-to-cell tau pathology propagation. Cell Rep. 11 (8), 1176−1183.

Carroll, J.C., Iba, M., Bangasser, D.A., Valentino, R.J., James, M.J., Brunden, K.R., Lee, V.M., Trojanowski, J.Q., 2011. Chronic stress exacerbates tau pathology, neurodegeneration, and cognitive performance through a corticotropin-releasing factor receptor-dependent mechanism in a transgenic mouse model of tauopathy. J. Neurosci. 31, 14436−14449.

Chan, S.J., San Segundo, B., McCormick, M.B., Steiner, D.F., 1986. Nucleotide and predicted amino acid sequences of cloned human and mouse preprocathepsin B cDNAs. Proc. Natl. Acad. Sci. U. S. A. 83 (20), 7721−7725.

Chan, F.K., Luz, N.F., Moriwaki, K., 2015. Programmed necrosis in the cross talk of cell death and inflammation. Annu. Rev. Immunol. 33, 79−106.

Chao, C.C., Hu, S., Molitor, T.W., Shaskan, E.G., Peterson, P.K., 1992. Activated microglia mediate neuronal cell injury via a nitric oxide mechanism. J. Immunol. 149 (8), 2736−2741.

Chen, J.S., Reddy, V., Chen, J.H., Shlykov, M.A., Zheng, W.H., Cho, J., Yen, M.R., Saier, M.H., 2011. Phylogenetic characterization of transport protein superfamilies: superiority of SuperfamilyTree programs over those based on multiple alignments. J. Mol. Microbiol. Biotechnol. 21 (3−4), 83−96.

Chevriaux, A., Pilot, T., Derangère, V., Simonin, H., Martine, P., Chalmin, F., Ghiringhelli, F., Rébé, C., 2020. Cathepsin B is required for NLRP3 inflammasome activation in macrophages, through NLRP3 interaction. Front. Cell Dev. Biol. 8.

Chiarini, A., et al., 2016. Calcium-sensing receptors of human neural cells play crucial roles in Alzheimer's disease. Front. Physiol. 7, 134.

Chiarini, A., Armato, U., Whitfield, J.F., Dal Pra, I., 2017. Targeting human astrocytes' calcium-sensing receptors for treatment of alzheimer's disease. Curr. Pharmaceut. Des. 23 (33), 4990−5000.

Chiozzi, P., Sarti, A.C., Sanz, J.M., Giuliani, A.L., Adinolfi, E., Vultaggio-Poma, V., Falzoni, S., Di Virgilio, F., 2019. Amyloid β-dependent mitochondrial toxicity in mouse microglia requires P2X7 receptor expression and is prevented by nimodipine. Sci. Rep. 9 (1), 6475.

Cho, H.J., Son, S.M., Jin, S.M., Hong, H.S., Shin, D.H., Kim, S.J., et al., 2009. RAGE regulates BACE1 and Abeta generation via NFAT1 activation in Alzheimer's disease animal model. FASEB. J. 23, 2639–2649.

Chuang, T.Y., et al., 2016. LRP1 expression in microglia is protective during CNS autoimmunity. Acta Neuropathol. Commun. 4 (1), 68.

Citron, M., Eckman, C.B., Diehl, T.S., Corcoran, C., Ostaszewski, B.L., Xia, W., Levesque, G., St George Hyslop, P., Younkin, S.G., Selkoe, D.J., 1998. Additive effects of PS1 and APP mutations on secretion of the 42-residue amyloid beta-protein. Neurobiol. Dis. 5 (2), 107–116.

Colton, C.A., Mott, R.T., Sharpe, H., et al., 2006. Expression profiles for macrophage alternative activation genes in AD and in mouse models of AD. J. Neuroinflam. 3, 27. https://doi.org/10.1186/1742-2094-3-27.

Coraci, I.S., Husemann, J., Berman, J.W., Hulette, C., Dufour, J.H., Campanella, G.K., Luster, A.D., Silverstein, S.C., El-Khoury, J.B., 2002. CD36, a class B scavenger receptor, is expressed on microglia in Alzheimer's disease brains and can mediate production of reactive oxygen species in response to beta-amyloid fibrils. Am. J. Pathol. 160 (1), 101–112.

Cotman, C.W., Qian, H.Y., Anderson, A.J., 2000. Cellular signaling pathways in neuronal apoptosis. Role in neurodegeneration and Alzheimer's disease. In: Reith, M.E.A. (Ed.), Cerebral Signal Transduction, From First to Fourth Messengers. Humana Press, Totowa, pp. 175–206.

Creasy, C.L., Ambrose, D.M., Chernoff, J., 1996. The Ste20-like protein kinase, Mst1, dimerizes and contains an inhibitory domain. J. Biol. Chem. 271 (35), 21049–21053.

Cribbs, D.H., Poon, W.W., Rissman, R.A., Blurton-Jones, M., 2004. Caspase-mediated degeneration in Alzheimer's disease. Am. J. Pathol. 165 (2), 353–355.

Cruts, M., Theuns, J., Van Broeckhoven, C., 2012. Locus-specific mutation databases for neurodegenerative brain diseases. Hum. Mutat. 33 (9), 1340–1344.

Dagbay, K.B., Bolik-Coulon, N., Savinov, S.N., Hardy, J.A., 2017. Caspase-6 undergoes a distinct helix-strand interconversion upon substrate binding. J. Biol. Chem. 292 (12), 4885–4897.

Dai, M.-H., Zheng, H., Zeng, L.-D., Zhang, Y., 2018. The genes associated with early-onset Alzheimer's disease. Oncotarget 9 (19), 15132–15143.

Davare, M.A., Hell, J.W., 2003. Increased phosphorylation of the neuronal L-type Ca^{2+} channel Cav1. 2 during aging. Proc. Natl. Acad. Sci. U. S. A. 100, 16018–16023.

de Calignon, A., Fox, L.M., Pitstick, R., Carlson, G.A., Bacskai, B.J., Spires-Jones, T.L., Hyman, B.T., 2010. Caspase activation precedes and leads to tangles. Nature 464, 1201–1204.

de la Monte, S.M., Sohn, Y.K., Wands, J.R., 1997. Correlates of p53- and Fas (CD95)-mediated apoptosis in Alzheimer's disease. J. Neurol. Sci. 152, 73–83.

de Rivero Vaccari, J.P., Lotocki, G., Marcillo, A.E., Dietrich, W.D., Keane, R.W., 2008. A molecular platform in neurons regulates inflammation after spinal cord injury. J. Neurosci. 28 (13), 3404–3414.

Delarasse, C., Auger, R., Gonnord, P., Fontaine, B., Kanellopoulos, J.M., 2011. The purinergic receptor P2X7 triggers alpha-secretase-dependent processing of the amyloid precursor protein. J. Biol. Chem. 286, 2596–2606.

Dineley, K.T., Westerman, M., Bui, D., Bell, K., Ashe, K.H., Sweatt, J.D., 2001. Beta-amyloid activates the mitogen-activated protein kinase cascade via hippocampal alpha7 nicotinic acetylcholine receptors: in vitro and in vivo mechanisms related to Alzheimer's disease. J. Neurosci. 21.

Di Virgilio, F., Dal Ben, D., Sarti, A.C., Giuliani, A.L., Falzoni, S., 2017. The P2X7 receptor in infection and inflammation. Immunity 47, 15−31.

Di Virgilio, F., Schmalzing, G., Markwardt, F., 2018. The elusive P2X7 macropore. Trends Cell Biol. 28, 392−404.

Dias, V., Junn, E., Mouradian, M.M., 2013. The role of oxidative stress in Parkinson's disease. J. Parkinsons Dis. 3, 461−491.

Díaz-Hernandez, M., del Puerto, A., Díaz-Hernandez, J.I., Diez-Zaera, M., Lucas, J.J., Garrido, J.J., Miras-Portugal, M.T., 2008. Inhibition of the ATP-gated P2X7 receptor promotes axonal growth and branching in cultured hippocampal neurons. J. Cell Sci. 121 (Pt 22), 3717−3728.

Diaz-Hernandez, J.I., Gomez-Villafuertes, R., Leon-Otegui, M., Hontecillas-Prieto, L., Del Puerto, A., Trejo, J.L., et al., 2012. In vivo P2X7 inhibition reduces amyloid plaques in Alzheimer's disease through GSK3beta and secretases. Neurobiol. Aging 33, 1816−1828.

DiMicco, J.A., Sarkar, S., Zaretskaia, M.V., Zaretsky, D.V., 2006. Stress-induced cardiac stimulation and fever: common hypothalamic origins and brainstem mechanisms. Auton. Neurosci. 126−127, 106−119.

Donmez, G., Outeiro, T.F., 2013. SIRT1 and SIRT2: emerging targets in neurodegeneration. EMBO Mol. Med. 5, 344−352.

Duan, W., Zhang, Z., Gash, D.M., Mattson, M.P., 1999. Participation of prostate apoptosis response-4 in degeneration of dopaminergic neurons in models of Parkinson's disease. Ann. Neurol. 46, 587−597.

Duncan, J.A., Bergstralh, D.T., Wang, Y., Willingham, S.B., Ye, Z., Zimmermann, A.G., Ting, J.P.-Y., 2007. Cryopyrin/NALP3 binds ATP/dATP, is an ATPase, and requires ATP binding to mediate inflammatory signaling. Proc. Natl. Acad. Sci. U. S. A. 104, 8041−8046.

D'Amelio, M., Cavallucci, V., Cecconi, F., 2010. Neuronal caspase-3 signaling: not only cell death. Cell Death Differ. 17, 1104−1114.

Eckman, C.B., Mehta, N.D., Crook, R., Perez-tur, J., Prihar, G., Pfeiffer, E., Graff-Radford, N., Hinder, P., Yager, D., Zenk, B., Refolo, L.M., Prada, C.M., Younkin, S.G., Hutton, M., Hardy, J., 1997. A new pathogenic mutation in the APP gene (I716V) increases the relative proportion of A beta 42(43). Hum. Mol. Genet. 6 (12), 2087−2089.

El Khoury, J., Toft, M., Hickman, S.E., Means, T.K., Terada, K., Geula, C., Luster, A.D., 2007. Ccr2 deficiency impairs microglial accumulation and accelerates progression of Alzheimer-like disease. Nat. Med. 13 (4), 432−438.

Ellis, R.J., Olichney, J.M., Thal, L.J., Mirra, S.S., Morris, J.C., Beekly, D., Heyman, A., 1996. Cerebral amyloid angiopathy in the brains of patients with Alzheimer's disease: the CERAD experience, Part XV. Neurology 46 (6), 1592−1596.

Enari, M., Sakahira, H., Yokoyama, H., Okawa, K., Iwamatsu, A., Nagata, S., 1998. A caspaseactivated DNase that degrades DNA during apoptosis, and its inhibitor ICAD. Nature 391, 43−50.

Evans, L.D., et al., 2018. Extracellular monomeric and aggregated tau efficiently enter human neurons through overlapping but distinct pathways. Cell Rep. 22 (13), 3612−3624.

Fellner, L., Irschick, R., Schanda, K., et al., 2013. Toll-like receptor 4 is required for α-synuclein dependent activation of microglia and astroglia. Glia 61 (3), 349−360.

Fernandes-alnemri, T., Litwack, G., Alnemri, E.S., 1995. Mch2, a new member of the apoptotic Ced-3/Ice cysteine protease gene family. Cancer Res. 55, 2737−2742.

Fernandes-Alnemri, T., Wu, J., Yu, J.W., Datta, P., Miller, B., Jankowski, W., Rosenberg, S., Zhang, J., Alnemri, E.S., 2007. The pyroptosome: a supramolecular assembly of ASC dimers mediating inflammatory cell death via caspase-1 activation. Cell Death Differ. 14 (9), 1590–1604.

Fernando, P., Megeney, L.A., 2007. Is caspase-dependent apoptosis only cell differentiation taken to the extreme? FASEB J. 21 (1), 8–17.

Ferrari, D., Wesselborg, S., Bauer, M.K., Schulze-Osthoff, K., 1997. Extracellular ATP activates transcription factor NF-kappaB through the P2Z purinoreceptor by selectively targeting NF-kappaB p65. J. Cell Biol. 139 (7), 1635–1643.

Franceschini, A., Capece, M., Chiozzi, P., Falzoni, S., Sanz, J.M., Sarti, A.C., Bonora, M., Pinton, P., Di Virgilio, F., 2015. The P2X7 receptor directly interacts with the NLRP3 inflammasome scaffold protein. FASEB. J. 29 (6), 2450–2461.

Franchi, L., Warner, N., Viani, K., Nuñez, G., 2009. Function of Nod-like receptors in microbial recognition and host defense. Immunol. Rev. 227 (1), 106–128.

Francistiová, L., Bianchi, C., Di Lauro, C., Sebastián-Serrano, Á., de Diego-García, L., Kobolák, J., Dinnyés, A., Díaz-Hernández, M., 2020. The role of P2X7 receptor in Alzheimer's disease. Front. Mol. Neurosci. 13, 94.

Frost, B., Jacks, R.L., Diamond, M.I., 2009. Propagation of tau misfolding from the outside to the inside of a cell. J. Biol. Chem. 284 (19), 12845–12852.

Frye, R.A., 1999. Characterization of five human cDNAs with homology to the yeast SIR2 gene: Sir2-like proteins (sirtuins) metabolize NAD and may have protein ADP-ribosyltransferase activity. Biochem. Biophys. Res. Commun. 260 (1), 273–279.

Fukumoto, H., Cheung, B.S., Hyman, B.T., Irizarry, M.C., 2002. Beta-secretase protein and activity are increased in the neocortex in Alzheimer disease. Arch. Neurol. 59 (9), 1381–1389.

Furukawa, K., Barger, S.W., Blalock, E.M., Mattson, M.P., 1996. Activation of K+ channels and suppression of neuronal activity by secreted beta-amyloid-precursor protein. Nature 379 (6560), 74–78.

Gamblin, T.C., Chen, F., Zambrano, A., Abraha, A., Lagalwar, S., Guillozet, A.L., Lu, M., Fu, Y., Garcia-Sierra, F., LaPointe, N., Miller, R., Berry, R.W., Binder, L.I., Cryns, V.L., 2003. Caspase cleavage of tau: linking amyloid and neurofibrillary tangles in Alzheimer's disease. Proc. Natl. Acad. Sci. 100 (17), 10032–10037.

Gastard, M.C., Troncoso, J.C., Koliatsos, V.E., 2003. Caspase activation in the limbic cortex of subjects with early Alzheimer's disease. Ann. Neurol. 54, 393–398.

Gavrieli, Y., Sherman, Y., Ben-Sasson, S.A., 1992. Identification of programmed cell death in situ via specific labeling of nuclear DNA fragmentation. J. Cell Biol. 119 (3), 493–501.

Gervais, F.G., Xu, D., Robertson, G.S., Vaillancourt, J.P., Zhu, Y., Huang, J., LeBlanc, A., Smith, D., Rigby, M., Shearman, M.S., Clarke, E.E., Zheng, H., Van Der Ploeg, L.H., Ruffolo, S.C., Thornberry, N.A., Xanthoudakis, S., Zamboni, R.J., Roy, S., Nicholson, D.W., 1999. Involvement of caspases in proteolytic cleavage of Alzheimer's amyloid-beta precursor protein and amyloidogenic A beta peptide formation. Cell 97 (3), 395–406.

Ghavami, S., Hashemi, M., Ande, S.R., Yeganeh, B., Xiao, W., Eshraghi, M., Bus, C.J., Kadkhoda, K., Wiechec, E., Halayko, A.J., Los, M., 2009. Apoptosis and cancer: mutations within caspase genes. J. Med. Genet. 46 (8), 497–510.

Ghosh, S., May, M.J., Kopp, E.B., 1998. NF-κB and Rel proteins: evolutionarily conserved mediators of immune responses. Annu. Rev. Immunol. 16, 225–260.

Giannakopoulos, P., Kovari, E., Savioz, A., De Bilabao, F., Dubois-Dauphin,¨M., Hof, P.R., Bouras, C., 1999. Differential distribution of presenilin-1, Bax, and Bcl-X in Alzheimer's disease and frontotemporal dementia. Acta Neuropathol. 98, 141–149.

Girod, J.P., Brotman, D.J., 2004. Does altered glucocorticoid homeostasis increase cardiovascular risk? Cardiovasc. Res. 64, 217–226.

Glenner, G.G., Wong, C.W., 1984. Alzheimer's disease and Down's syndrome: sharing of a unique cerebrovascular amyloid fibril protein. Biochem. Biophys. Res. Commun. 122 (3), 1131–1135.

Goate, A., Chartier-Harlin, M.C., Mullan, M., Brown, J., Crawford, F., Fidani, L., Giuffra, L., Haynes, A., Irving, N., James, L., 1991. Segregation of a missense mutation in the amyloid precursor protein gene with familial Alzheimer's disease. Nature 349 (6311), 704–706.

Goedert, M., Wischik, C.M., Crowther, R.A., Walker, J.E., Klug, A., 1988. Cloning and sequencing of the cDNA encoding a core protein of the paired helical filament of Alzheimer disease: identification as the microtubule-associated protein tau. Proc. Natl. Acad. Sci. U. S. A. 85 (11), 4051–4055.

Golde, T.E., Dickson, D., Hutton, M., 2006. Filling the gaps in the abeta cascade hypothesis of Alzheimer's disease. Curr. Alzheimer Res. 3 (5), 421–430.

Goldgaber, D., Lerman, M.I., McBride, O.W., Saffiotti, U., Gajdusek, D.C., 1987. Characterization and chromosomal localization of a cDNA encoding brain amyloid of Alzheimer's disease. Science 235 (4791), 877–880.

Goldstein, S.A., Bayliss, D.A., Kim, D., Lesage, F., Plant, L.D., Rajan, S., 2005. International Union of Pharmacology. LV. Nomenclature and molecular relationships of two-P potassium channels. Pharmacol. Rev. 57 (4), 527–540.

Gomez-Isla, T., Price, J., McKeel, D., Morris, J., Growdon, J., Hyman, B., 1997. Profound loss of layer II entorhinal cortex neurons occurs in very mild Alzheimer's disease. J. Neurosci. 16, 4491–4500.

Gonatas, N.K., Anderson, A., Evangelista, I., 1967. The contribution of altered synapses in the senile plaque. An electron microscopic study in Alzheimer dementia. J. Neuropathol. Exp. Neurol. 26, 25–39.

Gonfloni, S., Maiani, E., Di Bartolomeo, C., Diederich, M., Cesareni, G., 2012. Oxidative stress, DNA damage, and c-Abl signaling: at the crossroad in neurodegenerative diseases? Int. J. Cell Biol. 2012, 683097.

González-Scarano, F., Baltuch, G., 1999. Microglia as mediators of inflammatory and degenerative diseases. Annu. Rev. Neurosci. 22, 219–240.

Gorman, A.M., 2008. Cell death in neurodegenerative diseases: recurring themes around protein handling. J. Cell Mol. Med. 12 (6A), 2263–2280.

Goutte, C., Tsunozaki, M., Hale, V.A., Priess, J.R., 2002. APH-1 is a multipass membrane protein essential for the Notch signaling pathway in *Caenorhabditis elegans* embryos. Proc. Natl. Acad. Sci. U. S. A. 99 (2), 775–779.

Gratuze, M., Leyns, C.E.G., Holtzman, D.M., 2018. New insights into the role of TREM2 in Alzheimer's disease. Mol. Neurodegener. 13 (1), 66.

Graves, J.D., Draves, K.E., Gotoh, Y., Krebs, E.G., Clark, E.A., 2001. Both phosphorylation and caspase-mediated cleavage contribute to regulation of the Ste20-like protein kinase Mst1 during CD95/Fas-induced apoptosis. J. Biol. Chem. 276 (18), 14909–14915.

Gray, R., et al., 1996. Hippocampal synaptic transmission enhanced by low concentrations of nicotine. Nature 383 (6602), 713–716.

Guo, H., Albrecht, S., Bourdeau, M., Petzke, T., Bergeron, C., LeBlanc, A.C., 2004. Active caspase-6 and caspase-6 cleaved Tau in neuropil threads, neuritic plaques and neurofibrillary tangles of Alzheimer's disease. Am. J. Pathol. 165, 523–531.

Guo, H., et al., 2006. Caspase-1 activation of caspase-6 in human apoptotic neurons. Cell Death Differ. 13, 285–292.

Gustin, A., Kirchmeyer, M., Koncina, E., Felten, P., Losciuto, S., Heurtaux, T., Tardivel, A., Heuschling, P., Dostert, C., 2015. NLRP3 inflammasome is expressed and functional in mouse brain microglia but not in astrocytes. PLoS One 10, e0130624.

Guyant-Maréchal, L., Rovelet-Lecrux, A., Goumidi, L., Cousin, E., Hannequin, D., Raux, G., Penet, C., Ricard, S., Macé, S., Amouyel, P., Deleuze, J.-F., Frebourg, T., Brice, A., Lambert, J.-C.D., 2007. Variations in the APP gene promoter region and risk of Alzheimer disease. Camp. Neurol. 68 (9), 684–687.

Habermacher, C., Dunning, K., Chataigneau, T., Grutter, T., 2016. Molecular structure and function of P2X receptors. Neuropharmacology 104, 18–30.

Hardy, J.A., Higgins, G.A., 1992. Alzheimer's disease: the amyloid cascade hypothesis. Science 256 (5054), 184–185.

Hardy, J., Selkoe, D.J., 2002. The amyloid hypothesis of Alzheimer's disease: progress and problems on the road to therapeutics. Science 297, 353–356.

Harkat, M., Peverini, L., Cerdan, A.H., Dunning, K., Beudez, J., Martz, A., Calimet, N., Specht, A., Cecchini, M., 2017. On the permeation of large organic cations through the pore of ATP-gated P2X receptors. Proc. Natl. Acad. Sci. U. S. A. 114, E3786–E3795.

Hayward, J.H., Lee, S.J., 2014. A decade of research on tlr2 discovering its pivotal role in glial activation and neuroinflammation in neurodegenerative diseases. Exp. Neurobiol. 23, 138–147.

He, Y., Zeng, M.Y., Yang, D., Motro, B., Núñez, G., 2016. NEK7 is an essential mediator of NLRP3 activation downstream of potassium efflux. Nature 530 (7590), 354–357.

Helgason, E., Phung, Q.T., Dueber, E.C., 2013. Recent insights into the complexity of Tank-binding kinase 1 signaling networks: the emerging role of cellular localization in the activation and substrate specificity of TBK1. FEBS Lett. 587 (8), 1230–1237.

Hemmi, H., Takeuchi, O., Kawai, T., Kaisho, T., Sato, S., Sanjo, H., et al., 2000. A Toll-like receptor recognizes bacterial DNA. Nature 408 (6813), 740–745.

Henderson, B.W., Gentry, E.G., Rush, T., Troncoso, J.C., Thambisetty, M., Montine, T.J., Herskowitza, J.H., 2016. Rho-associated protein kinase 1 (ROCK1) is increased in Alzheimer's disease and ROCK1 depletion reduces amyloid-β levels in brain. J. Neurochem. 138 (4), 525–531.

Heneka, M.T., Kummer, M.P., Stutz, A., Delekate, A., Schwartz, S., Saecker, A., Griep, A., Axt, D., Remus, A., Tzeng, T.-C., Gelpi, E., Halle, A., Korte, M., Latz, E., Golenbock, D., 2013. NLRP3 is activated in Alzheimer's disease and contributes to pathology in APP/PS1 mice. Nature 493 (7434), 674–678.

Heneka, M.T., McManus, R.M., Latz, E., 2018. Inflammasome signalling in brain function and neurodegenerative disease. Nat. Rev. Neurosci. 19, 610–621.

Hengartner, M.O., 2000. The biochemistry of apoptosis. Nature 407, 770–776.

Herz, J., Hamann, U., Rogne, S., Myklebost, O., Gausepohl, H., Stanley, K.K., 1988. Surface location and high affinity for calcium of a 500-kd liver membrane protein closely related to the LDL-receptor suggest a physiological role as lipoprotein receptor. EMBO J. 7 (13), 4119–4127.

Hewinson, J., Mackenzie, A.B., 2007. P2X(7) receptor-mediated reactive oxygen and nitrogen species formation: from receptor to generators. Biochem. Soc. Trans. 35 (Pt 5), 1168–1170.

Hickman, S.E., Allison, E.K., El Khoury, J., 2008. Microglial dysfunction and defective β-amyloid clearance pathways in aging Alzheimer's disease mice. J. Neurosci. 28 (33), 8354–8360. https://doi.org/10.1523/JNEUROSCI.0616-08.2008.

Hiscott, J., Pitha, P., Genin, P., Nguyen, H., Heylbroeck, C., Mamane, Y., Algarte, M., Lin, R., 1999. Triggering the interferon response: the role of IRF-3 transcription factor. J. Interferon Cytokine Res. 19 (1), 1–13.

Hoebe, K., Du, X., Georgel, P., Janssen, E., Tabeta, K., Kim, S.O., et al., 2003. Identification of Lps2 as a key transducer of MyD88-independent TIR signalling. Nature 424 (6950), 743–748.

Hof, P.R., Bouras, C., Perl, D.P., Sparks, D.L., Mehta, N., Morrison, J.H., 1995. Age-related distribution of neuropathologic changes in the cerebral cortex of patients with Down's syndrome. Quantitative regional analysis and comparison with Alzheimer's disease. Arch. Neurol. 52 (4), 379–391.

Hofer, A.M., Brown, E.M., 2003. Extracellular calcium sensing and signalling. Nat. Rev. Mol. Cell Biol. 4 (7), 530–538.

Holtman, I.R., Raj, D.D., Miller, J.A., et al., 2015. Induction of a common microglia gene expression signature by aging and neurodegenerative conditions: a co-expression meta-analysis. Acta Neuropathol. Commun. 3 (1). https://doi.org/10.1186/s40478-015-0203-5.

Holtman, I.R., Skola, D., Glass, C.K., 2017. Transcriptional control of microglia phenotypes in health and disease. J. Clin. Invest. 127, 3220–3229.

Hook, V., Yoon, M., Mosier, C., Ito, G., Podvin, S., Head, B.P., Rissman, R., O'Donoghue, A.J., Hook, G., 2020. Cathepsin B in neurodegeneration of Alzheimer's disease, traumatic brain injury, and related brain disorders. Biochim. Biophys. Acta Protein Proteonomics 1868 (8), 140428.

Horowitz, P.M., et al., 2004. Early N-terminal changes and caspase-6 cleavage of tau in Alzheimer's disease. J. Neurosci. 24, 7895–7902.

Huang, K., et al., 2017. A common haplotype lowers PU. 1 expression in myeloid cells and delays onset of Alzheimer's disease. Nat. Neurosci. 20 (8), 1052–1061.

Illes, P., 2020. P2X7 receptors amplify CNS damage in neurodegenerative diseases. Int. J. Mol. Sci. 21, 5996. https://doi.org/10.3390/ijms21175996.

Illes, P., Rubini, P., Ulrich, H., Zhao, Y., Tang, Y., 2020. Regulation of microglial functions by purinergic mechanisms in the healthy and diseased CNS. Cells 9, 1108.

Jack Jr., C.R., et al., 2013. Tracking pathophysiological processes in alzheimer's disease: An updated hypothetical model of dynamic biomarkers. Te Lancet Neurol. 12, 207–216.

Jack Jr., C.R., Holtzman, D.M., 2013. Biomarker modeling of alzheimer's disease. Neuron 80, 1347–1358.

Jacobs, M.D., Harrison, S.C., 1998. Structure of an IκBα/NF-κB complex. Cell 95 (6), 749–758.

Jacobs, M., Hayakawa, K., Swenson, L., Bellon, S., Fleming, M., Taslimi, P., Doran, J., 2006. The structure of dimeric ROCK I reveals the mechanism for ligand selectivity. J. Biol. Chem. 281 (1), 260–268.

Jacobsen, T., Schubotz, R.I., Höfel, L., Cramon, D.Y.V., 2006. Brain correlates of aesthetic judgment of beauty. Neuroimage 29 (1), 276–285.

Jacobson, M.D., Weil, M., Raff, M.C., 1997. Programmed cell death in animal development. Cell 88 (3), 347–354.

Jakobsson, J., Pålsson, E., Sellgren, C., Rydberg, F., Ekman, A., Zetterberg, H., Blennow, K., Landen, M., 2016. CACNA1C polymorphism and altered phosphorylation of tau in bipolar disorder. Br. J. Psychiatry 208, 195–196.

Jan, A., Gokce, O., Luthi-Carter, R., Lashuel, H.A., 2008. The ratio of monomeric to aggregated forms of Abeta40 and Abeta42 is an important determinant of amyloid-beta aggregation, fibrillogenesis, and toxicity. J. Biol. Chem. 283 (42), 28176–28189.

Jayadev, S., Leverenz, J.B., Steinbart, E., Stahl, J., Klunk, W., Yu, C.E., Bird, T.D., 2010. Alzheimer's disease phenotypes and genotypes associated with mutations in presenilin 2. Brain 133 (Pt 4), 1143–1154.

Jellingera, K.A., Stadelmann, C., 2001. Problems of cell death in neurodegeneration and Alzheimer's Disease. J. Alzheimers Dis. 3, 31–40.

Jian, C., Lu, M., Zhang, Z., Liu, L., Li, X., Huang, F., Xu, N., Qin, L., Zhang, Q., Zou, D., 2017. miR-34a knockout attenuates cognitive deficits in APP/PS1 mice through inhibition of the amyloidogenic processing of APP. Life Sci. 182, 104–111.

Jog, N.R., Frisoni, L., Shi, Q., Monestier, M., Hernandez, S., Craft, J., Prak, E.T., Caricchio, R., 2012. Caspase-activated DNase is required for maintenance of tolerance to lupus nuclear autoantigens. Arthritis Rheum. 64 (4), 1247–1256.

Jones, A., Rådholm, K., Neal, B., 2018. Defining 'unhealthy': A systematic analysis of alignment between the Australian dietary guidelines and the health star rating system. Nutrients 10, 501.

Julia, T.C.W., Goate, A.M., 2017. Genetics of β-amyloid precursor protein in Alzheimer's disease. Cold Spring Harb. Perspect Med. 7 (6), a024539.

Kalia, M., 2003. Dysphagia and aspiration pneumonia in patients with Alzheimer's disease. Metabolism 52 (10 Suppl. 2), 36–38.

Kam, T.-I., Song, S., Gwon, Y., Park, H., Yan, J.-J., Im, I., Choi, J.-W., Choi, T.-Y., Kim, J., Song, D.-K., Takai, T., Kim, Y.-C., Kim, K.-S., Choi, S.-Y., Choi, S., Klein, W.L., Yuan, J., Jung, Y.-K., 2013. FcγRIIb mediates amyloid-β neurotoxicity and memory impairment in Alzheimer's disease. J. Clin. Invest. 123 (7), 2791–2802.

Kang, J., Lemaire, H.G., Unterbeck, A., Salbaum, J.M., Masters, C.L., Grzeschik, K.H., Multhaup, G., Beyreuther, K., Müller-Hill, B., 1987. The precursor of Alzheimer's disease amyloid A4 protein resembles a cell-surface receptor. Nature 325 (6106), 733–736.

Karin, M., 1999. How NF-kappaB is activated: the role of the IkappaB kinase (IKK) complex. Oncogene 18 (49), 6867–6874.

Katz, H.R., 2002. Inhibitory receptors and allergy. Curr. Opin. Immunol. 14 (6), 698–704.

Kawai, T., Akira, S., 2010. The role of pattern-recognition receptors in innate immunity: update on toll-like receptors. Nat. Immunol. 11 (5), 373–384.

Kawano, A., Tsukimoto, M., Noguchi, T., Hotta, N., Harada, H., Takenouchi, T., Kitani, H., Kojima, S., 2012. Involvement of P2X4 receptor in P2X7 receptor-dependent cell death of mouse macrophages. Biochem. Biophys. Res. Commun. 419 (2), 374–380.

Kegel, K.B., Kim, M., Sapp, E., et al., 2000. Huntingtin expression stimulates endosomal-lysosomal activity, endosome tubulation, and autophagy. J. Neurosci. 20, 7268–7278.

Kerr, J.F.R., Wyllie, A.H., Currie, A.R., 1972. Apoptosis: a basic biological phenomenon with wide-ranging implications in tissue kinetics. Br. J. Cancer 26, 239–257.

Khakh, B.S., Bao, X.R., Labarca, C., Lester, H.A., 1999. Neuronal P2X transmitter-gated cation channels change their ion selectivity in seconds. Nat. Neurosci. 2, 322—330.

Khan, M., Rutten, B.P.F., Kim, M.O., 2019. MST1 regulates neuronal cell death via JNK/ Casp3 signaling pathway in HFD mouse brain and HT22 cells. Int. J. Mol. Sci. 20 (10), 2504.

Kidd, M., 1963. Paired helical filaments in electron microscopy of Alzheimer's disease. Nature 197, 192—193.

Kim, T.W., Pettingell, W.H., Jung, Y.K., Kovacs, D.M., Tanzi, R.E., 1997. Alternative cleavage of Alzheimer-associated presenilins during apoptosis by a caspase-3 family protease. Science 277, 373—376.

Kim, S.Y., Moon, J.H., Lee, H.G., Kim, S.U., Lee, Y.B., 2007. ATP released from beta-amyloid-stimulated microglia induces reactive oxygen species production in an autocrine fashion. Exp. Mol. Med. 39, 820—827.

Kitamura, Y., Shimohama, S., Kamoshima, W., Ota, T., Matsuoka, Y., Nomura, Y., Smith, M.A., Perry, G., Whitehouse, P.J., Taniguchi, T., 1998. Alteration of proteins regulating apoptosis, Bcl-2, Bcl-x, Bax, Bak, Bad, ICH-1 and CPP32, in Alzheimer's disease. Brain Res. 780, 260—269.

Kitazawa, M., et al., 2005. Lipopolysaccharide-induced inflammation exacerbates tau pathology by a cyclin-dependent kinase 5-mediated pathway in a transgenic model of Alzheimer's disease. J. Neurosci. 25 (39), 8843—8853.

Klaiman, G., Champagne, N., LeBlanc, A.C., 2009. Self-activation of Caspase-6 in vitro and in vivo: caspase-6 activation does not induce cell death in HEK293T cells. Biochim. Biophys. Acta 1793, 592—601.

Kommaddi, R.P., Das, D., Karunakaran, S., Nanguneri, S., Bapat, D., Ray, A., Shaw, E., Bennett, D.A., Nair, D., Ravindranath, V., 2018. Aβ mediates F-actin disassembly in dendritic spines leading to cognitive deficits in Alzheimer's disease. J. Neurosci. 38 (5), 1085—1099.

Koonin, E.V., Aravind, L., 2000. The NACHT family - a new group of predicted NTPases implicated in apoptosis and MHC transcription activation. Trends Biochem. Sci. 25 (5), 223—224.

Korneyev, A.Y., 1998. Stress-induced tau phosphorylation in mouse strains with different brain Erk 1 + 2 immunoreactivity. Neurochem. Res. 23, 1539—1543.

Korvatska, O., et al., 2015. R47H variant of TREM2 associated with Alzheimer disease in a large late-onset family: clinical, genetic, and neuropathological study. JAMA Neurol. 72 (8), 920—927.

Kowal, R.C., et al., 1989. Low density lipoprotein receptor-related protein mediates uptake of cholesteryl esters derived from apoprotein E-enriched lipoproteins. Proc. Natl. Acad. Sci. U. S. A. 86 (15), 5810—5814.

Kuan, C.-Y., Roth, K.A., Flavell, R.A., Rakic, P., 2000. Mechanism of programmed cell death in the developing brain. Trends Neurosci. 23, 287—293.

Kuida, K., Lippke, J.A., Ku, G., Harding, M.W., Livingston, D.J., Su, M.S., Flavell, R.A., 1995. Altered cytokine export and apoptosis in mice deficient in interleukin-1 beta converting enzyme. Science 267, 2000—2003.

Kumar, A., Sivanandam, T.M., Thakur, M.K., 2016. Presenilin 2 overexpression is associated with apoptosis in Neuro2a cells. Transl. Neurosci. 7 (1), 71—75.

Lai, S.K., Wong, C.H., Lee, Y.P., Li, H.Y., 2011. Caspase-3-mediated degradation of condensin Cap-H regulates mitotic cell death. Cell Death Differ. 18 (6), 996–1004.

larke, P.G.H., 1990. Developmental cell death: morphological diversity and multiple mechanisms. Anat. Embryol. 181, 195–213.

Lavin, Y., Winter, D., Blecher-Gonen, R., David, E., Keren-Shaul, H., Merad, M., Jung, S., Amit, I., 2014. Tissue-resident macrophage enhancer landscapes are shaped by the local microenvironment. Cell 159, 1312–1326. https://doi.org/10.1016/j.cell.2014.11.018.

LeBlanc, A.C., 1995. Increased production of 4 kDa amyloid beta peptide in serum deprived human primary neuron cultures: possible involvement of apoptosis. J. Neurosci. 15, 7837–7846.

LeBlanc, A.C., 2013. Caspase-6 as a novel early target in the treatment of Alzheimer's disease. Eur. J. Neurosci. 37, 2005–2018.

LeBlanc, A., Liu, H., Goodyer, C., Bergeron, C., Hammond, J., 1999. Caspase-6 role in apoptosis of human neurons, amyloidogenesis, and Alzheimer's disease. J. Biol. Chem. 274 (33), 23426–23436.

Lee, Y., Jeon, K., Lee, J.T., Kim, S., Kim, V.N., 2002. MicroRNA maturation: stepwise processing and subcellular localization. EMBO J. 21, 4663–4670.

Lee, K.J., Moussa, C.E., Lee, Y., Sung, Y., Howell, B.W., Turner, R.S., Pak, D.T., Hoe, H.S., 2010. Beta amyloid-independent role of amyloid precursor protein in generation and maintenance of dendritic spines. Neuroscience 169, 344–356.

Lee, J., Sayed, N., Hunter, A., Au, K.F., Wong, W.H., Mocarski, E.S., et al., 2012. Activation of innate immunity is required for efficient nuclear reprogramming. Cell 151 (3), 547–558.

Lee, C.Y.D., et al., 2018. Elevated TREM2 gene dosage reprograms microglia responsivity and ameliorates pathological phenotypes in Alzheimer's disease models. Neuron 97 (5), 1032–10348 e5.

Lehnardt, S., Massillon, L., Follett, P., et al., 2003. Activation of innate immunity in the CNS triggers neurodegeneration through a Toll-like receptor 4-dependent pathway. Proc. Natl. Acad. Sci. U. S. A. 100 (14), 8514–8519.

Lei, P., Ayton, S., Finkelstein, D.I., Adlard, P.A., Masters, C.L., Bush, A.I., 2010. Tau protein: relevance to Parkinson's disease. Int. J. Biochem. Cell Biol. 42 (11), 1775–1778.

Leon-Otegui, M., Gomez-Villafuertes, R., Diaz-Hernandez, J.I., Diaz-Hernandez, M., Miras-Portugal, M.T., Gualix, J., 2011. Opposite effects of P2X7 and P2Y2 nucleotide receptors on alpha-secretase-dependent APP processing in Neuro-2a cells. FEBS Lett. 585, 2255–2262.

Levy-Lahad, E., Wijsman, E.M., Nemens, E., Anderson, L., Goddard, K.A., Weber, J.L., Bird, T.D., Schellenberg, G.D., 1995. A familial Alzheimer's disease locus on chromosome 1. Science 269 (5226), 970–973.

Li, X.H., Lv, B.L., Xie, J.Z., Liu, J., Zhou, X.W., Wang, J.Z., 2012. AGEs induce Alzheimer-like tau pathology and memory deficit via RAGE-mediated GSK-3 activation. Neurobiol. Aging 33, 1400–1410.

Lill, C.M., et al., 2015. The role of TREM2 R47H as a risk factor for Alzheimer's disease, frontotemporal lobar degeneration, amyotrophic lateral sclerosis, and Parkinson's disease. Alzheimers Dement 11 (12), 1407–1416.

Lin, M.T., Beal, M.F., 2006. Mitochondrial dysfunction and oxidative stress in neurodegenerative diseases. Nature 443 (7113), 787–795.

Ling, Y., Morgan, K., Kalsheker, N., 2003. Amyloid precursor protein (APP) and the biology of proteolytic processing: relevance to Alzheimer's disease. Int. J. Biochem. Cell Biol. 35 (11), 1505–1535.

Lippa, C.F., Swearer, J.M., Kane, K.J., Nochlin, D., Bird, T.D., Ghetti, B., Nee, L.E., St George-Hyslop, P., Pollen, D.A., Drachman, D.A., 2000. Familial Alzheimer's disease: site of mutation influences clinical phenotype. Ann. Neurol. 48 (3), 376–379.

Liu, P.-P., Xie, Y., Meng, X.-Y., Kang, J.-S., 2019. History and progress of hypotheses and clinical trials for Alzheimer's disease. Signal Transduct. Target Ther. 4, 29.

Liu, X., Zou, H., Slaughter, C., Wang, X., 1997. DFF, a heterodimeric protein that functions downstream of caspase-3 to trigger DNA fragmentation during apoptosis. Cell 89 (2), 175–184.

Liu, C.C., et al., 2017. Astrocytic LRP1 mediates brain Abeta clearance and impacts amyloid deposition. J. Neurosci. 37 (15), 4023–4031.

Loetscher, H., Deuschle, U., Brockhaus, M., Reinhardt, D., Nelboeck, P., Mous, J., et al., 1997. Presenilins are processed by caspase-type proteases. J. Biol. Chem. 272, 20655–20659.

Long, J.M., et al., 2019. Novel upregulation of amyloid-beta precursor protein (APP) by microRNA-346 via targeting of APP mRNA 5′-untranslated region: implications in Alzheimer's disease. Mol. Psychiatr. 24 (3), 345–363.

Louis, C., Burns, C., Wicks, I., 2018. TANK-binding kinase 1-dependent responses in health and autoimmunity. Front. Immunol. 9, 434.

Lue, L.F., Walker, D.G., Brachova, L., Beach, T.G., Rogers, J., Schmidt, A.M., et al., 2001. Involvement of microglial receptor for advanced glycation endproducts (RAGE) in Alzheimer's disease: identification of a cellular activation mechanism. Exp. Neurol. 171, 29–45.

Luo, W., et al., 2015. Microglial internalization and degradation of pathological tau is enhanced by an anti-tau monoclonal antibody. Sci. Rep. 5, 11161.

Luse, S.A., Smith Jr., K.R., 1964. The ultrastructure of senile plaques. Am. J. Pathol. 44, 553–563.

MacGibbon, G.A., Lawlor, P.A., Walton, M., Sirimanne, E., Faull, R.L.M., Synek, B., Mee, E., Connor, B., Dragunow, M., 1997. Expression of Fos, Jun, and Krox family proteins in Alzheimer's disease. Exp. Neurol. 147, 316–332.

Mackenzie, I.R., Neumann, M., Baborie, A., Sampathu, D.M., Du Plessis, D., Jaros, E., Perry, R.H., Trojanowski, J.Q., Mann, D.M., Lee, V.M., 2011. A harmonized classification system for FTLD-TDP pathology. Acta Neuropathol. 122 (1), 111–113.

Maejima, Y., Kyoi, S., Zhai, P., Liu, T., Li, H., Ivessa, A., Sciarretta, S., Del Re, D.P., Zablocki, D.K., Hsu, C.P., Lim, D.S., Isobe, M., Sadoshima, J., 2013. Mst1 inhibits autophagy by promoting the interaction between Beclin1 and Bcl-2. Nat. Med. 19 (11), 1478–1488.

Mahla, R.S., Reddy, M.C., Prasad, D.V., Kumar, H., 2013. Sweeten PAMPs: role of sugar complexed PAMPs in innate immunity and vaccine biology. Front. Immunol. 4, 248.

Majno, G., Joris, I., 1995. Apoptosis, oncosis, and necrosis. An overview of cell death. Am. J. Pathol. 146, 3–15.

Mak, K.S., Funnell, A.P.W., Pearson, R.C.M., Crossley, M., 2011. PU.1 and haematopoietic cell fate: dosage matters. Int. J. Cell Biol. 2011 (808524).

Mangan, M.S.J., Olhava, E.J., Roush, W.R., Seidel, H.M., Glick, G.D., Latz, E., 2018. Targeting the NLRP3 inflammasome in inflammatory diseases. Nat. Rev. Drug Discov. 17 (8), 588–606.

Mariathasan, S., Newton, K., Monack, D.M., Vucic, D., French, D.M., Lee, W.P., Roose-Girma, M., Erickson, S., Dixit, V.M., 2004. Differential activation of the inflammasome by caspase-1 adaptors ASC and Ipaf. Nature 430 (6996), 213–218. https://doi.org/10.1038/nature02664.

Martin, L.J., 2010. Mitochondrial and cell death mechanisms in neurodegenerative diseases. Pharmaceuticals 3 (4), 839–915.

Martin, E., Amar, M., Dalle, C., Youssef, I., Boucher, C., Le Duigou, C., Brückner, M., Prigent, A., Sazdovitch, V., Halle, A., et al., 2019. New role of P2X7 receptor in an Alzheimer's disease mouse model. Mol. Psychiatr. 24, 108–125.

Martinon, F., Tschopp, J., 2005. NLRs join TLRs as innate sensors of pathogens. Trends Immunol. 26, 447–454.

Martinon, F., Tschopp, J., 2007. Inflammatory caspases and inflammasomes: master switches of inflammation. Cell Death Differ. 14 (1), 10–22.

Marwarha, G., Raza, S., Meiers, C., Ghribi, O., 2014. Leptin attenuates BACE1 expression and amyloid-β genesis via the activation of SIRT1 signaling pathway. Biochim. Biophys. Acta (BBA) 1842, 1587–1595.

Matthews-Roberson, T.A., Quintanilla, R.A., Ding, H., Johnson, G.V.W., 2008. Immortalized cortical neurons expressing caspase-cleaved tau are sensitized to endoplasmic reticulum stress induced cell death. Brain Res. 1234, 206–212.

Mattson, M.P., 2000. Apoptosis in neurodegenerative disorders. Nat. Rev. Mol. Cell Biol. 1, 120–129.

Mawal-Dewan, M., Henley, J., Van de Voorde, A., Trojanowski, J.Q., Lee, V.M., 1994. The phosphorylation state of tau in the developing rat brain is regulated by phosphoprotein phosphatases. J. Biol. Chem. 269 (49), 30981–30987.

May, P., et al., 2004. Neuronal LRP1 functionally associates with postsynaptic proteins and is required for normal motor function in mice. Mol. Cell Biol. 24 (20), 8872–8883.

McCarthy, A.E., Yoshioka, C., Mansoor, S.E., 2019. Full-length P2X7 structures reveal how palmitoylation prevents channel desensitization. Cell 179, 659–670.

McCarty, J.S., Toh, S.Y., Li, P., 1999. Study of DFF45 in its role of chaperone and inhibitor: two independent inhibitory domains of DFF40 nuclease activity. Biochem. Biophys. Res. Commun. 264 (1), 176–180.

McLarnon, J.G., Ryu, J.K., Walker, D.G., Choi, H.B., 2006. Upregulated expression of purinergic P2X(7) receptor in Alzheimer disease and amyloid-beta peptide-treated microglia and in peptide-injected rat hippocampus. J. Neuropathol. Exp. Neurol. 65, 1090–1097.

Meda, L., Cassatella, M.A., Szendrei, G.I., Otvos Jr., L., Baron, P., Villalba, M., Ferrari, D., Rossi, F., 1995. Activation of microglial cells by beta-amyloid protein and interferon-gamma. Nature 374 (6523), 647–650.

Medzhitov, R., Preston-Hurlburt, P., Janeway, C.A., 1997. A human homologue of the Drosophila Toll protein signals activation of adaptive immunity. Nature 388 (6640), 394–397.

Miller, M.C., Tavares, R., Johanson, C.E., Hovanesian, V., Donahue, J.E., Gonzalez, L., et al., 2008. Hippocampal RAGE immunoreactivity in early and advanced Alzheimer's disease. Brain Res. 1230, 273–280.

Minoretti, P., Gazzaruso, C., Vito, C.D., Emanuele, E., Bianchi, M., et al., 2006. Effect of the functional toll-like receptor 4 Asp299Gly polymorphism on susceptibility to late-onset alzheimer's disease. Neurosci. Lett. 391, 147–149.

Mirra, S.S., Heyman, A., McKeel, D., Sumi, S.M., Crain, B.J., Brownlee, L.M., Vogel, F.S., Hughes, J.P., van Belle, G., Berg, L., 1991. The consortium to establish a registry for

Alzheimer's disease (CERAD). Part II. Standardization of the neuropathological assessment of Alzheimer's disease. Neurology 41, 479–486.

Mohandas, E., Rajmohan, V., Raghunath, B., 2009. Neurobiology of Alzheimer's disease. Indian J. Psychiatr. 51 (1), 55–61.

Moore, K.J., El Khoury, J., Medeiros, L.A., Terada, K., Geula, C., Luster, A.D., Freeman, M.W., 2002. A CD36-initiated signaling cascade mediates inflammatory effects of -amyloid. J. Biol. Chem. 277 (49), 47373–47379.

Mosconi, L., Berti, V., Glodzik, L., Pupi, A., De Santi, S., de Leon, M.J., 2010. Pre-clinical detection of Alzheimer's disease using FDG-PET, with or without amyloid imaging. J. Alzheimer's Dis. 20 (3), 843–854.

Mullan, M., Crawford, F., Axelman, K., Houlden, H., Lilius, L., Winblad, B., Lannfelt, L., 1992. A pathogenic mutation for probable Alzheimer's disease in the APP gene at the N-terminus of beta-amyloid. Nat. Genet. 1 (5), 345–347.

Nakagawa, O., Fujisawa, K., Ishizaki, T., Saito, Y., Nakao, K., Narumiya, S., 1996. ROCK-I and ROCK-II, two isoforms of Rho-associated coiled-coil forming protein serine/threonine kinase in mice. FEBS Lett. 392 (2), 189–193.

Nakanishi, H., 2003. Neuronal and microglial cathepsins in aging and age-related diseases. Ageing Res. Rev. 2 (4), 367–381.

Nakanishi, H., Wu, Z., 2009. Microglia-aging: roles of microglial lysosome- and mitochondria-derived reactive oxygen species in brain aging. Behav. Brain Res. 201 (1), 1–7.

National Center for Biotechnology Information (NCBI), 2020. CASP6 Caspase 6. Available at:https://www.ncbi.nlm.nih.gov/gene/839.

NCBI, 2021. P2RX7 Purinergic Receptor P2X 7 [*Homo sapiens* (Human)]. Available online at: https://www.ncbi.nlm.nih.gov/gene?Db=gene&Cmd=DetailsSearch&Term=5027.

Neve, R.L., Harris, P., Kosik, K.S., Kurnit, D.M., Donlon, T.A., 1986. Identification of cDNA clones for the human microtubule-associated protein tau and chromosomal localization of the genes for tau and microtubule-associated protein 2. Brain Res. 387 (3), 271–280.

Nicholson, D.W., 1999. Caspase structure, proteilytic substrates, and function during apoptotic cell death. Cell Death Differ. 6, 1028–1042.

Nicholson, D.W., 2000. From bench to clinic with apoptosis-based therapeutic agents. Nature 407, 810–816.

Nicke, A., 2008. Homotrimeric complexes are the dominant assembly state of native P2X7 subunits. Biochem. Biophys. Res. Commun. 377, 803–808.

Nicke, A., Bäumert, H.G., Rettinger, J., Eichele, A., Lambrecht, G., Mutschler, E., Schmalzing, G., 1998. P2X1 and P2X3 receptors form stable trimers: a novel structural motif of ligand-gated ion channels. EMBO J. 17 (11), 3016–3028.

Nimmerjahn, A., Kirchhoff, F., Helmchen, F., 2005. Resting microglial cells are highly dynamic surveillants of brain parenchyma in vivo. Science 308 (5726), 1314–1318.

North, R.A., 2002. Molecular physiology of P2X receptors. Physiol. Rev. 82 (4), 1013–1067.

Nuttle, L.C., Dubyak, G.R., 1994. Differential activation of cation channels and non-selective pores by macrophage P2z purinergic receptors expressed in *Xenopus* oocytes. J. Biol. Chem. 269, 13988–13996.

Obulesu, M., Lakshmi, M.J., 2014. Apoptosis in Alzheimer's disease: an understanding of the physiology, pathology and therapeutic avenues. Neurochem. Res. 39 (12), 2301–2312.

Ojala, J., Alafuzoff, I., Herukka, S.-K., van Groen, T., Tanila, H., Pirttilä, T., 2009. Expression of interleukin-18 is increased in the brains of Alzheimer's disease patients. Neurobiol. Aging 30, 198–209.

Okawa, Y., Ishiguro, K., Fujita, S.C., 2003. Stress-induced hyperphosphorylation of tau in the mouse brain. FEBS Lett. 535, 183−189.

Ou, S.H., Wu, F., Harrich, D., García-Martínez, L.F., Gaynor, R.B., 1995. Cloning and characterization of a novel cellular protein, TDP-43, that binds to human immunodeficiency virus type 1 TAR DNA sequence motifs. J. Virol. 69 (6), 3584−3596.

Oyama, F., Cairns, N.J., Shimada, H., Oyama, R., Titani, K., Ihara, Y., 1994. Down's syndrome: up-regulation of beta-amyloid protein precursor and tau mRNAs and their defective coordination. J. Neurochem. 62 (3), 1062−1066.

Parvathenani, L.K., Tertyshnikova, S., Greco, C.R., Roberts, S.B., Robertson, B., Posmantur, R., 2003. P2X7 mediates superoxide production in primary microglia and is up-regulated in a transgenic mouse model of Alzheimer's disease. J. Biol. Chem. 278, 13309−13317.

Peng, Q., et al., 2010. TREM2- and DAP12-dependent activation of PI3K requires DAP10 and is inhibited by SHIP1. Sci. Signal. 3 (122), ra38.

Pétrilli, V., Papin, S., Dostert, C., Mayor, A., Martinon, F., Tschopp, J., 2007. Activation of the NALP3 inflammasome is triggered by low intracellular potassium concentration. Cell Death Differ. 14 (9), 1583−1589.

Pfeifer, L.A., White, L.R., Ross, G.W., Petrovitch, H., Launer, L.J., 2002. Cerebral amyloid angiopathy and cognitive function: the HAAS autopsy study. Neurology 58 (11), 1629−1634.

Pimenova, A.A., Herbinet, M., Gupta, I., Machlovi, S.I., Bowles, K.R., Marcora, E., Goate, A.M., 2021. Alzheimer's-associated PU.1 expression levels regulate microglial inflammatory response. Neurobiol. Dis. 148, 105217.

Pirttimaki, T.M., et al., 2013. alpha7 Nicotinic receptor-mediated astrocytic gliotransmitter release: abeta effects in a preclinical Alzheimer's mouse model. PLoS One 8 (11), e81828.

Prinz, M., Priller, J., Sisodia, S.S., Ransohoff, R.M., 2011. Heterogeneity of CNS myeloid cells and their roles in neurodegeneration. Nat. Neurosci. 14 (10), 1227−1235.

Pritchard, N.R., Smith, K.G., 2003. B cell inhibitory receptors and autoimmunity. Immunology 108 (3), 263−273.

Qi, Y., Sun, D., Yang, W., Xu, B., Lv, D., Han, Y., Sun, M., Jiang, S., Hu, W., Yang, Y., 2020. Mammalian sterile 20-like kinase (MST) 1/2: crucial players in nervous and immune system and neurological disorders. J. Mol. Biol. 432 (10), 3177−3190.

Qin, J., Zhang, X., Wang, Z., Li, J., Zhang, Z., Gao, L., Ren, H., Qian, M., Du, B., 2017. Presenilin 2 deficiency facilitates Aβ-induced neuroinflammation and injury by upregulating P2X7 expression. Sci. China Life Sci. 60 (2), 189−201.

Qu, J., Zhao, H., Li, Q., Pan, P., Ma, K., Liu, X., Feng, H., Chen, Y., 2018. MST1 suppression reduces early brain injury by inhibiting the NF-κB/MMP-9 pathway after subarachnoid hemorrhage in mice. Behav. Neurol. 2018, 6470957.

Quintanilla, R.A., Matthews-Roberson, T.A., Dolan, P.J., Johnson, G.V.W., 2009. Caspase-cleaved tau expression induces mitochondrial dysfunction in immortalized cortical neurons: implications for the pathogenesis of Alzheimer disease. J. Biol. Chem. 284 (28), 18754−18766.

Rath, N., Olson, M.F., 2012. Rho-associated kinases in tumorigenesis: re-considering ROCK inhibition for cancer therapy. EMBO Rep. 13 (10), 900−908.

Reddy, P.H., Tonk, S., Kumar, S., Vijayan, M., Kandimalla, R., Kuruva, C.S., Reddy, A.P., 2017. A critical evaluation of neuroprotective and neurodegenerative MicroRNAs in Alzheimer's disease. Biochem. Biophys. Res. Commun. 483, 1156−1165.

Reed, J.C., 2000. Mechanism of apoptosis. Am. J. Pathol. 157, 1415−1430.

Reiche, E.M., Nunes, S.O., Morimoto, H.K., 2004. Stress, depression, the immune system, and cancer. Lancet Oncol. 5, 617−625.

Reitz, C., 2012. Alzheimer's disease and the amyloid cascade hypothesis: a critical review. Int. J. Alzheimer's Dis. 2012, 11. Article ID 369808.

Riccardi, D., Kemp, P.J., 2012. The calcium-sensing receptor beyond extracellular calcium homeostasis: conception, development, adult physiology, and disease. Annu. Rev. Physiol. 74, 271−297.

Ricciarelli, R., Fedele, E., 2017. The amyloid cascade hypothesis in Alzheimer's disease: it's time to change our mind. Curr. Neuropharmacol. 15 (6), 926−935.

Roberson, E.D., Scearce-Levie, K., Palop, J.J., Yan, F., Cheng, I.H., Wu, T., Gerstein, H., Yu, G.Q., Mucke, L., 2007. Reducing endogenous tau ameliorates amyloid beta-induced deficits in an Alzheimer's disease mouse model. Science 316 (5825), 750−754.

Roberts, G.W., 1988. Immunocytochemistry of neurofibrillary tangles in dementia pugilistica and Alzheimer's disease: evidence for common genesis. Lancet 2 (8626−8627), 1456−1458.

Rogaev, E.I., Sherrington, R., Rogaeva, E.A., Levesque, G., Ikeda, M., Liang, Y., Chi, H., Lin, C., Holman, K., Tsuda, T., 1995. Familial Alzheimer's disease in kindreds with missense mutations in a gene on chromosome 1 related to the Alzheimer's disease type 3 gene. Nature 376 (6543), 775−778.

Rohn, T.T., 2008. Caspase-cleaved TAR DNA-binding protein-43 is a major pathological finding in Alzheimer's disease. Brain Res. 1228, 189−198.

Roman, A.Y., Devred, F., Byrne, D., La Rocca, R., Ninkina, N.N., Peyrot, V., Tsvetkov, P.O., 2019. Zinc induces temperature-dependent reversible self-assembly of tau. J. Mol. Biol. 431 (4), 687−695.

Rosa, M.L., Guimaraes, F.S., de Oliveira, R.M., Padovan, C.M., Pearson, R.C., Del Bel, E.A., 2005. Restraint stress induces beta-amyloid precursor protein mRNA expression in the rat basolateral amygdala. Brain Res. Bull. 65, 69−75.

Roth, K.A., 2001. Caspases, apoptosis, and Alzheimer disease: causation, correlation, and confusion. J. Neuropathol. Exp. Neurol. 60 (9), 829−838.

Rovelet-Lecrux, A., Hannequin, D., Raux, G., Le Meur, N., Laquerrière, A., Vital, A., Dumanchin, C., Feuillette, S., Brice, A., Vercelletto, M., Dubas, F., 2006. APP locus duplication causes autosomal dominant early-onset Alzheimer disease with cerebral amyloid angiopathy. Nat. Genet. 38 (1), 24−26.

Sadowski, M., Wisniewski, H.M., Tarnawski, M., Kozlowski, P.B., Lach, B., Wegiel, J., 1999. Entorhinal cortex of aged subjects with Down's syndrome shows severe neuronal loss caused by neurofibrillary pathology. Acta Neuropathol. 97 (2), 156−164.

Sakahira, H., Enari, M., Nagata, S., 1999. Functional differences of two forms of the inhibitor of caspase-activated DNase, ICAD-L, and ICAD-S. J. Biol. Chem. 274 (22), 15740−15744.

Salminen, A., Ojala, J., Suuronen, T., Kaarniranta, K., Kauppinen, A., 2008. Amyloid-β oligomers set fire to inflammasomes and induce Alzheimer's pathology. J. Cell Mol. Med. 12 (6a), 2255−2262.

Sandbrink, R., Masters, C.L., Beyreuther, K., 1996. APP gene family. Alternative splicing generates functionally related isoforms. Ann. N. Y. Acad. Sci. 777, 281−287.

Sannerud, R., Esselens, C., Ejsmont, P., Mattera, R., Rochin, L., Tharkeshwar, A.K., De Baets, G., De Wever, V., Habets, R., Baert, V., Vermeire, W., Michiels, C., Groot, A.J., Wouters, R., Dillen, K., Vints, K., Baatsen, P., Munck, S., Derua, R., Waelkens, E., Basi, G.S., Mercken, M., Vooijs, M., Bollen, M., Schymkowitz, J., Rousseau, F.,

Bonifacino, J.S., Van Niel, G., De Strooper, B., Annaert, W., 2016. Restricted location of PSEN2/γ-secretase determines substrate specificity and generates an intracellular aβ pool. Cell 166 (1), 193–208.

Sanz, J.M., Chiozzi, P., Ferrari, D., Colaianna, M., Idzko, M., Falzoni, S., et al., 2009. Activation of microglia by amyloid {beta} requires P2X7 receptor expression. J. Immunol. 182, 4378–4385.

Sapolsky, R.M., Krey, L.C., McEwen, B.S., 1985. Prolonged glucocorticoid exposure reduces hippocampal neuron number: implications for aging. J. Neurosci. 5 (5), 1222–1227.

Sasaki, N., Toki, S., Chowei, H., Saito, T., Nakano, N., Hayashi, Y., et al., 2001. Immunohistochemical distribution of the receptor for advanced glycation end products in neurons and astrocytes in Alzheimer's disease. Brain Res. 888, 256–262.

Satoh, J.-I., Asahina, N., Kitano, S., Kino, Y., 2014. A comprehensive profile of ChIP-seq-based PU.1/Spi1 target genes in microglia. Gene Regul. Syst. Biol. 8, 127–139.

Schmidt, A.M., Hori, O., Brett, J., Yan, S.D., Wautier, J.L., Stern, D., 1994. Cellular receptors for advanced glycation end products. Implications for induction of oxidant stress and cellular dysfunction in the pathogenesis of vascular lesions. sscler. Thromb. J. Vasc. Biol. 14 (10), 1521–1528. https://doi.org/10.1161/01.atv.14.10.1521.

Schmid, C.D., et al., 2002. Heterogeneous expression of the triggering receptor expressed on myeloid cells-2 on adult murine microglia. J. Neurochem. 83 (6), 1309–1320.

Schwartz, L.M., Milligan, C.E., 1996. Cold thoughts of death : the role of ICE proteases in neuronal cell death. Trends Neurosci. 19, 555–562.

Sebastian-Serrano, A., de Diego-Garcia, L., di Lauro, C., Bianchi, C., Diaz-Hernandez, M., 2019. Nucleotides regulate the common molecular mechanisms that underlie neurodegenerative diseases; therapeutic implications. Brain Res. Bull. 151, 84–91.

Sebbagh, M., Renvoizé, C., Hamelin, J., Riché, N., Bertoglio, J., Bréard, J., 2001. Caspase-3-mediated cleavage of ROCK I induces MLC phosphorylation and apoptotic membrane blebbing. Nat. Cell Biol. 3 (4), 346–352.

Selkoe, D.J., 2001. Alzheimer's disease: genes, proteins, and therapy. Physiol. Rev. 81, 741–766.

Selye, H., 1950. Stress and the general adaptation syndrome. Br. Med. J. 1, 1383–1392.

Sergeant, N., Delacourte, A., Buée, L., 2005. Tau protein as a differential biomarker of tauopathies. Biochim. Biophys. Acta (BBA) - Mol. Basis Dis. 1739 (2–3), 179–197.

Shaw, M.H., Reimer, T., Kim, Y.G., Nuñez, G., 2008. NOD-like receptors (NLRs): bona fide intracellular microbial sensors. Curr. Opin. Immunol. 20 (4), 377–382.

Shi, H., Murray, A., Beutler, B., 2016. Reconstruction of the mouse inflammasome system in HEK293T cells. Bio Protoc. 6 (21).

Shimohama, S., 2000. Apoptosis in Alzheimer's disease–an update. Apoptosis 5 (1), 9–16.

Shinohara, M., et al., 2017. Role of LRP1 in the pathogenesis of Alzheimer's disease: evidence from clinical and preclinical studies. J. Lipid Res. 58 (7), 1267–1281.

Sierra, A., Encinas, J.M., Deudero, J.J., Chancey, J.H., Enikolopov, G., Overstreet-Wadiche, L.S., Tsirka, S.E., Maletic-Savatic, M., 2010. Microglia shape adult hippocampal neurogenesis through apoptosis-coupled phagocytosis. Cell Stem Cell 7 (4), 483–495.

Simon, D.J., et al., 2012. A caspase cascade regulating developmental axon degeneration. J. Neurosci. 32, 17540–17553.

Skaper, S.D., Facci, L., Culbert, A.A., Evans, N.A., Chessell, I., Davis, J.B., Richardson, J.C., 2006. Glia 54 (3), 234–242. https://doi.org/10.1002/glia.20379.

Sleegers, K., Brouwers, N., Gijselinck, I., Theuns, J., Goossens, D., Wauters, J., Del-Favero, J., Cruts, M., van Duijn, C.M., Van Broeckhoven, C., 2006. APP duplication is sufficient to cause early onset Alzheimer's dementia with cerebral amyloid angiopathy. Brain 129 (Pt 11), 2977−2983.

Sluyter, R., 2017. The P2X7 receptor. Adv. Exp. Med. Biol. 1051, 17−53.

Smale, G., Nichols, N.R., Brady, D.R., Finch, C.E., Horton, W.E.J., 1995. Evidence for apoptotic cell death in Alzheimer's disease. Exp. Neurol. 133, 225−230.

Smith, S.M., Vale, W.W., 2006. The role of the hypothalamic-pituitary-adrenal axis in neuroendocrine responses to stress. Dialogues Clin. Neurosci. 8 (4), 383−395.

Smith, E.M., Gregg, M., Hashemi, F., Schott, L., Hughes, T.K., 2006. Corticotropin releasing factor (CRF) activation of NF-κB-directed transcription in leukocytes. Cell. Mol. Neurobiol. 26 (4−6), 1021−1036.

Smith, A.M., et al., 2013. The transcription factor PU.1 is critical for viability and function of human brain microglia. Glia 61 (6), 929−942.

Sperandio, S., de Belle, I., Bredesen, D.E., 2000. An alternative, nonapoptotic form of programmed cell death. Proc. Natl. Acad. Sci. U. S. A. 97, 14376−14381.

Stadelmann, C., Brück, W., Bancher, C., Jellinger, K., Lassmann, H., 1998. Alzheimer disease: DNA fragmentation indicates increased neuronal vulnerability, but not apoptosis. J. Neuropathol. Exp. Neurol. 57, 456−464.

Stancu, I.C., et al., 2019. Aggregated tau activates NLRP3-ASC inflammasome exacerbating exogenously seeded and non-exogenously seeded tau pathology in vivo. Acta Neuropathol. 137 (4), 599−617.

Stockley, J.H., O'Neill, C., 2008. Understanding BACE1: essential protease for amyloid-beta production in Alzheimer's disease. Cell. Mol. Life Sci. 65, 3265−3289.

Stutz, A., Golenbock, D.T., Latz, E., 2009. Inflammasomes: too big to miss. J. Clin. Invest. 119, 3502−3511.

Su, J.H., Anderson, A.J., Cummings, B.J., Cotman, C.W., 1994. Immunohistochemical evidence for apoptosis in Alzheimer's disease. Neuroreport 5, 2529−2533.

Sun, L., Wu, Z., Hayashi, Y., Peters, C., Tsuda, M., Inoue, K., Nakanishi, H., 2012. Microglial cathepsin B contributes to the initiation of peripheral inflammation-induced chronic pain. J. Neurosci. 32, 11330−11342.

Sun, L., Zhou, R., Yang, G., Shi, Y., 2017. Analysis of 138 pathogenic mutations in presenilin-1 on the in vitro production of Aβ42 and Aβ40 peptides by γ-secretase. Proc. Natl. Acad. Sci. U. S. A. 114 (4), E476−E485.

Sutinen, E.M., Pirttilä, T., Anderson, G., Salminen, A., Ojala, J.O., 2012. Pro-inflammatory interleukin-18 increases Alzheimer's disease-associated amyloid-β production in human neuron-like cells. J. Neuroinflammation 9, 199.

Suzuki, T., Nishiyama, K., Yamamoto, A., et al., 2000. Molecular cloning of a novel apoptosis-related gene, human Nap1 (NCKAP1), and its possible relation to Alzheimer disease. Genomics 63, 246−254.

Takeda, K., Akira, S., 2004. TLR signaling pathways. Semin. Immunol. 16, 3−9.

Takeda, K., Akira, S., 2005. Toll-like receptors in innate immunity. Int. Immunol. 17 (1), 1−14.

Taniguchi, T., Kawamata, T., Mukai, H., Hasegawa, H., Isagawa, T., Yasuda, M., et al., 2001. Phosphorylation of tau is regulated by PKN. J. Biol. Chem. 276 (13), 10025−10031.

Tarawneh, R., Holtzman, D.M., 2012. The clinical problem of symptomatic Alzheimer disease and mild cognitive impairment. Cold Spring Harb. Perspect Med. 2 (5), a006148.

Terry, R.D., 2000. Cell death or synaptic loss in alzheimer disease. J. Neuropathol. Exp. Neurol. 59 (12), 1118−1119.

Terry, R.D., Peck, A., DeTeresa, R., et al., 1981. Some morphometric aspects of the brain in senile dementia of the Alzheimer type. Ann. Neurol. 10, 184–192.

Terry, R.D., Masliah, E., Salmon, D.P., Butters, N., DeTeresa, R., Hill, R., Hansen, L.A., Katzman, R., 1991. Physical basis of cognitive alterations in Alzheimer's disease: synapse loss is the major correlate of cognitive impairment. Ann. Neurol. 30, 572–580.

Tesco, G., Koh, Y.H., Kang, E.L., Cameron, A.N., Das, S., Sena-Esteves, M., et al., 2007. Depletion of GGA3 stabilizes BACE and enhances beta-secretase activity. Neuron 54, 721–737.

Timmer, J.C., Salvesen, G.S., 2007. Caspase substrates. Cell Death Differ. 14, 66–72.

Tiso, N., Pallavicini, A., Muraro, T., Zimbello, R., Apolloni, E., Valle, G., Lanfranchi, G., Danieli, G.A., 1996. Chromosomal localization of the human genes, CPP32, Mch2, Mch3, and Ich-1, involved in cellular apoptosis. Biochem. Biophys. Res. Commun. 225 (3), 983–989.

Toné, S., Sugimoto, K., Tanda, K., Suda, T., Uehira, K., Kanouchi, H., Samejima, K., Minatogawa, Y., Earnshaw, W.C., 2007. Three distinct stages of apoptotic nuclear condensation revealed by time-lapse imaging, biochemical and electron microscopy analysis of cell-free apoptosis. Exp. Cell Res. 313 (16), 3635–3644.

Tortosa, A., Lopez, W., Ferrer, I., 1998. Bcl-2 and Bax protein expression in Alzheimer's disease. Acta Neuropathol. 95, 407–412.

Tremblay, C., St-Amour, I., Schneider, J., Bennett, D.A., Calon, F., 2011. Accumulation of transactive response DNA binding protein 43 in mild cognitive impairment and Alzheimer disease. J Neuropathol. Exp. Neurol. 70 (9), 788–798.

Tyler, S.J., Dawbarn, D., Wilcock, G.K., Allen, S.J., 2002. alpha- and beta-secretase: profound changes in Alzheimer's disease. Biochem. Biophys. Res. Commun. 299, 373–376.

Ura, S., Masuyama, N., Graves, J.D., Gotoh, Y., 2001. Caspase cleavage of MST1 promotes nuclear translocation and chromatin condensation. Proc. Natl. Acad. Sci. U. S. A. 98, 10148–10153.

Valenzuela, S.M., Mazzanti, M., Tonini, R., Qiu, M.R., Warton, K., Musgrove, E.A., Campbell, T.J., Breit, S.N., 2000. The nuclear chloride ion channel NCC27 is involved in regulation of the cell cycle. J. Physiol. 529 (Pt 3), 541–552.

Vassar, R., 2005. β-Secretase, APP and aβ in Alzheimer's disease. In: Harris, J.R., Fahrenholz, F. (Eds.), Alzheimer's Disease. Subcellular Biochemistry, vol. 38. Springer, Boston, MA. https://doi.org/10.1007/0-387-23226-5_4.

Vega, M.V., Nigro, A., Luti, S., Capitini, C., Fani, G., Gonnelli, L., Boscaro, F., Chiti, F., 2019. Isolation and characterization of soluble human full-length TDP-43 associated with neurodegeneration. FASEB. J. 33 (10), 10780–10793.

Velez-Fort, M., Audinat, E., Angulo, M.C., 2009. Functional alpha 7-containing nicotinic receptors of NG2-expressing cells in the hippocampus. Glia 57 (10), 1104–1114.

Venegas, C., Heneka, M.T., 2017. Danger-associated molecular patterns in Alzheimer's disease. J. Leukoc. Biol. 101, 87–98.

Venigalla, M., Sonego, S., Gyengesi, E., Sharman, M.J., Munch, G., 2016. Novel promising therapeutics against chronic neuroinflammation and neurodegeneration in Alzheimer's disease. Neurochem. Int. 95, 63–74.

Virginio, C., MacKenzie, A., Rassendren, F.A., North, R.A., Surprenant, A., 1999. Pore dilation of neuronal P2X receptor channels. Nat. Neurosci. 2 (4), 315–321.

Wahrborg, P., 1998. Mental stress and ischaemic heart disease: an underestimated connection. Eur. Heart J. 19 (Suppl. O), O20–O23.

Walker, E.S., Martinez, M., Brunkan, A.L., Goate, A., 2005. Presenilin 2 familial Alzheimer's disease mutations result in partial loss of function and dramatic changes in Abeta 42/40 ratios. J. Neurochem. 92 (2), 294−301.

Walters, J., Pop, C., Scott, F.L., Drag, M., Swartz, P., Mattos, C., Salvesen, G.S., Clark, A.C., 2009. A constitutively active and uninhibitable caspase-3 zymogen efficiently induces apoptosis. Biochem. J. 424 (3), 335−345.

Wang, H.Y., Lee, D.H., D'Andrea, M.R., Peterson, P.A., Shank, R.P., et al., 2000. beta-Amyloid(1-42) binds to alpha7 nicotinic acetylcholine receptor with high affinity. Implications for Alzheimer's disease pathology. J. Biol. Chem. 275, 5626−5632.

Wang, X.J., et al., 2010. Crystal structures of human caspase 6 reveal a new mechanism for intramolecular cleavage self-activation. EMBO Rep. 11, 841−847.

Wang, L.Z., Yu, J.T., Miao, D., Wu, Z.C., Zong, Y., et al., 2011. Genetic association of TLR4/11367 polymorphism with late-onset alzheimer's disease in a han Chinese population. Brain Res. 1381, 202−207.

Wang, L., et al., 2013. Expression of Tau40 induces activation of cultured rat microglial cells. PLoS One 8 (10), e76057.

Wang, Y., et al., 2015. TREM2 lipid sensing sustains the microglial response in an Alzheimer's disease model. Cell 160 (6), 1061−1071.

Wang, P.F., Xu, D.Y., Zhang, Y., Liu, X.B., Xia, Y., Zhou, P.Y., Fu, Q.G., Xu, S.G., 2017. Deletion of mammalian sterile 20-like kinase 1 attenuates neuronal loss and improves locomotor function in a mouse model of spinal cord trauma. Mol. Cell. Biochem. 431 (1−2), 11−20.

Wellington, C.L., Hayden, M.R., 2000. Caspases and neurodegeneration: on the cutting edge of new therapeutic approaches. Clin. Genet. 57, 1−10.

Wewers, M.D., Sarkar, A., 2009. P2X7 receptor and macrophage function. Purinergic Signal. 5 (2), 189−195.

Wiltfang, J., Esselmann, H., Bibl, M., Smirnov, A., Otto, M., Paul, S., Schmidt, B., Klafki, H.W., Maler, M., Dyrks, T., Bienert, M., Beyermann, M., Rüther, E., Kornhuber, J., 2002. Highly conserved and disease-specific patterns of carboxyterminally truncated Abeta peptides 1-37/38/39 in addition to 1-40/42 in Alzheimer's disease and in patients with chronic neuroinflammation. J. Neurochem. 81 (3), 481−496.

Wisniewski, K.E., Wisniewski, H.M., Wen, G.Y., 1985. Occurrence of neuropathological changes and dementia of Alzheimer's disease in Down's syndrome. Ann. Neurol. 17 (3), 278−282.

Wolfe, M.S., Guénette, S.Y., 2007. APP at a glance. J. Cell Sci. 120 (Pt 18), 3157−3161.

Wu, Du C., Jackson-Lewis, V., Vila, M., Tieu, K., Teismann, P., Vadseth, C., Choi, D.-K., Ischiropoulos, H., Przedborski, S., 2002. Blockade of microglial activation is neuroprotective in the 1-methyl-4-phenyl-1,2,3,6-tetrahydropyridine mouse model of Parkinson disease. J. Neurosci. 22 (5), 1763−1771. https://doi.org/10.1523/JNEUROSCI.22-05-01763.2002.

Wyss-Coray, T., et al., 2003. Adult mouse astrocytes degrade amyloid-beta in vitro and in situ. Nat. Med. 9 (4), 453−457.

Xie, Y., Kaufmann, D., Moulton, M.J., Panahi, S., Gaynes, J.A., Watters, H.N., Zhou, D., Xue, H.H., Fung, C.M., Levine, E.M., Letsou, A., Brennan, K.C., Dorsky, R.I., 2017. Lef1-dependent hypothalamic neurogenesis inhibits anxiety. PLoS Biol. 15 (8), e2002257.

Xing, J., Titus, A.R., Humphrey, M.B., 2015. The TREM2-DAP12 signaling pathway in Nasu-Hakola disease: a molecular genetics perspective. Res. Rep. Biochem. 5, 89−100.

Yamamoto, M., Sato, S., Hemmi, H., Uematsu, S., Hoshino, K., Kaisho, T., et al., 2003. TRAM is specifically involved in the Toll-like receptor 4-mediated MyD88-independent signaling pathway. Nat. Immunol. 4 (11), 1144−1150.

Yan, S.D., Chen, X., Fu, J., Chen, M., Zhu, H., Roher, A., et al., 1996. RAGE and amyloid-beta peptide neurotoxicity in Alzheimer's disease. Nature 382, 685−691.

Yang, L., et al., 2016. LRP1 modulates the microglial immune response via regulation of JNK and NF-kappaB signaling pathways. J. Neuroinflammation 13 (1), 304.

Yankner, B.A., Duffy, L.K., Kirschner, D.A., 1990. Neurotrophic and neurotoxic effects of amyloid beta protein: reversal by tachykinin neuropeptides. Science 250 (4978), 279−282.

Yano, S., Brown, E.M., Chattopadhyay, N., 2004. Calcium-sensing receptor in the brain. Cell Calcium 35 (3), 257−264.

Yu, S.P., Farhangrazi, Z.S., Ying, H.S., Yeh, C.H., Choi, D.W., 1998. Enhancement of outward potassium current may participate in beta-amyloid peptide-induced cortical neuronal death. Neurobiol. Dis. 5 (2), 81−88.

Yu, H.B., Li, Z.B., Zhang, H.X., Wang, X.L., 2006. Role of potassium channels in Abeta(1-40)-activated apoptotic pathway in cultured cortical neurons. J. Neurosci. Res. 84 (7), 1475−1484.

Yuan, J., Shaham, S., Ledoux, S., Ellis, H.M., Horvitz, H.R., 1993. TheCElegans cell death gene ced-3 encodes a protein similar to mammalian interleukin-1beta-converting enzyme. Cell 75, 641−652.

Yun, H.J., Yoon, J.H., Lee, J.K., Noh, K.T., Yoon, K.W., Oh, S.P., Oh, H.J., Chae, J.S., Hwang, S.G., Kim, E.H., Maul, G.G., Lim, D.S., Choi, E.J., 2011. Daxx mediates activation-induced cell death in microglia by triggering MST1 signalling. EMBO J. 30 (12), 2465−2476.

Yuste, V.J., Sánchez-López, I., Solé, C., Moubarak, R.S., Bayascas, J.R., Dolcet, X., et al., 2005. The contribution of apoptosis-inducing factor, caspase-activated DNase, and inhibitor of caspase-activated DNase to the nuclear phenotype and DNA degradation during apoptosis. J. Biol. Chem. 280 (42), 35670−35683.

Zatti, G., Ghidoni, R., Barbiero, L., Binetti, G., Pozzan, T., Fasolato, C., Pizzo, P., 2004. The presenilin 2 M239I mutation associated with familial Alzheimer's disease reduces Ca^{2+} release from intracellular stores. Neurobiol. Dis. 15 (2), 269−278.

Zatti, G., Burgo, A., Giacomello, M., Barbiero, L., Ghidoni, R., Sinigaglia, G., Florean, C., Bagnoli, S., Binetti, G., Sorbi, S., Pizzo, P., Fasolato, C., 2006. Presenilin mutations linked to familial Alzheimer's disease reduce endoplasmic reticulum and Golgi apparatus calcium levels. Cell Calcium 39, 539−550.

Zhang, Y., Goodyer, C., LeBlanc, A., 2000. Selective and protracted apoptosis in human primary neurons microinjected with active caspase-3,-6,-7, and-8. J. Neurosci. 20 (22), 8384−8389.

Zhang, Q.S., Liu, W., Lu, G.X., 2017. miR-200a-3p promotes b-Amyloid-induced neuronal apoptosis through down-regulation of SIRT1 in Alzheimer's disease. J. Biosci. 42 (3), 397−404.

Zhao, Y., et al., 2018. TREM2 is a receptor for beta-amyloid that mediates microglial function. Neuron 97 (5), 1023−10231 e7.

Zhong, L., et al., 2018. Amyloid-beta modulates microglial responses by binding to the triggering receptor expressed on myeloid cells 2 (TREM2). Mol. Neurodegener. 13 (1), 15.

Zubenko, G.S., et al., 2003. A collaborative study of the emergence and clinical features of the major depressive syndrome of Alzheimer's disease. Am. J. Psychiatr. 160 (5), 857−866.

Zuo, L., Zhou, T., Pannell, B.K., Ziegler, A.C., Best, T.M., 2015. Biological and physiological role of reactive oxygen species—the good, the bad and the ugly. Acta Physiol. 214, 329−348.

Role of caspases and apoptosis in Parkinson's disease

Introduction

Parkinson's disease (PD) belongs to a group of chronic and progressive neurodegenerative disorders. In 1817, Dr. James Parkinson, an English surgeon was credited for his work on "Shaking Palsy" (Parkinson, 1817) in which he described "Paralysis Agitans" that was later named as Parkinson's disease by Jean-Martin Charcot (French neurologist and anatomist).

It is mainly characterized by motor manifestations followed by the development of nonmotor symptoms in the affected persons. Parkinson's disease is marked with rigidity, bradykinesia, postural instability, and tremor at rest as the cardinal motor symptoms (Kurlawala et al., 2021). The nonmotor characteristics of Parkinson's disease include cognitive dysfunction, sleep disorder, mood changes, impaired cardiovascular function, disturbed bowel movement, and reproductive function (Poewe et al., 2017). These prodromal symptoms appear before the manifestations of motor symptoms (Schrag et al., 2000).

Parkinson's disease has multifactorial etiopathogenesis including genetic susceptibility, involvement of environmental factors, and coupled with age as the predilection factor with 60 years of age as the median age for the onset of disease. (Lees et al., 2009).

In persons between 70 and 79 years of age group, the incidence of Parkinson's disease increases to nearly 93.1% (per 100,000 person-years) (de Rijk et al., 1995). Geographical and cultural variations affect the prevalence with a higher percentage of diseases estimated in North America, South America, and European countries than African, Asian, and Arabic countries (Kalia and Lang, 2015).

Several studies focus on the specific pattern of involvement of brain regions in Parkinson's disease. In the early phase, PD affects the medulla involving the dorsal nucleus of the vagus nerve (cluster of neurons in medulla occupying the ventral position to the floor of fourth ventricle), and olfactory bulbs involving the locus coeruleus (chief site in the brain for noradrenaline synthesis), and subsequently, PD affects the neurons in the substantia nigra pars compacta (Kwan and Whitehill, 2011).

The neurons in the cortical region of the brain are damaged in the later stage of the disease that results in multiple facets of disease manifestations in terms of motor dysfunction as well as loss of cognitive function and psychiatric manifestations in the affected persons (Melissa and Michael, 2020; Kwan and Whitehill, 2011).

Dopamine is a neurotransmitter in the brain synthesized from the decarboxylation of L-DOPA. It belongs to a group of catecholamines neurotransmitters (Michel, 2018). The group of neurons in the brain involved in the synthesis and release of dopamine is called as dopaminergic neurons. The cell body of each dopaminergic neuron synthesizes and release dopamine and transmits it through the axon to the presynaptic membrane where dopamine is stored in the storage vesicle (Kravitz et al., 2010). The influx of calcium ions triggers the release of dopamine into the synapse, and it binds with the dopamine receptors located on the postsynaptic membrane. Surplus and free dopamine in the synapse is degraded by monoamine oxidase (Edmondson et al., 2004) and catechol-O-methyl transferase (Tai and Wu, 2002).

Particularly, dopaminergic neurons are abundant in the substantia nigra pars compacta, ventral tegmental area, and the retrorubral field of the brain. These regions are rich in iron content and neuromelanin (Chinta and Andersen, 2005).

Dopaminergic neurons serve multiple functions including reward, motivation, voluntary movements, mood, addiction, stress, and learning (Kravitz et al., 2010).

The nigrostriatal pathway (dopaminergic neurons) connects the substantia nigra pars compacta with the caudate nucleus and the putamen (dorsal striatum). The pathway controls motor function, reward, and motivation.

The mesolimbic pathway of dopaminergic neurons connects the ventral tegmental area with the ventral striatum (nucleus accumbens and olfactory tubercle), while the mesocortical pathway of dopaminergic neurons connects the ventral tegmental area with the prefrontal cortex.

Dysfunction in the dopaminergic pathway leads to neuropsychiatric disorders, while selective **degeneration of dopaminergic neurons in substantia nigra pars compacta results in Parkinson's disease** (Damier et al., 1999).

Although slow and small degeneration of dopaminergic neurons is reported with the advancing age without manifestation of motor symptoms, in the idiopathic Parkinson's disease, the degeneration of dopaminergic neurons is slow and gradual with the appearance of motor symptoms in the affected persons (Cookson, 2009), while in the early onset of PD, rapid degeneration of dopaminergic neurons coupled with the manifestation of motor symptoms are reported much earlier in life in comparison to the idiopathic PD. Another case of in utero environmental factor-induced or genetic factor-mediated marked low count of dopaminergic neurons at birth is reported with high susceptibility to the development of Parkinson's disease (Haas et al., 2012).

Thus, degeneration of dopaminergic neurons in the basal ganglia (caudate, putamen, and globus pallidus, substantia nigra, nucleus accumbens, and subthalamic nucleus) (Davie, 2008), affecting maximally the neurons in pars compacta in substantia nigra leads to a significant decline in the synthesis of dopamine, especially the pars compacta (part of substantia nigra in the midbrain) region is associated with the rise in excitatory activity, impaired motor activity, and disordered voluntary control of skeletal muscles in the pathogenesis of PD (Gasparini et al., 2013).

The etiopathology of Parkinson's disease is still inconclusive, but the role of genetic factors and environmental factors and their interplay are attributed to the pathology of Parkinson's disease. The significant loss of dopaminergic neurons is the cardinal feature of the pathology of Parkinson's disease.

The chapter deciphers the role of a cascade of events and signaling pathways implicated in the pathogenesis of Parkinson's disease.

Neuroinflammation is the typical sign associated with neurodegenerative disease including Parkinson's disease. However, it is true to seek justification whether neuroinflammation contributes to the pathogenesis of Parkinson's disease or the outcome of the disease. Wanga et al. (2016) provided evidence for the role of neuroinflammation in the cause of Parkinson's disease. After the outbreak of the 1918 influenza pandemic, several survivors were affected with manifestations of postencephalitic Parkinsonism characterized by stooped posture, masklike faces, rigidity, and resting tremors suggesting the role of neuroinflammation in Parkinson's disease.

Role of caspase 1 in pathology of Parkinson's disease

The aggregate of α-synuclein marks the important pathological feature of Parkinson's disease. Wanga et al. (2016) posited the caspase-1 mediated aggregation ability of α-synuclein in PD.

Multiple posttranslational modifications of α-synuclein exhibit the potential to enhance the tendency to form aggregates of α-synuclein. Oueslati et al. (2010) described main posttranslational modifications including phosphorylation, truncation, and ubiquitination in α-synuclein associated with Parkinson's disease. These increase the aggregation potential and toxicity of α-synuclein implicated in the pathogenesis of PD).

Caspase-1 mediated truncation of α-synuclein and neuroinflammation in Parkinson's disease

Truncation of α-synuclein is an irreversible process, and truncated α-synuclein is reported in the structure of Lewy bodies potentially linked with the pathology of PD.

Michell et al. (2007) studied the effect of truncated α-synuclein on dopaminergic cell count.

In vitro study by Michell et al. (2007) showed a specific decline in the number of dopaminergic neurons count in transgenic embryonic ventral mesencephalic cell cultures.

Michell et al. (2007) suggested **that α-synuclein (1−120) increases the tendency of dopaminergic neurons to oxidative stress explaining the mechanism of truncated protein in inducing neuronal apoptosis leading to pathology of sporadic PD.**

Another study by Periquet et al. (2007) showed that aggregated α-synuclein can induce dopaminergic neurotoxicity in the brain in PD.

Periquet et al. (2007) expressed mutant α-synuclein (deleted amino acid residues from 71 to 82) in the transgenic Drosophila model of PD. The mutant α-synuclein exhibited an absence of aggregation or oligomerization in the transgenic Drosophila model.

In vivo study by Periquet et al. (2007) expressed truncated α-synuclein that exhibited enhanced aggregation potential to form inclusions bodies coupled with a higher tendency to neurotoxicity over dopaminergic neurons.

Thus, aggregation of α-synuclein has neurotoxic potential to dopaminergic neurons.

The inflammasomes are the multiprotein oligomeric elements of the immune cells including microglial cells. The pathogen-associated molecular patterns and damage-associated molecular patterns are recognized by pattern recognition receptors like TLR-4 on the microglial cells (Bsibsi et al., 2002). These are activated and initiate the cascade of events involving NF-κB nuclear translocation, synthesis of proinflammatory cytokine, and NLRP3. The latter is involved in the formation of NLRP3 inflammasome through its binding with adaptor protein named as apoptosis-associated speck-like protein containing a caspase activation and recruitment domain (CARD). The activated NLRP3 inflammasome binds with the procaspase-1 either through its CARD or by linking to CARD localized in the associated adaptor protein, ASC.

The procaspase-1 is cleaved into active caspase 1 (Yamin et al., 1996). The active caspase 1 is made up of two heterodimers of which each contains p20 and p10 subunits (total two p20 and two p10 subunits in active caspase 1).

Caspase-1 further activates the proinflammatory cytokine (ProIL-1B) into active form leading to neuroinflammation (Ma et al., 2018). Caspase 1 also splits the full-length α-Syn and produces truncated α-Syn (α-Syn121), C-terminus 19 amino acid residues fragment. Caspase I cleaves after amino acid residue 121 in the α-synuclein.

Ma et al. (2018) explored the neurotoxic potential of truncated α-Syn in the pathology of PD. The α-Syn121 showed accelerated potential in the formation of aggregates characterized by amorphous aggregates containing random coil structures in place of β-sheet structure.

Furthermore, α-Syn121 expressed higher neurotoxicity and lesser membrane disruption potential. Surprisingly, authors (Ma et al., 2018) reported that aggregates of α-Syn121, in turn, could activate the caspase-1-induced cleavage of α-Syn full-length into α-Syn121 implicating via vicious cycle leading to rise in endogenous levels of α-Syn121 and intracellular S129 phosphorylated α-Syn inclusion bodies.

Thus, activate caspase-1 has the enormous potential to truncate α-Syn full length and generate α-Syn121 fragment with intense neurotoxic and neuroinflammatory potential leading to neuroinflammation in the regions of the brain implicated in the initiation and progression of PD as evidenced from studies by authors (Harris et al., 2012; Rail et al., 1981; Gao et al., 2011; Liu et al., 2003).

Caspase-1 mediated dopaminergic neuronal death in Parkinson's disease

The Caspase-1 also called as Interleukin-1 converting enzyme is the evolutionarily conserved protease. It activates proinflammatory cytokines like IL-1B and IL-18 into their active forms (Thornberry et al., 1992). Active caspase −1 is also involved in the death of dopaminergic neurons implicated in the pathology of Parkinson's disease.

A study by Qiao et al. (2017) described the effect of caspase 1 deficiency on the dopaminergic neuronal death in 1-methyl-4-phenyl-1,2,3,6-tetrahydropyridine (MPTP)/p mouse model of PD.

Qiao et al. (2017) reported that **the association between caspase-1 in the neuro-inflammation and Parkinson disease is established, while the involvement of caspase 1 in inducing neuronal degeneration of dopaminergic neurons in the PD is still uncertain.**

In the in vivo study by Qiao et al. (2017), the MPTP/p mouse model of Parkinson's disease was used. This represents the most commonly used model of Parkinson's disease with the use of toxin as MPTP to elicit nigral neuronal loss, striatal dopaminergic loss, coupled with alteration in behavior in mice model. The motor symptoms in the MPTP mice model do not simulate as reported in PD.

The caspase-1 knockout MPTP model of mice showed a reduction in dopaminergic neuronal loss and dyskinesia.

Qiao et al. (2017) reported that deficiency of caspase-1 led to suppressed cleavage and activation of caspase-7 followed by restricted nuclear translocation of poly (ADP-ribose) polymerase 1 leading to inhibition in the synthesis and release of apoptosis-inducing factors.

The apoptosis-inducing factor is involved in the caspase-independent apoptosis of cells and neurons. Another study by Yoo et al. (2017) described the role of apoptosis-inducing factors in inducing dopaminergic neuron death in the caspase-independent pathway in the 6-hydroxydopamine treated mice model.

Thus, the novel role of caspase-1 in controlling dopaminergic neuronal death is implicated in the pathology of PD in mice models through activated caspase-7/PARP1/AIF cascade of events. The findings suggest drug targeting of caspase 1 in the treatment of PD.

Role of caspase 3 in pathology of Parkinson's disease

Caspase 3 is the executioner caspase, and it remains inactive until cleaved and activated by the initiator caspases in response to the apoptotic signaling cascade (Walters et al., 2009).

Active caspase-3 is formed from 32 kDa zymogen. It is split into 17 and 12 kDa subunits. The active caspase 3 is heterotetramer. Each heterodimer contains four antiparallel β-sheets from subunit p17 and two antiparallel β-sheets from the p12 subunit. Thus, two heterodimers are associated with each other via hydrophobic interactions to form complete 12-stranded β-sheet composition guarded by α-helices (Salvesen, 2002; Porter and Jänicke, 1999).

The caspase-3 is the precarious factor in dopaminergic neuronal death in the substantia nigra implicated in the pathology of Parkinson's disease.

Caspase-3 serves as an executioner of apoptosis in experimental animal models of Parkinson's disease. Hartmann et al. (2000) studied the isolated tissues from human brain autopsies with PD. The authors reported a higher proportion of active caspase-3-positive neurons in dopaminergic neurons in the brains regions of patients with Parkinson's disease in comparison to control subjects.

On the basis of immunohistochemical studies (Hartmann et al., 2000), in vitro studies and findings from electron microscopy in the human brains collectively revealed that activation of caspase-3 occurs before the apoptosis of dopaminergic neurons in Parkinson's disease. **The neurons with active caspase-3 undergo pathological changes more rapidly than neurons without the expression of active caspase-3 rendering caspase-3 a susceptibility factor to the cell death of dopaminergic neurons in Parkinson's disease.**

Caspase-3 and parkin protein in Parkinson's disease

Parkin represents the E3 ubiquitin ligase composed of 465 amino acid residues. It is helpful in the covalent bonding of mutant proteins with ubiquitin (Ubiquitination) leading to their degradation by proteasomes (Pickart and Eddins, 2004; Seirafi et al., 2015).

The parkin protein is essential for the survival of the neurons that undergo apoptosis in Parkinson's disease.

Parkin protein suppresses the mitochondria-independent and mitochondria-dependent apoptosis of neurons, and thus it is essential in enhancing the survival of neurons that undergo apoptosis in substantia nigra in the brain in Parkinson's disease (Dawson and Dawson, 2014).

Parkin protein-mediated survival of neurons is regulated via **activated NF-kB-mediated intracellular signaling in the neurons.** After the oxidative injury to neurons, parkin protein stimulates the HOIP subunit of another linear ubiquitin chain assembly complex (LUBAC). Subsequently, activated HOIP subunit of LUBAC initiates the aggregation of ubiquitin polymers on the NF-κB essential modulator (NEMO) that is an inhibitor of nuclear factor kappa-B kinase subunit **gamma** (IKK-γ) and activates the nuclear translocation of NF-κB and ultimately activates the transcription of OPA1 (mitochondrial GTPase) (Aleksaniants, 2013).

The OPA1 protein inhibits the release of cytochrome c from the mitochondria. **Thus, parkin is involved in the suppression of caspase-mediated neuronal apoptosis and the promotion of neuronal survival that undergo death in the brain in Parkinson's disease** (Müller-Rischart et al., 2013).

Kahns et al. (2002) reported that compounds as okadaic acid (toxin from species of dinoflagellates), staurosporine (derived from *Streptomyces staurosporeus*), and camptothecin (topoisomerase toxin) induce apoptosis of neurons and are associated with proteolytic cleavage of parkin protein during apoptosis. Parkin is split into 38-kDa C-terminal fragment and a 12-kDa N-terminal fragment (Kahns et al., 2002). It is identified that cleavage of parkin during apoptosis is suppressed by caspase inhibitors suggesting the role of caspases in inducing cleavage of parkin protein.

Thus, concluded that caspase-mediated parkin cleavage might contribute to the pathology of Parkinson's disease. Cell survival function of cleaved parkin is compromised depressing the neuronal oxidative stress threshold and exaggerated activation of caspases mediating in a vicious cycle of events in the pathology of sporadic Parkinson's disease.

Another study (Kahns et al., 2003) posited that mutations in the parkin gene are responsible for the loss of dopaminergic neurons involved in the pathology of early-onset Parkinson's disease and further augment the cell survival function of parkin protein as was described in the aforementioned study.

A study (Kahns et al., 2003) involving transient coexpression of caspases and wild-type parkin in HEK-293 cells reported that prominently caspase-1 and another caspase as 3, and 8 mediate parkin cleavage and render it inactive. The study identified that the main site of cleavage is located after Asp126 residue in parkin protein.

It can be inferred that both caspase-1 and caspase-8 can cleave parkin protein rendering it inactive and subsequent activation of death receptor and inflammatory stress with loss of E3 ubiquitin ligase role of parkin leading to accumulation of toxic cleaved parkin fragments promoting the death of dopaminergic neurons in the pathology of Parkinson's disease.

Role of caspase 8 in pathology of Parkinson's disease
Activation of caspase 8 in apoptosis

Caspase-8 belongs to the family of cysteine-aspartic proteases. Its synthesis is controlled by the *CASP8* gene. The caspase-8 belongs to the initiator caspases and is involved in apoptosis. It is synthesized in a single polypeptide chain that remains inactive form called as procaspase or zymogen. Procaspase-8 is activated into active caspase-8 through dimerization and proteolytic cleavage (Chang et al., 2003).

The activation of procaspase-8 is started by the dimerization of procaspase-8 with another molecule of procaspase-8. The dimerization is facilitated by the binding of procaspase-8 with the adaptor protein via protein–protein interaction motifs. These are called as "death folds," which are found in the prodomain (structural domain of caspases) of the procaspase 8 (Yigong, 2004).

Single "death fold" is present in the prodomain of the initiator caspases involved in the intrinsic pathway of apoptosis and is called as "caspase recruitment domain" (CARD), whereas the initiator caspases involved in the extrinsic pathway of apoptosis contain two death folds called as "death effector domains" (DEDs) (Kumar, 2006). Caspase-8 contains two DED). (Andersen et al., 2006).

The dimer of procaspase-8 undergoes cleavage by autocatalysis. This results in the elimination of the prodomains and the splitting of the linker region between the large and small subunits. There is the formation of heterotetramer that is active caspases 8 (Riedl and Shi, 2004).

The **FasL** is also designated as CD95L. The Fas ligand is the homotrimer belonging to the type II transmembrane protein of the tumor necrosis factor family.

The Fas receptor belongs to the family of death receptors expressed on the surface of cells. The binding of homotrimer FasL with FasR induces the trimerization of death receptors leading to the formation of "death-inducing signaling complex" that controls the apoptosis.

The activated Fas receptor recruits adapter protein, Fas-associated death domain (FADD). This activity brings about binding between FADD with the death domain of Fas receptor.

Near the amino terminus, the FADD, additionally contains death effector domain that binds with DED of procaspase-8. It undergoes autocatalysis and conversion into active caspase8 and is released into the cytosol (Sheikh and Fornace, 2000).

The activated caspase-8, in turn, activates the downstream caspase 3 and mediates apoptosis via the extrinsic pathway. The activated caspase 8 can also be interlinked to the intrinsic pathway of apoptosis via inducing cleavage of proapoptotic protein of the Bcl-2 family, **BH3 interacting-domain death agonist** (BiD) to form truncated tBiD (Wang et al., 1996).

The truncated BiD facilitates the release of apoptogenic proteins like cytochrome c from the mitochondria and triggers the activation of caspase-9 in a complex with dATP and Apaf-1. The activation of caspase-9 subsequently activates downstream caspases and induces apoptosis in the intrinsic pathway caspase-8 (Kruidering and Evan, 2000).

Suggested that active caspase-8 is essential for apoptosis via **FasL, and death receptors Fas-mediated signaling pathway; additionally, active caspase 8 is involved in interlinking of intrinsic as well as extrinsic apoptotic pathways.**

Caspase 8 as effector in apoptosis of dopaminergic neurons in Parkinson's disease

The apoptosis plays a central role in the dopaminergic neurons death in the substantia nigra pars compacta in patients with Parkinson's disease.

It is activated by the tumor necrosis factor receptor-induced signaling pathway involving active caspase 8 as the effector molecule in apoptosis.

The facts are supported by a study (Boka et al., 1994) in which TNFR1-positive dopaminergic neurons in the substantia nigra pars compacta in both the control subjects and persons with Parkinsonian disorder, while the density of tumor necrosis factor α-positive glial cells was higher in the substantia nigra pars compacta in the persons with Parkinson's disease than the control subjects suggesting the role of TNF receptor/TNF-α-mediated apoptosis of dopaminergic neurons in Parkinson's disease (Boka et al., 1994).

The ligand, tumor necrosis factor α activates the trimerization of the TNF receptor leading to its activation and recruitment of adaptor protein. The interaction between tumor necrosis factor receptor-associated death domain and Fas-associated death domain induces autocatalysis of procaspase 8 into active caspase 8 (Schulze-Osthoff et al., 1998).

Activated caspase 8 can activate the effector caspase 3 in the apoptosis of neurons. Additionally, activated caspase 8 can mediate mitochondrial translocation of proapoptotic protein, BID belonging to the family of Bcl proteins. It induces the release of cytochrome c from the intermembrane space of mitochondria into the cytosol (Green, 2005). The release of cytochrome c initiates the activation of caspase c involved in the death of dopaminergic neurons in the substantia nigra pars compacta (Hartmann et al., 2000) in Parkinson's disease.

Thus, caspase-8 serves as the final effector of apoptosis either via death receptor-induced signaling or BID induced release of cytochrome c and associated cascade of events.

Caspase 8 and activation of microglial cells

The microglia are the resident immune cells in the brain and play essential roles in the physiology of the brain. Moreover, impaired, altered, and uncontrolled activation of microglial cells in the brain is incriminated in the neurotoxicity due to release of proinflammatory cytokine as interleukin-1b, activation of tumor necrosis factor α (Block et al., 2007), and release of nitric oxide exhibiting neuroinflammation and neurotoxicity. Activated microglial cells are involved in the pathology of Parkinson disease, Alzheimer disease, and neurodegenerative diseases (Hanisch and Kettenmann, 2007).

Caspase 8 controls the activation of caspases 3 and 7, which are the prime executioners of the activation of microglial cells.

Thus, a cascade of events leading to serial activation of caspase-8, caspase 3, and caspase 7 are implicated in the activation of microglia, release of proinflammatory cytokines and apoptotic death via activation of protein kinase C-δ-dependent signaling pathway (Burguillos et al., 2011). Further, knockdown of these caspases suppressed the activation of microglia and reduced neurotoxicity.

Burguillos et al. (2011) identified these caspases in activated microglia in the ventral mesencephalon in brains in patients with Parkinson's disease. Another study delineates the role of caspase 8 in the activation of microglial cells in Parkinson's disease. The density of microglial cells is high in the substantia nigra in the brain region. In Parkinson's disease, activated and clustered microglial cells are identified in the substantia nigra in brains in Parkinson's disease.

Role of apoptosis in Parkinson's disease

Apoptosis is critically implicated in the dopaminergic neuronal death in Parkinson's disease that is supported by reporting of DNA fragments and chromatin changes in the brain regions as revealed through a postmortem of brains in Parkinson's disease.

A study (Tompkins et al., 1997) emphasized the dopaminergic neuronal death in the pars compacta in the substantia nigra region of the brain.

Tompkins et al. (1997) compared the brain tissues from substantia nigra of Parkinson's disease, Alzheimer's disease, and healthy controls utilizing in situ end-labeling to identify DNA fragments.

Authors reported the presence of chromatin condensation, DNA fragments, and apoptotic-like bodies in the substantia nigra in brains in healthy persons and with Alzheimer's disease and Parkinson's disease (Tompkins et al., 1997). **But the amount of apoptotic bodies was much higher in the brain regions in Parkinson's disease than in healthy controls suggesting the potential role of apoptosis in the death of neurons in the substantia nigra pars compacta**.

Initial evidence for the role of apoptosis in cell death in Parkinson's was provided by in vitro study involving use of 1-methyl-4-phenylpyridinium (neurotoxin that can destroy dopaminergic neurons in the nigrostriatal tract causing the expression of the parkinsonian syndrome) (Dipasquale et al., 1991) and cultures of cerebellar granule neurons.

Dipasquale et al. (1991) reported suppression of MPP(+)-mediated neuronal apoptosis in the cerebellar granule neurons with cycloheximide inhibited gene expression.

It is suggested that neurodegenerative diseases including Parkinson's disease is the outcome of an impaired apoptotic signaling pathway in the dopaminergic neurons in the brain regions.

Several studies described the presence of chromatin condensation, DNA fragments, and apoptotic like bodies in the brain regions in the postmortem brain in the patients confirming the role of apoptosis in dopaminergic neuronal death in Parkinson's disease (Mochizuki et al., 1996; Anglade et al., 1997), while the studies by Kosel et al. (1997), Banati et al. (1998), Wullner et al. (1999) failed to identify the apoptotic changes in the neurons in the brain regions in Parkinson's disease.

Thus, the role of apoptosis in the neuronal death and pathology of Parkinson's is still contradictory.

The discrepancy in the findings of those who favor the role of apoptosis in the pathology of Parkinson's disease and with those who failed to detect apoptosis-related observations in the postmortem brain in Parkinson's disease might be attributed to several limitations including the small number of dopaminergic neuron subjected to apoptotic death in brain regions (Elmore, 2007), variations in the experimental design, and methodology to identify DNA fragments by utilizing terminal deoxynucleotidyl transferase dUTP nick-end labeling (TUNEL) technique (Tatton, 2000; Tatton et al., 2003), coupled with ambiguity in criteria in describing apoptosis (Tatton et al., 2003).

Other studies described the role of additional techniques in supporting the role of apoptosis in Parkinson's disease. These studies utilized the TUNEL technique coupled with the use of fluorescent DNA binding dyes to identify the DNA fragments in the same neurons (Tatton et al., 1998; Tatton, 2000). Also, studies

(Hartmann et al., 2000; Tatton, 2000) identified the raised activity of caspase-9, caspase-3, and caspase-8 in dopaminergic neurons in substantia nigra pars compacta.

Thus, the role of apoptosis in the pathology of Parkinson's disease is beset with contradictions but its potential in dopaminergic neuronal death in Parkinson's disease cannot be refuted.

Pathways of apoptosis in neuronal death in Parkinson's disease

Two pathways are involved in the apoptotic neuronal death in Parkinson's disease as well as another neurodegenerative disease.

Extrinsic pathway of apoptosis (death receptor)

The death receptors on the surface of cells belong to the tumor necrosis factor family of receptors including Fas, TNF receptor 1, and TRAIL receptors. After binding with death ligands, the death receptors undergo trimerization and initiate the extrinsic pathway of apoptosis (Venderova and Park, 2012).

Several studies (Boka et al., 1994; de la Monte et al., 1998; Mogi et al., 2000; Hayley et al., 2004; Ho et al., 2009; Simunovic et al., 2009; Fu et al., 2011) emphasize the role of the death receptor pathway in apoptosis of neurons in Parkinson's disease.

The death receptors are transmembrane proteins containing the death domain located at their cytoplasmic regions and serve as sensors of the internal environment of the cell. The trimerization of the death receptor results in the conformational alteration in the death domains and recruitment of adaptor protein containing death effector domain (TRADD or FADD). The FADD in turn bind with procaspase 8 leading to the formation of death-inducing signaling complex and activation of caspase 8 resulting in apoptosis of neurons (Venderova and Park, 2012).

Intrinsic pathway of apoptosis (mitochondrial pathway)

The mitochondrial pathway or also named as intrinsic pathway represents the major pathway implicated in the death of dopaminergic neurons.

Intrinsic pathway is insinuated in the activation of caspases and cascade of molecular events leading to manifestations of Parkinson's disease.

The mitochondria serve as the site of ATP synthesis through the coordinated activity of series of enzymes located on the inner mitochondrial membrane constituted as electron transport chain.

Cardiolipin is the important phospholipid in the inner mitochondrial membrane. It forms about 20% of the total lipids in mitochondria and serves as the key molecule in integrating several proteins and enzymes in the mitochondria and helps in maintaining the integrity of mitochondrial membranes and patency of mitochondrial function (Osellame et al., 2012).

Cardiolipin is essential for the stabilization of respiratory chain complexes and attaches the cytochrome C molecule to the inner mitochondrial membrane (Mejia and Hatch, 2016).

Reactive oxygen species and mutated misfolded proteins can activate the intrinsic pathway of apoptosis (Bossy-Wetzel et al., 2003), possibly via interacting with cardiolipin molecules.

Alpha-synuclein also plays a causative role in the pathology of Parkinson's disease via its oligomers that contribute to dopaminergic neuronal apoptosis (Ghio et al., 2016).

The α-synuclein oligomers exert a neurotoxic effect on the mitochondrial membranes through binding with the components of mitochondrial membranes, primarily, the negatively charged molecules like Cardiolipin (Ghio et al., 2016). Thus, α-synuclein oligomers can bind and disrupt the membranes of mitochondria characterized by pore formation, leakage, and lysis of the mitochondrial membranes.

One of the important molecular events is the release of cytochrome C from the inner mitochondrial membrane during the preliminary stage of lysis of mitochondria in the dopaminergic neurons.

The cytochrome c is localized to the inner mitochondrial membrane. It is bound with negatively charged phospholipids, mainly, the Cardiolipin through "loosely bound" and "tightly bound" conformation states.

In a loosely bound conformation state between Cardiolipin and cytochrome c, there is ionic bond formation between the anionic phosphate groups present in the cardiolipin molecule and cationic lysine amino acid residues present in the cytochrome c molecule (Nicholls, 1974).

In a tightly bound conformation between Cardiolipin and cytochrome c, the hydrophobic interaction exists between Cardiolipin acyl chain and cytochrome c resulting in the embedding of cytochrome c into the inner mitochondrial membrane (Chan, 2006).

Thus, it was posited that two separate pools of cytochrome c can be solubilized based on the nature of stress stimuli.

The first type of stress stimulus can disrupt the ionic bonds between cytochrome c and Cardiolipin resulting in solubilization of cytochrome c and thus confirming the presence of a loosely bound conformation state (Chan, 2006).

While another stimulus can induce lipid peroxidation of Cardiolipin in mitochondria and result in the mobilization of cytochrome c confirming the existence of a tightly bound conformation state (Chan, 2006).

After the solubilization of cytochrome c in the inner mitochondrial membrane, Bax protein is activated. The **BAX**, protein is also named as **bcl-2-like protein 4**, which represents the regulator of apoptosis. It is coded by the *BAX* gene in humans and belongs to the super family of the Bcl-2 gene (Oltvai et al., 1993).

Under normal physiological conditions, BAX protein is localized in the cytosol. Apoptotic stimuli can induce a conformational change in the BAX protein and are associated with the outer mitochondrial membrane (Gross et al., 1998).

According to one hypothesis, activated BAX interacts with mitochondrial voltage-dependent anion channel, also named as mitochondrial porins. The Bax protein induces the opening of mitochondrial voltage-dependent anion channel that plays role in the apoptosis of neurons (Lemasters and Holmuhamedov, 2006).

The mitochondrial voltage-dependent anion channel alters the mitochondrial permeability transition pore(mitochondrial protein located on the inner membrane) resulting in the release of cytochrome c (Lemasters and Holmuhamedov, 2006).

Additionally, activated BAX protein can form a **mitochondrial apoptosis-induced channel** (**MAC**) that serves as a biomarker of the apoptosis (Buytaert et al., 2006).

The **mitochondrial apoptosis-induced channel** is formed on the outer membrane and permits the extrusion of cytochrome c into the cytosol.

Cytosolic localization of cytochrome c

Cytochrome c is released into extra-mitochondrial space, and it serves as a key molecule in triggering apoptosis via series of events (Wang and Youle, 2009).

Cytosolic Cyt c in the presence of ATP, induces the allosteric activation of apoptosis-protease activating factor 1 (Apaf-1) resulting in its oligomerization and formation of apoptosome complex.

Each apoptosome (quaternary protein structure) recruits seven dimers of procaspase-9 leading to auto-proteolysis and formation of active caspase 9 (Acehan et al., 2002). The activated caspase 9 subsequently activates the procaspase 3 and mediates apoptosis (Venderova and Park, 2012).

An additional feature of the intrinsic pathway of apoptosis is related to the impaired mitochondrion fission process in the synaptic terminals.

These regions of the dopaminergic neurons harbor a large number of mitochondria whose activity is determined by the ratio between mitochondrial fusion to mitochondrial fission.

Proapoptotic stimuli inhibit the mitochondrion fission resulting in an increase in the size of mitochondria but the reduction in the number of mitochondria in synaptic terminals leads to depleted oxidative phosphorylation, reduced ATP, and increased formation of ROS causing disruption of mitochondrial membrane potential (Jalmar et al., 2013) and impaired mitochondrial function. These events contribute to reduced ubiquitination, activation of procaspases 1, 3, and 8 (Williams et al., 2005), and initiation of apoptosis.

It can be concluded that Apoptosis is critically involved in the dopaminergic neuronal death implicated pathology of Parkinson's disease based on the post-mortem studies and in vitro studies.

Role of caspase 2 in the pathology of Parkinson's disease

Caspase 2 is encoded by *CASP2* gene (Kumar et al., 1995).

It belongs to the Ich-1 subfamily. It is evolutionary conserved protein (Krumschnabel et al., 2009) and contains amino acid sequence similar to caspase 1, caspase 4, caspase 5, and caspase 9 (initiator caspases). It is released in the inactive form containing prodomain similar to prodomain of caspase 9 and CARD domain. It also contains subunit p19 and subunit p12 (Krumschnabel et al., 2009).

Active caspase 2 interacts with several proapoptotic proteins through its CARD domain. The caspase 2 can form PIDDosome (Tinel and Tschopp, 2004; Zhivotovsky and Orrenius, 2005).

The PIDDosome is the complex involved in the activation of caspase 2. It contains p53-induced protein with a death domain (PIDD) (expression and synthesis controlled by p53 protein) and the RAIDD (adaptor protein). The overexpression of PIDD is linked to rapid activation of caspase-2 Tinel and Tschopp (2004).

Another study revealed that 6-hydroxy dopamine-mediated activation of caspases-2 resulted in the activation of caspase-3 and apoptosis of dopaminergic neurons.

Furthermore, the study revealed that 6-hydroxy dopamine-mediated neurotoxicity induced caspase-mediated apoptosis of dopaminergic neurons.

The fact was supplemented by the findings provided by several animal models of Parkinson's disease. The activated caspase 2 can in turn activate the caspase-3 in Parkinson's disease and might be the result of the nuclear factor kappa B activation pathway.

Certainly, a therapeutic intervention targeting the activation of caspase 2 can be helpful in mitigating the manifestations of Parkinson's disease.

Caspase-2 and Δtau314 in Parkinson's disease

Tau is the soluble protein in the neurons and is made up of the N-terminal projection domain, the C-terminal tail, proline-rich domain, and microtubule-binding domain.

Tau protein can interact with several biomolecules as enzymes, motor proteins, chaperones, lipids, DNA, and RNA.

Under normal physiological conditions, the modified tau protein is involved in controlling cell differentiation, neurite polarity and plasticity, cargo transport along axons, synapse plasticity, and genome stability (Arendt et al., 2016).

Additionally, oxidative stress and proapoptotic stimuli can induce hyperphosphorylation of τ and could impair its binding with microtubule protein leading to surplus accumulation of τ proteins in the dendritic spines and nuclei.

Proapoptotic stimuli can activate caspase 2 and in turn result in generation of caspase 2 mediated site-specific cleaved tau protein in the neurons associated with cognitive impairment.

The Δtau314 is the fragment of tau protein derived from the brain regions in patients with cognitive impairment. This soluble tau protein fragment contains intact N-terminal but truncated C-terminal by caspase-2 mediated cleavage at aspartate 314 residue (Zhao et al., 2016) and was named as Δtau314. The activated caspase 2 cleaved Δtau314 fragment was found to accumulate in dendritic spines and contributed to closing excitatory postsynaptic transmission partially (Hoover et al., 2010).

The animal models of Parkinson's correlated the reduced synaptic activity with the early dysfunction of neurons before the apparent manifestation of apoptosis of dopaminergic neurons and Parkinson's disease.

The latest study (Zhao et al., 2016) identified the presence of Δtau31416 in the rTg4510 mouse that expresses human tau protein with P301L mutation (Ramsden et al., 2005; Santacruz et al., 2005) and was considered as animal model for study of tauopathies.

A study by Zhao et al., (2016) described the proline-to-leucine mutation at amino acid residue 301 (P301L) linked to frontotemporal dementia and parkinsonism. The genetic mutation was associated with human chromosome 17 (Hutton et al., 1998; Dumanchin et al., 1998).

Furthermore, the study described that reduction in the levels of caspase 2 in the body was associated with the decline in the generation of Δtau314 tau fragment leading to amelioration of the cognitive impairment in the rTg4510 mice. Additionally, point mutation at D314 (D314E) leading to replacement of aspartate by glutamate residue renders tau noncleavable by caspase 2, and its localization in the dendritic spines is inhibited coupled with betterment in the synaptic transmission in cultured primary neurons expressing the τ P301L mutant (Zhao et al., 2016).

The above-mentioned findings were additionally supplemented by a transduction experiment involving mice transduced with adeno-associated virus infected with τ D314E mutation. It was identified that mice showed normal cognitive function. Possibly, τ protein with a point mutation at D314 prevented the expression of Δtau314 τ fragment and its lethal neurotoxic effects on the cognitive function in mice.

The aforementioned findings could be helpful in concluding the role of caspase 2-mediated cleavage of τ protein into the generation of neurotoxic Δtau314 τ fragment and its ill effects on the cognitive function in an animal model of Parkinson's disease.

References

Acehan, D., Jiang, X., Morgan, D.G., Heuser, J.E., Wang, X., Akey, C.W., 2002. Three-dimensional structure of the apoptosome: implications for assembly, procaspase-9 binding, and activation. Mol. Cell 9, 423–432.

Aleksaniants, G.D., 2013. Use of balneo-, peloid- and centimeter-wave therapy in the complex treatment of patients with circumscribed scleroderma. Vestn. Dermatol. i Venerol. 32 (6), 58–60.

Andersen, M.H., Schrama, D., Thor Straten, P., Becker, J.C., 2006. Cytotoxic T cells. J. Invest. Dermatol. 126 (1), 32–41.

Anglade, P., Vyas, S., Javoy-Agid, F., Herrero, M.T., Michel, P.P., Marquez, J., Mouatt-Prigent, A., Ruberg, M., Hirsch, E.C., Agid, Y., 1997. Apoptosis and autophagy in nigral neurons of patients with Parkinson's disease. Histol. Histopathol. 12, 25–31.

Arendt, T., Stieler, J.T., Holzer, M., 2016. Tau and tauopathies. Brain Res. Bull. 126, 238–292.

Banati, R.B., Daniel, S.E., Blunt, S.B., 1998. Glial pathology but absence of apoptotic nigral neurons in long-standing Parkinson's disease. Mov. Disord. 13, 221–227.

Block, M.L., Zecca, L., Hong, J.S., 2007. Microglia-mediated neurotoxicity: uncovering the molecular mechanisms. Nat. Rev. Neurosci. 8, 57–69.

Boka, G., Anglade, P., Wallach, D., Javoy-Agid, F., Agid, Y., Hirsch, E.C., 1994. Immunocytochemical analysis of tumor necrosis factor and its receptors in Parkinson's disease. Neurosci. Lett. 172, 151–154.

Bossy-Wetzel, E., Barsoum, M.J., Godzik, A., Schwarzenbacher, R., Lipton, S.A., 2003. Mitochondrial fission in apoptosis, neurodegeneration and aging. Curr. Opin. Cell Biol. 15, 706–716.

Bsibsi, M., Ravid, R., Gveric, D., van Noort, J.M., 2002. Broad expression of toll-like receptors in the human central nervous system. J. Neuropathol. Exp. Neurol. 61, 1013–1021. https://doi.org/10.1093/jnen/61.11.1013.

Burguillos, M.A., Deierborg, T., Kavanagh, E., Persson, A., Hajji, N., Garcia-Quintanilla, A., Cano, J., Brundin, P., Englund, E., Venero, J.L., Joseph, B., 2011. Caspase signalling controls microglia activation and neurotoxicity. Nature 472 (7343), 319–324.

Buytaert, E., Callewaert, G., Vandenheede, J.R., Agostinis, P., 2006. Deficiency in apoptotic effectors Bax and Bak reveals an autophagic cell death pathway initiated by photodamage to the endoplasmic reticulum. Autophagy 2 (3), 238–240.

Chan, D.C., 2006. Mitochondria: dynamic organelles in disease, aging, and development. Cell 125, 1241–1252.

Chang, D.W., Xing, Z., Capacio, V.L., Peter, M.E., Yang, X., 2003. Interdimer processing mechanism of procaspase-8 activation. EMBO J. 22, 4132–4142.

Chinta, S.J., Andersen, J.K., 2005. Dopaminergic neurons. Int. J. Biochem. Cell Biol. 37 (5), 942–946.

Cookson, M.R., 2009. α-Synuclein and neuronal cell death. Mol. Neurodegener. 4, 9. Retrieved February 23, 2013 from: http://www.molecularneurodegeneration.com/content/4/1/9.

Damier, P., Hirsch, E.C., Agid, Y., Graybiel, A.M., 1999. The substantia nigra of the human brain. II. Patterns of loss of dopamine-containing neurons in Parkinson's disease. Brain 122 (Pt 8), 1437–1448.

Davie, C.A., 2008. A review of Parkinson's disease. Br. Med. Bull. 86 (1), 109–127.

Dawson, T.M., Dawson, V.L., 2014. The role of parkin in familial and sporadic Parkinson's disease. Mov. Disord. 25 (Suppl. 1), S32–S39.

de la Monte, S.M., Sohn, Y.K., Ganju, N., Wands, J.R., 1998. P53- and CD95-associated apoptosis in neurodegenerative diseases. Lab. Invest. 78, 401–411.

de Rijk, M.C., Breteler, M.M., Graveland, G.A., Ott, A., Grobbee, D.E., van der Meché, F.G., et al., 1995. Prevalence of Parkinson's disease in the elderly: the Rotterdam study. Neurology 45 (12), 2143–2146.

Dipasquale, B., Marini, A.M., Youle, R.J., 1991. Apoptosis and DNA degradation induced by 1-methyl-4-phenylpyridinium in neurons. Biochem. Biophys. Res. Commun. 181, 1442–1448.

Dumanchin, C., et al., 1998. Segregation of a missense mutation in the microtubule-associated protein tau gene with familial frontotemporal dementia and parkinsonism. Hum. Mol. Genet. 7, 1825–1829.

Edmondson, D.E., Mattevi, A., Binda, C., Li, M., Hubálek, F., 2004. Structure and mechanism of monoamine oxidase. Curr. Med. Chem. 11 (15), 1983–1993.

Elmore, S., 2007. Apoptosis: a review of programmed cell death. Toxicol. Pathol. 35, 495—516.

Fu, K., Ren, H., Wang, Y., Fei, E., Wang, H., Wang, G., 2011. DJ-1 inhibits TRAIL-induced apoptosis by blocking pro-caspase-8 recruitment to FADD. Oncogene 31, 1311—1322.

Gao, H.-M., et al., 2011. Neuroinflammation and α-synuclein dysfunction potentiate each other, driving chronic progression of neurodegeneration in a mouse model of Parkinson's disease. Environ. Health Perspect. 119 (6), 807—814.

Gasparini, F., Di Paolo, T., Gomez-Mancilla, B., 2013. Metabotropic glutamate receptors for Parkinson's disease therapy. Parkinson's Dis. 2013. Article ID 196028.

Ghio, S., Kamp, F., Cauchi, R., Giese, A., Vassallo, N., 2016. Interaction of α-synuclein with biomembranes in Parkinson's disease—role of cardiolipin. Prog. Lipid Res. 61, 73—82.

Green, D.R., 2005. Apoptotic pathways: ten minutes to dead. Cell 121, 671—674.

Gross, A., Jockel, J., Wei, M.C., Korsmeyer, S.J., 1998. Enforced dimerization of BAX results in its translocation, mitochondrial dysfunction and apoptosis. EMBO J. 17 (14), 3878—3885.

Hanisch, U.K., Kettenmann, H., 2007. Microglia: active sensor and versatile effector cells in the normal and pathologic brain. Nat. Neurosci. 10, 1387—1394.

Harris, M.A., Tsui, J.K., Marion, S.A., Shen, H., Teschke, K., 2012. Association of Parkinson's disease with infections and occupational exposure to possible vectors. Mov. Disord. 27 (9), 1111—1117.

Hartmann, A., Hunot, S., Michel, P.P., Muriel, M.P., Vyas, S., Faucheux, B.A., Mouatt-Prigent, A., Turmel, H., Srinivasan, A., Ruberg, M., et al., 2000. Caspase-3: a vulnerability factor and final effector in apoptotic death of dopaminergic neurons in Parkinson's disease. Proc. Natl. Acad. Sci. U. S. A. 97, 2875—2880.

Hass, C.J., Malczak, P., Nocera, J., et al., 2012. Quantitative normative gait data in a large cohort of ambulatory persons with Parkinson's disease. PLoS One 7 (8), e42337.

Hayley, S., Crocker, S.J., Smith, P.D., Shree, T., Jackson-Lewis, V., Przedborski, S., Mount, M., Slack, R., Anisman, H., Park, D.S., 2004. Regulation of dopaminergic loss by Fas in a 1-methyl-4-phenyl-1,2,3,6-tetrahydropyridine model of Parkinson's disease. J. Neurosci. 24, 2045—2053.

Ho, C.C., Rideout, H.J., Ribe, E., Troy, C.M., Dauer, W.T., 2009. The Parkinson disease protein leucine-rich repeat kinase 2 transduces death signals via Fas-associated protein with death domain and caspase-8 in a cellular model of neurodegeneration. J. Neurosci. 29, 1011—1016.

Hoover, B.R., Reed, M.N., Su, J., Penrod, R.D., Kotilinek, L.A., Grant, M.K., Pitstick, R., Carlson, G.A., Lanier, L.M., Yuan, L.L., Ashe, K.H., Liao, D., 2010. Tau mislocalization to dendritic spines mediates synaptic dysfunction independently of neurodegeneration. Neuron 68, 1067—1081.

Hutton, M., et al., 1998. Association of missense and 5′-splice-site mutations in tau with the inherited dementia FTDP-17. Nature 393, 702—705. https://doi.org/10.1038/31508.

Jalmar, O., François-Moutal, L., García-Sáez, A.J., Perry, M., Granjon, T., Gonzalvez, F., et al., 2013. Caspase-8 binding to cardiolipin in giant unilamellar vesicles provides a functional docking platform for bid. PLoS One 8, e55250.

Kahns, S., Lykkebo, S., Jakobsen, L.D., Nielsen, M.S., Jensen, P.H., 2002. Caspase-mediated parkin cleavage in apoptotic cell death. Biol. Chem. 277, 15303—15308.

Kahns, S., Kalai, M., Jakobsen, L.D., Clark, B.F.C., Vandenabeele, P., Jensen, P.H., 2003. Caspase-1 and caspase-8 cleave and inactivate cellular parkin. J. Biol. Chem. 278 (26), 23376—23380.

Kalia, L.V., Lang, A.E., 2015. Parkinson's disease. Lancet 386 (9996), 896−912. https://doi.org/10.1016/S0140-6736(14)61393-3.

Kosel, S., Egensperger, R., von Eitzen, U., Mehraein, P., Graeber, M.B., 1997. On the question of apoptosis in the parkinsonian substantia nigra. Acta Neuropathol. 93, 105−108.

Kravitz, A.V., Freeze, B.S., Parker, P.R., Kay, K., Thwin, M.T., Deisseroth, K., Kreitzer, A.C., 2010. Regulation of parkinsonian motor behaviours by optogenetic control of basal ganglia circuitry. Nature 466, 622−626.

Kruidering, M., Evan, G.I., 2000. Caspase-8 in apoptosis: the beginning of "the end"? IUBMB Life 50 (2), 85−90.

Krumschnabel, G., Manzl, C., Villunger, A., 2009. Caspase-2: killer, savior and safeguard−emerging versatile roles for an ill-defined caspase. Oncogene 28 (35), 3093.

Kumar, S., 2006. Caspase function in programmed cell death. Cell Death Differ. 14 (1), 32−43.

Kumar, S., White, D.L., Takai, S., Turczynowicz, S., Juttner, C.A., Hughes, T.P., 1995. Apoptosis regulatory gene NEDD2 maps to human chromosome segment 7q34-35, a region frequently affected in haematological neoplasms. Hum. Genet. 95 (6), 641−644.

Kurlawala, Z., Shadowen, P.H., McMillan, J.D., Beverly, L.J., Friedland, R.P., 2021. Progression of nonmotor symptoms in Parkinson's disease by sex and motor laterality. Parkinson's Dis. 2021. Article ID 8898887, 12 pages.

Kwan, L.C., Whitehill, T.L., 2011. Perception of speech by Individuals with Parkinson's disease: a review. Parkinson's Dis. 2011. Article ID 389767, 11 pp.

Lees, A.J., Hardy, J., Revesz, T., 2009. Parkinson's disease. Lancet 373 (9680), 2055−2066. https://doi.org/10.1016/S0140-6736(09)60492-X.

Lemasters, J.J., Holmuhamedov, E., 2006. Voltage-dependent anion channel (VDAC) as mitochondrial governor−thinking outside the box. Biochim. Biophys. Acta 1762 (2), 181−190.

Liu, B., Gao, H.M., Hong, J.S., 2003. Parkinson's disease and exposure to infectious agents and pesticides and the occurrence of brain injuries: role of neuroinflammation. Environ. Health Perspect. 111 (8), 1065−1073.

Ma, L., Yang, C., Zhang, X., Li, Y., Wang, S., Zheng, L., Huang, K., 2018. C-terminal truncation exacerbates the aggregation and cytotoxicity of α-synuclein: a vicious cycle in Parkinson's disease. Biochim. Biophys. Acta Mol. Basis Dis. 1864, 3714−3725.

Mejia, E.M., Hatch, G.M., 2016. Mitochondrial phospholipids: role in mitochondrial function. J. Bioenerg. Biomembr. 48, 99−112.

Melissa, J., Michael, S., 2020. Diagnosis and treatment of Parkinson disease: a review. J. Am. Med. Assoc. 323 (6), 548−560.

Michel, Le M., February 5, 2018. Mesocorticolimbic Dopaminergic Neurons. Neuropsychopharmacology: The Fifth Generation of Progress.

Michell, A.W., et al., 2007. The effect of truncated human alpha-synuclein (1-120) on dopaminergic cells in a transgenic mouse model of Parkinson's disease. Cell Transplant. 16 (5), 461−474.

Mochizuki, H., Goto, K., Mori, H., Mizuno, Y., 1996. Histochemical detection of apoptosis in Parkinson's disease. J. Neurol. Sci. 137, 120−123.

Mogi, M., Togari, A., Kondo, T., Mizuno, Y., Komure, O., Kuno, S., Ichinose, H., Nagatsu, T., 2000. Caspase activities and tumor necrosis factor receptor R1 (p55) level are elevated in the substantia nigra from parkinsonian brain. J. Neural. Transm. 107, 335−341.

Müller-Rischart, A.K., Pilsl, A., Beaudette, P., Patra, M., Hadian, K., Funke, M., Peis, R., Deinlein, A., Schweimer, C., Kuhn, P.H., Lichtenthaler, S.F., Motori, E., Hrelia, S.,

Wurst, W., Trümbach, D., Langer, T., Krappmann, D., Dittmar, G., Tatzelt, J., Winklhofer, K.F., 2013. The E3 ligase parkin maintains mitochondrial integrity by increasing linear ubiquitination of NEMO. Mol. Cell 49 (5), 908–921.

Nicholls, P., 1974. Cytochrome c binding to enzymes and membranes. Biochim. Biophys. Acta 346 (3–4), 261–310.

Oltvai, Z.N., Milliman, C.L., Korsmeyer, S.J., 1993. Bcl-2 heterodimerizes in vivo with a conserved homolog, Bax, that accelerates programmed cell death. Cell 74 (4), 609–619.

Osellame, L.D., Blacker, T.S., Duchen, M.R., 2012. Cellular and molecular mechanisms of mitochondrial function. Best Pract. Res. Clin. Endocrinol. Metabol. 26, 711–723.

Oueslati, A., Fournier, M., Lashuel, H.A., 2010. Role of post-translational modifications in modulating the structure, function and toxicity of alpha-synuclein: implications for Parkinson's disease pathogenesis and therapies. Prog. Brain Res. 183, 115–145.

Parkinson, J., 1817. An Essay on the Shaking Palsy. Sherwood Neely and Jones, London.

Periquet, M., Fulga, T., Myllykangas, L., Schlossmacher, M.G., Feany, M.B., 2007. Aggregated α-synuclein mediates dopaminergic neurotoxicity in vivo. J. Neurosci. 27 (12), 3338–3346. https://doi.org/10.1523/JNEUROSCI.0285-07.2007.

Pickart, C.M., Eddins, M.J., 2004. Ubiquitin: structures, functions, mechanisms. Biochim. Biophys. Acta Mol. Cell Res. 1695 (1–3), 55–72.

Poewe, W., Seppi, K., Tanner, C.M., et al., 2017. Parkinson disease. Nat. Rev. Dis. Prim. 3. Article ID 17013.

Porter, A.G., Jänicke, R.U., 1999. Emerging roles of caspase-3 in apoptosis. Cell Death Differ. 6 (2), 99–104.

Qiao, C., Zhang, L.-X., Sun, X.-Y., Ding, J.-H., Lu, M., Hu, G., 2017. Caspase-1 deficiency alleviates dopaminergic neuronal death via inhibiting caspase-7/AIF pathway in MPTP/p mouse model of Parkinson's disease. Mol. Neurobiol. 54 (6), 4292–4302.

Rail, D., Scholtz, C., Swash, M., 1981. Post-encephalitic parkinsonism: current experience. J. Neurol. Neurosurg. Psychiatr. 44 (8), 670–676.

Ramsden, M., Kotilinek, L., Forster, C., Paulson, J., McGowan, E., Santacruz, K., Guimaraes, A., Yue, M., Lewis, J., Carlson, G., Hutton, M., Ashe, K.H., 2005. Age-dependent neurofibrillary tangle formation, neuron loss, and memory impairment in a mouse model of human tauopathy (P301L). J. Neurosci. 25 (46), 10637–10647.

Riedl, S.J., Shi, Y., 2004. Molecular mechanisms of caspase regulation during apoptosis. Nature reviews Mol. Cell Biol. 5 (11), 897–907. https://doi.org/10.1038/nrm1496.

Salvesen, G.S., 2002. Caspases: opening the boxes and interpreting the arrows. Cell Death Differ. 9 (1), 3–5.

Santacruz, K., Lewis, J., Spires, T., Paulson, J., Kotilinek, L., Ingelsson, M., Guimaraes, A., DeTure, M., Ramsden, M., McGowan, E., Forster, C., Yue, M., Orne, J., Janus, C., Mariash, A., Kuskowski, M., Hyman, B., Hutton, M., Ashe, K.H., 2005. Tau suppression in a neurodegenerative mouse model improves memory function. Science 309 (5733), 476–481.

Schrag, A., Jahanshahi, M., Quinn, N., 2000. What contributes to quality of life in patients with Parkinson's disease? J. Neurol. Neurosurg. Psychiatr. 69 (3), 308–312.

Schulze-Osthoff, K., Ferrari, D., Los, M., Wesselborg, S., Peter, M.E., 1998. Apoptosis signaling by death receptors. Eur. J. Biochem. 254 (3), 439–459. https://doi.org/10.1046/j.1432-1327.1998.2540439.x.

Seirafi, M., Kozlov, G., Gehring, K., 2015. Parkin structure and function. FEBS J. 282 (11), 2076–2088.

Sheikh, M.S., Fornace, A.J., 2000. Death and decoy receptors and p53-mediated apoptosis. Leukemia 14 (8), 1509–1513.

Simunovic, F., Yi, M., Wang, Y., Macey, L., Brown, L.T., Krichevsky, A.M., Andersen, S.L., Stephens, R.M., Benes, F.M., Sonntag, K.C., 2009. Gene expression profiling of substantia nigra dopamine neurons: further insights into Parkinson's disease pathology. Brain J. Neurol. 132, 1795—1809.

Tai, C.H., Wu, R.M., 2002. Catechol-*O*-methyltransferase and Parkinson's disease. Acta Med. Okayama 56 (1), 1—6.

Tatton, N.A., 2000. Increased caspase 3 and Bax immunoreactivity accompany nuclear GAPDH translocation and neuronal apoptosis in Parkinson's disease. Exp. Neurol. 166, 29—43.

Tatton, N.A., Maclean-Fraser, A., Tatton, W.G., Perl, D.P., Olanow, C.W., 1998. A fluorescent double-labeling method to detect and confirm apoptotic nuclei in Parkinson's disease. Ann. Neurol. 44, S142—S148.

Tatton, W.G., Chalmers-Redman, R., Brown, D., Tatton, N., 2003. Apoptosis in Parkinson's disease: signals for neuronal degradation. Ann. Neurol. 53, S61—S70 discussion S70—62.

Thornberry, N.A., Bull, H.G., Calaycay, J.R., Chapman, K.T., Howard, A.D., Kostura, M.J., Miller, D.K., Molineaux, S.M., Weidner, J.R., Aunins, J., 1992. A novel heterodimeric cysteine protease is required for interleukin-1 beta processing in monocytes. Nature 356 (6372), 768—774.

Tinel, A., Tschopp, J., 2004. The PIDDosome, a protein complex implicated in activation of caspase-2 in response to genotoxic stress. Science 304 (5672), 843—846.

Tompkins, M.M., Basgall, E.J., Zamrini, E., Hill, W.D., 1997. Apoptotic-like changes in Lewy-body-associated disorders and normal aging in substantia nigral neurons. Am. J. Pathol. 150 (1), 119—131.

Venderova, K., Park, D.S., 2012. Programmed cell death in Parkinson's disease. Cold Spring Harb. Perspect. Med. 2, a009365.

Walters, J., Pop, C., Scott, F.L., Drag, M., Swartz, P., Mattos, C., Salvesen, G.S., Clark, A.C., 2009. A constitutively active and uninhibitable caspase-3 zymogen efficiently induces apoptosis. Biochem. J. 424 (3), 335—345.

Wang, K., Yin, X.M., Chao, D.T., Milliman, C.L., Korsmeyer, S.J., 1996. BID: a novel BH3 domain-only death agonist. Genes Dev. 10 (22), 2859—2869.

Wang, C., Youle, R.J., 2009. The role of mitochondria in apoptosis. Ann. Rev. Genet. 43, 95—118.

Wanga, W., Nguyena, L.T.T., Burlak, C., Chegini, F., Guo, F., Chataway, T., Ju, S., Fishere, O.S., Miller, D.W., Datta, D., Wu, F., Wu, C.-X., Landeru, A., Wellsh, J.A., Cooksong, M.R., Boxerj, M.B., Thomasj, C.J., Gaid, W.P., Ringee, D., Petskoe, G.A., Hoang, Q.Q., 2016. Caspase-1 causes truncation and aggregation of the Parkinson's disease-associated protein α-synuclein. Proc. Natl. Acad. Sci. U. S. A. 113.

Williams, A.C., Cartwright, L.S., Ramsden, D.B., 2005. Parkinson's disease: the first common neurological disease due to auto-intoxication? QJM 98, 215—226.

Wullner, U., Kornhuber, J., Weller, M., Schulz, J.B., Loschmann, P.A., Riederer, P., Klockgether, T., 1999. Cell death and apoptosis regulating proteins in Parkinson's disease—a cautionary note. Acta Neuropathol. 97, 408—412.

Yamin, T.T., Ayala, J.M., Miller, D.K., 1996. Activation of the native 45-kDa precursor form of interleukin-1 converting enzyme. J. Biol. Chem. 271 (22), 13273—13282.

Yigong, S., 2004. Caspase activation. Cell 117 (7), 855—858.

Yoo, H.I., Ahn, G.Y., Lee, E.J., et al., 2017. 6-Hydroxydopamine induces nuclear translocation of apoptosis inducing factor in nigral dopaminergic neurons in rat. Mol. Cell. Toxicol. 13, 305—315.

Zhao, X., Kotilinek, L.A., Smith, B., Hlynialuk, C., Zahs, K., Ramsden, M., Cleary, J., Ashe, K.H., 2016. Caspase-2 cleavage of tau reversibly impairs memory. Nat. Med. 22, 1268–1276.

Zhivotovsky, B., Orrenius, S., 2005. Caspase-2 function in response to DNA damage. Biochem. Biophys. Res. Commun. 331 (3), 859–867.

Illustrated etiopathogenesis of Huntington's disease

Introduction

Huntington's disease is the inherited neurodegenerative disease with autosomal dominant pattern of inheritance. It is manifested as the disorders in the movement of the affected persons with decline in cognitive function. The disorders in mobility include chorea and impaired coordination. The obsessive—compulsive disorder and depression are the common psychiatric manifestations linked with the Huntington disease (Rosenblatt, 2007).

The Huntington disease is linked with neurodegeneration in the striatum coupled with neuronal apoptosis of medium spiny neurons (special type of GABAergic inhibitory neurons constituting around 95% of the neurons in the striatum) (Reiner et al., 1988).

Additionally, regionally specific thinning of the cortical ribbon was associated with the Huntington disease. The cortical ribboning is related to appearance of magnetic resonance imaging characterized by hyper-intense signal changes in cerebral cortex and basal ganglia.

The MRI image is related to the intense loss of neurons in several disease (Tschampa et al., 2007).

There is progressive loss of cortical mass and it starts from the posterior cortical region to anterior cortical region and is considered as early clinical sign of Huntington disease (Rosas et al., 2002).

Huntington disease is caused by mutations in the *HTT* gene that in humans encodes huntingtin protein. The cytosine-adenine-guanine (CAG) trinucleotide repeats in the exon-1 of gene *HTT* encodes the polyglutamine tract in huntingtin protein.

The wild HTT gene contains maximum 35 CAG repeats while there is expansion of CAG repeats more than 36 in patients with Huntington disease (Rubinsztein et al., 1996).

For the complete penetrance of the Huntington disease, it has been reported that CAG repeats of more than or equal to ($\geq$42) are implicated while the CAG repeats between 36 and 41 exhibits clinical manifestations of the disease in their life span (Rubinsztein et al., 1996; Brinkman et al., 1997).

Human Caspases and Neuronal Apoptosis in Neurodegenerative Diseases. https://doi.org/10.1016/B978-0-12-820122-0.00002-9

Furthermore, number of CAG repeats and the age of onset of disease are inversely related with higher CAG repeats are linked with early onset of Huntington disease (Andrew et al., 1993).

Structure of Huntingtin

Huntingtin protein is encoded by HTT gene (The Huntington's Disease Collaborative Research Group, 1993). The HTT gene is mapped to short arm of human chromosome 4 at position 16.3. it spans 3,074,510 to 3,243,960 base pairs (NCBI, 2021). The sequence of CAG is present at the $5'$ end of the HTT gene and codes for glutamine residue. This domain is trinucleotide repeat with normal CAG repeat is between 6 and 35 (NCBI, 2021).

The HTT gene possesses 67 exons (coding regions) and contains mRNA transcript of 10,366 bp and another mRNA transcript of 13,711 bp (Lin et al., 1993). The second mRNA transcript contains $3'$ untranslated region with sequence of 3360 bp and is predominant in brain regions (Saudou and Humbert, 2016).

The structure of the Huntingtin protein is highly variable due to genetic polymorphisms and variability in the number of glutamine amino acid residues in its structure. Huntingtin protein is highly conserved protein from nonchordates to vertebrates.

Wild type huntingtin protein contains 35 glutamine residues with molecular weight 350 kDa. The N-terminal of huntingtin protein is extensively studied.

The N-terminal starts with the presence of 17 amino acid residues followed by expandable polyglutamine tract and proline-rich domain (Li et al., 2006), where the latter two domain are polymorphic.

The N-terminal 17 amino acid residues are evolutionary conserved in humans (Tartari et al., 2008). These amino acid residues are organized into an amphipathic α-helix (containing hydrophilic and hydrophobic amino acid residues forming two facets of the helix).

The amphipathic α-helix serves as nuclear export signal and can be modified by posttranslational modifications including acetylation at lysine sixth and ubiquitination at lysine 15th coupled with phosphorylation at serine residues 13th and 16th. The posttranslational modifications of NES in amphipathic α-helix can in turn influence the cellular trafficking of huntingtin protein (Desmond et al., 2012; Saudou and Humbert, 2016).

The polyglutamine stretch, in comparison to echinoderms and deuterostomes is longer in mammals including humans. It is polymorphic in humans. The precise role of variability in the polyQ tract on the function of wild type huntingtin is uncertain, but the polyQ stretch knockout model of mice showed increased longevity (Zheng et al., 2010; Saudou and Humbert, 2016).

Proline-rich region in huntingtin protein at the amino terminal is reported in mammals including humans indicating the evolutionary changes in the huntingtin protein (Tartari et al., 2008; Saudou and Humbert, 2016).

The Proline-rich region is essential for interaction with cellular proteins and has role in controlling protein—protein interaction. This region has modulating effect on the structure of polyglutamine tract.

Neveklovska et al. (2012) produced knock-in allele (knock-in) describes one-for-one substitution of DNA sequence) of the mouse with Huntington's disease (HD) that expresses full-length wild type huntingtin without proline-rich region.

The motor functions and motor learning were similar in homozygous mutants and wild-type control mice (Saudou and Humbert, 2016).

The findings suggested that proline-rich domain was not critical for huntingtin activity in the embryonic development. Furthermore, the proline-rich region exhibits proline—proline helix with kinked conformation (Saudou and Humbert, 2016). The proline—proline helix helps stabilize the structure of polyglutamine stretch.

The remaining huntingtin protein is made up of amino acid residues 69 to amino acid residues 3144 constituting nearly 97.8% of the protein and is encoded by 66 exons.

The amino acid residues from 69 to 3144 contain multiple "HEAT repeats" involved in protein—protein interaction (Palidwor et al., 2009). **The HEAT repeat stands for four proteins namely huntingtin,** Elongation factor 3, protein phosphatase 2A, and TOR1 where these repeats are localized (Palidwor et al., 2009).

The HEAT repeats are made up of two antiparallel α-helices linked by nonhelical region. Analysis of huntingtin protein reported the occurrence of HEAT repeats between 16 and 36 which clustered to form 3 to 5 macro sized α-rod domains linked by disordered domains (Palidwor et al., 2009). These are involved in cytosolic trafficking of molecules.

Huntingtin protein is enormously expressed in the striatal large neurons and cortico-striatal neurons. It is located in the cytoplasm of neurons and vesicle membranes (DiFiglia et al., 1995; Cornett et al., 2005).

Ubiquitous nature of Huntingtin

HTT protein is ubiquitously expressed in the human body tissues including its highest levels in the nervous system and its maximum peripheral distribution in the testes (Li et al., 1993; Marques Sousa and Humbert, 2013). The HTT distribution is much higher in nervous system than other tissues of body (Saudou and Humbert, 2016) **and reported in neurons, and glial cells** (Landwehrmeyer et al., 1995).

The HTT is localized in several types of striatal projection neurons, interneurons (central nodes of neural circuits helping interneuron communication), in the cerebellum, cortex, and hippocampus.

Mutated huntingtin in Huntington disease

Mutation in the HTT gene results into mutant Huntingtin protein responsible for the manifestations of Huntington disease (Choi et al., 2014).

The mutation in HTT gene is involved in the anomalous expansion of a polyglutamine tracts located at the amino terminus of universally expressed huntingtin protein (The Huntington's Disease Collaborative Research Group, 1993). The mutation of HTT results into formation of an erroneous huntingtin protein named as **mutant huntingtin (mHtt)**.

Additionally, mutant huntingtin protein interacts with other cellular proteins primarily expressed in the brain regions as striatum and cortex. Important proteins namely huntingtin protein-associated protein, calmodulin, huntingtin protein-interacting proteins (HIP-1) and huntingtin protein-interacting proteins (HIP-2) and glyceraldehyde-3-phosphate dehydrogenase mainly interact with the mutant huntingtin protein and might be implicated in the progression of Huntington disease (Wang et al., 2010). The mutant huntingtin protein may exert altered effect on the neuronal functions. The mutant huntingtin protein can also influence the transcriptional activity of transcription proteins (Petersén et al., 1999; Wu and Zhou 2009).

The proteolytic cleavage of mutant polyQ huntingtin generates N-terminal pathogenic fragments of mutant Htt that aggregating to form neuronal intranuclear inclusions within the neurons. According to one hypothesis, presence of neuronal intranuclear inclusions was confirmed as the basis of the neurological dysfunction and neuronal apoptosis in the in HD animal model of transgenic mice (Davies et al., 1997).

Furthermore, presence of mutant huntingtin aggregates in the brain regions is the hallmark of the manifestations of the Huntington disease. In vitro study revealed that huntingtin aggregates starts by the nucleation-based activity leading to polyglutamine repeats organizing in the form of β-sheet fibrillar structure stabilized by interchain hydrogen bonds (Perutz et al., 1994).

Contrarily, studies by (Saudou et al., 1998; Arrasate et al., 2004) **negated the neurotoxic potential of neuronal intranuclear inclusions. These studies described that inclusion bodies in the neurons offered protection against the neuronal apoptosis.**

Thus, precise role of neuronal intranuclear inclusions in the etiopathogenesis of Huntington disease is still elusive.

Pathogenic role of exon 1 HTT protein

The exon 1 HTT protein exerts pathogenic potential in the brains of patients with Huntington disease and it was assumed to be formed by incomplete splicing of mRNA in Huntington's disease.

Study was conducted by Neueder et al. (2017) involving HD knock-in mouse models and identified the presence of incompletely spliced mRNA of Htt (Sathasivam et al., 2013).

Further, incomplete splicing of mRNA of Htt led to formation of short HTTexon1 mRNA that was made up of Htt exon 1 and the 5′ intron 1 (Neueder et al., 2017) and generation of the exon 1 HTT protein.

The R6/2 mouse cell line (R6/2 transgenic mouse expresses the 5′ end of human HTT with variable CAG repeats) (Mangiarini et al., 1996) expresses exon 1 HTT protein and the study concluded the high neurotoxicity attributed to this fragment (Barbaro et al., 2015).

The aforementioned facts were supplemented by additional studies (DiFiglia et al., 1997; Lunkes et al., 2002) involving human postmortem of brain tissues exhibiting the presence of small N-terminal fragments of HTT.

Therapeutic intervention blocking the production of pathogenic exon 1 HTT would be helpful in mitigating the ill effects of huntingtin aggregates in brain regions of patients with Huntington disease.

It can be concluded that mutant huntingtin protein might be implicated in triggering cascade of molecular events involved in the neuronal apoptosis including the medium spiny neurons and pathology of Huntington disease.

Clinical relevance of length of expanded polyglutamine tract

The abnormal expansion of polyglutamine tract occurs beyond the 35 CAG repeats (Huntingtin polyglutamine domain in normal population) and is closely associated with the manifestations of Huntington disease (Myers, 2004), while suffers with HD possess expansion of polyglutamine tract $\geq$36 CAG repeats (Myers, 2004).

The length of polyglutamine tract is inversely correlated to the age of onset of disease.

The polyglutamine tract having CAG repeats between 40 and 50 are associated with adult-onset of HD while polyglutamine tract with CAG repeats between 50 and 120 are characteristic of early onset of Huntington disease (Squitieri et al., 2002).

Introduction to etiopathogenesis of Huntington' disease

The abnormal expansion of CAG trinucleotide repeats results into abnormal expression of polyglutamine tracts that is implicated in the etiopathogenesis of Huntington disease. The abnormal expansion polyglutamine repeats is representative of toxic gain-of-function mutation and is specifically injurious to the neurons in the regions in brain affected with such polyglutamine tracts. Thus, a short fragment of protein derived from the cleavage of full-length protein exhibits toxic potential due to pathogenic conformation assumed by the fragment of protein.

Microglia activation in pathology of Huntington's disease

Microglial cells represent subtype of neuroglia (nonneuronal cells) distributed in the central nervous system (Fields et al., 2014). Microglial cells constitute around 15% of the total cells in the brain (Lawson et al., 1992). Microglial cells are the tissue-specific macrophages in central nervous system (Bushong et al., 2002) and are the principal phagocytes in the CNS.

Microglial cells are highly sensitive to the minute changes in the brain (Dissing-Olesen et al., 2007) and scavenges the dead proteins, dead neurons, plaques, and pathogens (Gehrmann et al., 1995).

In normal healthy conditions, microglial cells have sharp sensing ability to the changes in the microenvironment of brain.

Thus, microglial cells as resident macrophages in brain are is most essential in maintaining the health of neurons and synapses and thus microglial cells act as protective and supportive nonneuronal cells in brain and spinal cord.

Process of microglia activation

The process involves transformation of resident microglia into the reactive and activated form in response to pathogenic stimuli in brain tissues has been identified and mentioned in several research publications (Hanisch and Kettenmann, 2007). The microglial activation involves highly complex and diverse cascade of molecular events at the transcriptional and nontranscriptional levels with myriad of consequences including the **manifestations of several neurodegenerative disease** (Yang et al., 2017).

In the presence of pathological stimuli in brain, transformation of resident microglial cells involves metamorphosis of microglial cells exhibiting small cell bodies with fine radiating axons and dendrites into the retraction of processes and presence of enlarged cell bodies (Moller, 2010; Hanisch and Kettenmann, 2007).

Activated microglial cells exhibit enlarged cell bodies with ameba like movements in the brain tissues The fines processes in the microglial cells either disappear after activation or transforms into short and thick processes in the patients with Huntington disease (Sapp et al., 2001; Yang et al., 2017).

Sapp et al. (2001) conducted study to identify early accumulation of activated microglia in brain regions in the patients with Huntington disease.

The thymosin β-4 (Tβ4) is the small peptide in humans involved in the wound healing angiogenesis, and increased metastatic activity in tumor cells. it is also associated with activated microglial cells in brain regions (Gómez-Márquez et al., 1989).

The early accumulation of activated microglial cells in Huntington brains was traced through the tracing of thymosin β-4 whose concentrations in the activated microglia were elevated.

Sapp et al. (2001) reported the higher accumulation of activated microglial cells in the cortex, neostriatum, and globus pallidus and white matter of the brains in patients with Huntington disease while presence of activated microglial cells in the brain regions of controls were not detected.

Sapp et al. (2001) reported the occurrence of activated microglial cells coupled with gradual rise in the content of reactive microglia in striatum and cortex. Furthermore, authors reported significant difference in the morphological features of activated microglial cells in the striatum and cortex.

In the brain regions affected with Huntington disease, the processes of reactive microglia showed specific patterns in relation to the severity of disease. In low-grade Huntington disease, the processes of reactive microglial cells were prominent, while in the grade 2 and grade 3 Huntington disease, colocalization of reactive microglia processes with the apical dendrites of pyramidal neurons was reported by authors Sapp et al. (2001).

The huntingtin-positive nuclear inclusions were seen in few reactive microglia that were aligned with degenerating pyramidal neurons (Sapp et al., 2001).

It can be concluded that **premature and perpetual rise in density of reactive microglia coupled with their colocalization with degenerating neurons in the striatum and cortex regions of brain in Huntington disease are characteristic features of the implication of the activated microglia in the pathology of Huntington disease.**

Additional study involving the brains of patients ($n = 9$) affected with Huntington's disease and controls ($n = 3$) was conducted by Singhrao et al. (1999). The enhanced **complement biosynthesis by activated microglial cells and activation of complement in neurons in Huntington's disease** were identified by Singhrao et al. (1999) and served as biomarkers for the microgliosis and astrogliosis in the caudate and internal capsule in Huntington disease, while these regions in normal brains exhibited normal population of microglia and astrocytes (Singhrao et al., 1999).

Positron emission tomography technique was utilized by several workers to identify the changes in the microglia population in the brain regions affected with Huntington disease.

Pavese et al. (2006) and Politis et al. (2011) delineated the existence of microglial activation in brain regions inflicted with Huntington disease and further posited that intensity of microglial activation was consistent with the severity of Huntington disease.

Tai et al. (2007) described the occurrence of activated microglia in the brain regions preceding clinical manifestation of Huntington disease.

Aforementioned studies provide ample evidences to correlated the presence of activated microglia in brain regions with Huntington disease that might be implicated in the pathology of the disease.

The medium spiny neurons constitute around 95% of the neurons in striatum in human brain. The GABAergic inhibitory neurons in the central nervous system exert a critical role in the pathology of several neurodegenerative diseases including gamma oscillations, neural activity of cortical and hippocampal neurons in brain. **Altered functioning of GABAergic inhibitory neurons can disturb a balance between excitation to inhibition in the neurons that might be implicated in the pathogenesis of Huntington disease** (Xu and Wong, 2018).

Furthermore, mutant huntingtin protein largely accumulate and form aggregate in neuronal processes and axonal terminals (Li et al., 2000).

The GABAergic inhibitory neurons in the striatum are highly sensitive to the neurotoxicity induced by accumulated mutant huntingtin protein.

The microglia are the active sensor of microenvironment of brain. In response to accumulated mHTT in several regions of brain, resident microglia are activated into reactive microglia (Hanisch and Kettenmann, 2007).

These facts were further supplemented in the in vitro primary neuronal cultures (Yang et al., 2017) and cortico-striatal brain slice (Yang et al., 2017) study that expression of mHTT in the neurons in the cortex and striatum mediated inflammatory response leading to metamorphosis of resident microglia into reactive activated form of microglia coupled with rise in number of reactive activated microglial via increase in the proliferation ability of microglia (Yang et al., 2017).

Reactive and activated microglia released proinflammatory cytokines mediating inflammatory changes in the neurons in brain regions and possibly have a role in the pathology of Huntington disease. Thus, activated and reactive microglia are construed as neurotoxic.

Therapeutic intervention targeting at reducing the expression of mHTT in the neurons located in the cortex and hippocampal regions of brain could mitigate the severity of Huntington disease via **reduction in the activation of resident microglia in the brain.**

Autonomous activation of microglia

The microglia as resident macrophages in the brain have potent immune function via phagocytosis and scavenger action and these cells remove the mHTT from brain regions. Thus, accumulation of mHTT in the microglia is minimized due to presence of proteasome subunit (Orre et al., 2013) and autophagosomes (Su et al., 2016) in microglia.

The mHTT impaired the activity of microglia as was demonstrated in the R6/2 transgenic mice model of Huntington disease (Bjorkqvist et al., 2008). In response to mHTT expression, authors identified the presence of highly reactive and transformed microglia in brain regions.

The mHTT in microglia led to expression of cell-autonomous proinflammatory genes that is subsequently regulated by **transcription factor PU.1 and transcriptional factor CCAAT-enhancer-binding protein** (Yang et al., 2017).

Role of phenotypes of microglia in pathology of Huntington's disease

Two functional phenotypes of microglial cells in the brain constitute a critical element in the pathophysiology of neuroinflammation and progress of neurodegeneration diseases including Huntington disease (Jha et al., 2016; Yang et al., 2017).

In response to the duration and severity of pathological stimuli in brain regions, microglial cells exist in the two phenotypes namely M1 phenotype (classically activated microglial cells) and M2 phenotype (alternatively activated microglial cells) (Yang et al., 2017).

Classically activated microglia

These cells secrete proinflammatory cytokines namely inducible nitric oxide synthase, IL-8, IL-1β, IL-6, and tumor necrosis factor α (Yang et al., 2017; Kroner et al., 2014).

Further these microglial cells express surface receptors namely CD32, CD16, CD36, CD86 and CD68 (Kroner et al., 2014; Yang et al., 2017).

Microglial cells isolated from the R6/2 transgenic mice model of Huntington showed higher levels of inducible nitric oxide synthase activity, elevated levels of cytokines like TNF-α IL-1β, IL-6 (Hsiao et al., 2014; Yang et al., 2017).

Furthermore, isolated microglial cells form the transgenic porcine model of HD (Valekova et al., 2016; Yang et al., 2017) **showed the presence of elevated levels of cytokines like IL-1β and IL-8 implying the role of these proinflammatory cytokines and activated microglial cells in the pathology of HD.**

Bjorkqvist et al. (2008), Chang et al. (2015), Yang et al. (2017) **identified the higher levels of cytokines TNF-α, IL-6, IL-1β, and IL-8 from the striatum and cerebrospinal fluid in brain and in the blood circulation in patients affected with HD.**

Thus, it can be inferred that M1 phenotype of classically activated microglial cells with their potential to release proinflammatory cytokines in the regions in brain might be implicated in the pathology of HD.

Alternately activated microglia

These cells secrete antiinflammatory cytokines namely, interleukin 10 (IL-10), transforming growth factor beta (TGF-β), CD206, arginase 1 (Arg1), chitinase like protein 3 in humans (Ym1) (Franco and Fernandez-Suarez, 2015; Yang et al., 2017).

The M2 phenotype of microglial cells are involved in the repair of neural tissues and phagocytosis of pathogens and damaged proteins and tissues in the brain (Miron et al., 2013).

The antiinflammatory cytokine interleukin 10 secreted by M2 microglial cells exhibit a crucial role in the mediation of phagocytosis of apoptotic neurons in brain (Chhor et al., 2013).

Furthermore, existence of M2 microglial biomarkers in the brain regions in HD patients were poorly reported except the presence of transforming growth factor β and vascular endothelial growth factor in association with the biomarkers of M1 phenotype of microglial cells in the postmortem brain tissues of patients with HD (Di Pardo et al., 2013; Chang et al., 2015).

Thus, it can be concluded that M1 and M2 phenotypes of microglia possess contradictory functions and specific biomarkers.

Nuclear factor kappa B mediated microglia activation in Huntington's disease

Toll-like receptors are abundantly expressed on the surface of microglial cells. these receptors can recognize the pathological stimuli and mediate downstream cascade of molecular events leading to secretion of proinflammatory cytokines predominantly tumor necrosis factor alpha (TNF-α), IL-12, IL-10, IL-6, and **C-X-C motif chemokine ligand 10** (CXCL-10) (Jack et al., 2005).

Predominantly, TLR-2, TLR-3, and TLR-4 are widely expressed on the microglial cells. the toll like receptor 2 is the abundant cell surface receptor in microglial cells and is closely related with the release of IL-6 and IL-10 in response to the pathological stimuli.

Thus, toll like receptors are the novel receptors that modulate and mediate innate immune response in the microglial cells in humans. The activity of innate immunity via involvement of toll like receptors in brain regions is subjected to the microenvironment of brain regions.

The nuclear factor kappa B is localized in the cytoplasm in the cells. its activity is restricted by the inhibitory proteins, I-κBs (Ghosh et al., 1998). After phosphorylation of I-κBs that is catalyzed by enzyme I-κB kinase, the nuclear factor kappa B is detached from I-κB proteins and is translocated to the nucleus (Ghosh and Karin, 2002). Nuclear localization of nuclear factor kappa B regulates the expression of gene involved in the synthesis of proinflammatory cytokines.

Impaired transcriptional activity induced by the mutant huntingtin protein (mHTT) is insinuated in the pathology of HD.

In vitro study by Khoshnan et al. (2004) involved the cultured cells and cells from the striatum region of brain in transgenic mice model of HD.

The culture cells expressing mutant HTT and striatal cells from transgenic mice model of HD showed the raised activity of nuclear factor-kappa B (Khoshnan et al., 2004). The striatal cells exhibited nuclear localization of nuclear factor-kappa B in neurons in the brains of transgenic mice in HD.

In HD transgenic mice model, mutant HTT induce activation of IkappaB kinase complex that regulates the NF-kappaB (Khoshnan et al., 2004). The expanded PolyQ tract and proline-rich motifs of the mutant huntingtin directly interact with the regulatory subunit of IKK named as IKK-γ **resulting into activation of** IkappaB kinase complex (Khoshnan et al., 2004).

The activity of IKK-γ **enhances the** mutant Htt exon-1aggregation and its localization in the nucleus (Khoshnan et al., 2004).

Khoshnan et al. (2004) demonstrated that amino-terminally truncated type IKK-γ **can suppress the activity of** IkappaB kinase complex leading to inhibition of mutant Htt-induced neurotoxicity in medium-sized spiny neurons.

Thus, it can be inferred that mutant huntingtin protein induced dysregulation of the NF-kappa-B signaling pathway might be implicated in the pathology of Huntington disease.

The mutant huntingtin induced dysregulation in NFκB signaling pathway can be aborted by reducing the expression of mutant huntingtin.

Study was conducted by Träger et al. (2014) utilizing the targeting delivery of small interfering RNA in primary human cells to reduce the expression of huntingtin. The method used the encapsulated small interfering RNA particles targeted at immune cells, macrophages in the Huntington's disease that led to decline in the levels of cytokine production (Träger et al., 2014).

The findings substantiate the role of mHTT in altering the NFκB signaling pathway, activation of microglial cells, expression of proinflammatory cytokines and neurodegeneration in the Huntington disease.

Furthermore, targeted delivery of small interfering RNA particles in reducing the expression of mutant HTT might be a therapeutic intervention in reducing the severity of clinical manifestations of Huntington disease.

Kynurenine signaling pathway in microglia activation in Huntington's disease

The kynurenine pathway is involved in the synthesis of NAD+ from the L-tryptophan amino acid. Kynurenine is a metabolite of the tryptophan metabolism (Savitz, 2020). Altered kynurenine pathway is involved in several neurological and psychiatric disorders.

The enzyme indoleamine 2,3-dioxygenase (first enzyme in kynurenine pathway) is predominantly expressed in the microglia (Heyes et al., 1996). Study involving transgenic N171-82Q mice model of HD showed that indoleamine 2,3-dioxygenase enzymatic activity was raised in the transgenic mice with HD in comparison to wild type mice (Donley et al., 2016).

Enzyme indoleamine 2,3-dioxygenase is the first enzyme and rate limiting enzyme in the L-tryptophan catabolic pathway (Lilla et al., 2018).

The enzyme converts the L-tryptophan into N-formylkynurenine that is rapidly converted into kynurenine. Subsequently, kynurenine is converted into active metabolites namely hydroxykynurenine, anthranilic acid, kynurenic acid, 3-hydroxyanthranilic acid, picolinic acid and **quinolinic acid** (Lilla et al., 2018).

Raised activity of enzyme indoleamine 2,3-dioxygenase is linked with surplus accumulation of neurotoxic metabolites of kynurenine pathway, predominantly the quinolinic acid in the brain regions.

In vitro study by Heyes et al. (1996) involved the cultures of human fetal brain cells, microglial cells and peripheral blood macrophages. These cells were incubated with L-tryptophan in the absence or presence of soluble interferon, interferon γ.

Heyes et al. (1996) identified the quinolinic acid gas chromatography and electron-capture negative-chemical ionization mass spectrometry.

Heyes et al. (1996) reported the formation of L-kynurenine and quinolinic acid from the cultures of microglial cells and peripheral macrophages without the stimulation by Interferon γ.

The interferon γ stimulates the activity of indoleamine 2,3-dioxygenase leading to rise in the formation of L-kynurenine from microglia, peripheral macrophages and cultures of human fetal brain cells.

The concentration of L-kynurenine was identified to be more than 40 microM (Heyes et al., 1996).

The significant increased levels of quinolinic acid were reported from human microglia and peripheral macrophages (Heyes et al., 1996), while formation of quinolinic acid from the cultures of human fetal brain cells was negligible (up to 2 nM).

Thus, it can be inferred that infiltrated peripheral macrophages in brain and activated microglial cells might be involved in the surplus formation of neurotoxic quinolinic acid in brain regions due to altered metabolism of L-tryptophan through dysregulated kynurenine signaling pathway in the Huntington disease.

Microglial activation is associated with the clinical manifestation of Huntington disease and has been supported by publications past several years. the mutant HTT induced neurotoxicity is intensively linked with the microglial activation that is crucial element in the neuroinflammation associated with the HD.

But the precise role of microglial activation in the pathogenesis of HD is still to be elucidated.

Role of caspase 2 activation in Huntington disease

Activation of caspases and their involvement in the breakdown of substrates including the mutant huntingtin have been proposed as the etiopathogenesis of neurodegenerative disease covering the Huntington disease.

Caspases are the group of cysteine proteases and caspase belongs to the family of Ich-1 subfamily of proteins. Caspase 2 is the highly conserved protein in the animal kingdom (Krumschnabel et al., 2009).

Transcription of caspase 2 in the striatum in the yeast artificial chromosome (YAC72) mice model of Huntington disease was reported to be elevated (Hermel et al., 2004; Carroll et al., 2011) leading to overall rise in the immuno-reactivity of caspase 2 in the striatum of the patients with Huntington disease (Hermel et al., 2004; Carroll et al., 2011).

The YAC 128 mouse model of HD manifests the cognitive impairment, motor impairments and other characteristic neurological involvements similar to the progression of HD in humans (Ghilan et al., 2014).

Caspase 2 has a role in the pathogenesis of HD (Hermel et al., 2004). It was posited that binding of caspase-2 to the huntingtin is the primary step in the cascade of events leading to neuronal death in Huntington disease (Carroll et al., 2011).

Hermel et al. (2004) and Carroll et al. (2011) revealed that activation of caspase 2 is critical in the apoptosis of primary striatal cells of transgenic mouse model (YAC72) of HD that expressed full-length huntingtin.

Hermel et al. (2004) and Carroll et al. (2011) described that dominant-negative caspase-2 expression led to inhibition of the apoptosis of the primary striatal cells in mice model (YAC72) of HD.

Hermel et al. (2004) and Carroll et al. (2011) involved the histochemcial analysis of the tissues derived from the postmortem of brain of patients with HD and tissues from the brain of mice model (YAC72) of HD. Hermel et al. (2004) and Carroll et al. (2011) identified high level of immuno-reactivity of caspase-2 in medium spiny neurons in the striatum and the cortical projection neurons in (Hermel et al., 2004; Carroll et al., 2011) patients with HD in comparison to the control subjects.

The brain-derived neurotrophic factor (BDNF) has neuro-protective role in the human brain.it promotes the survival of neurons, differentiation of neurons and growth of new synapses in the brain (Acheson et al., 1995).

The enhanced transcription of caspase-2 has been correlated with the downregulation in the expression of brain-derived neurotrophic factor in the striatum and cortex regions of brain in mice model (YAC72) of HD (Hermel et al., 2004; Carroll et al., 2011).

Thus, caspase 2 activation is involved in the neuronal apoptosis in selected regions of brain and is associated with the pathology of Huntington disease.

Oxidative stress, mitochondrial dysfunction and Huntington's disease

The genetic mutation in HTT gene is responsible for the manifestations of Huntington disease.

The CAG repeats expansion sequence was detected in the DNA mutant huntingtin protein leading to abnormal expansion of polyglutamine tract in the expressed mutant huntingtin protein (MacDonald et al., 1993).

The abnormally expanded polyQ tract might initiate the misfolding of mutant huntingtin protein (DiFiglia et al., 1997; Penney et al., 1997).

The misfolded mutant huntingtin is further subjected to proteolysis by proteases at N-terminal to form polyQ expanded Htt fragments with neurotoxicity, cytotoxicity and high potential to form aggregates. The cleavage of misfolded mutant huntingtin is construed as the prime cellular event in the etiopathogenesis of Huntington's disease (Lunkes et al., 2002).

The N-terminal mutant Htt fragment contains 17 amino acids sequence (N17). This fragment serves as nuclear import signal and promotes the formation of mutant huntingtin aggregates inside the nucleus of neurons (Lunkes et al., 2002).

Cytoplasmic organelles like mitochondria and endoplasmic reticulum generative high reactive oxygen species during the metabolic activities. The reactive oxygen species are removed by antioxidant system of the body and it's a natural defense mechanism of body. Failure to scavenge and/or surplus generation of reactive oxygen species impair the antioxidant system of body. The imbalance between prooxidant mechanism and antioxidant system manifest as oxidative stress.

Surplus accumulation of reactive oxygen species results into structural damage of lipids, proteins, and DNA that is the basis of pathogenesis of several diseases (Sies, 1991; Yu, 1994; Weidinger and Kozlov, 2015).

Neurons in the brain regions have high energy demand that is supplied by the mitochondria present in the axons (Chan, 2006).

The reduced coenzymes namely NADH2 and FADH2 transfer the protons to the mitochondrial matrix and subsequent transfer of electrons along the electron transport chain located on the inner membrane of mitochondria generate ATP for the various activities in the brain regions.

Occasionally, electrons during the oxidative phosphorylation process in mitochondria also interact with the molecular oxygen to generate superoxide.

Around 2% of the total mitochondrial oxygen reacts with electron to generate superoxide (ROS) and remaining 98% of the oxygen is utilized in ATP generation in mitochondria (Chance et al., 1979).

The mutant huntingtin aggregates are lipophilic and bind with the mitochondrial membranes resulting into inhibition of **succinate-CoQ reductase** (complexes II) and **CoQH$_2$-cytochrome c reductase** (complex III) and impaired electron transport and oxidative phosphorylation in mitochondria. The cytosolic ATP store is depleted. The accumulated electrons react with the molecular oxygen and generate surplus reactive oxygen species (Jenkins et al., 1993).

The microenvironment of the regions of brain becomes highly toxic, thus, ROS inflicts oxidative damage on the mitochondrial DNA (mtDNA). The ROS are also released into the cytosol (Polidori et al., 1999). The oxidative lesions of mtDNA are directly linked with pathogenesis of HD (Polidori et al., 1999).

Overall, depleted ATP, elevated ROS levels coupled with damaged mtDNA have been reported in the mutant huntingtin expressing cells in striatum in brain (Zheng et al., 2018; Siddiqui et al., 2012). **The oxidative damage of mtDNA is the key finding in the oxidative stress mediated by mHTT in the mitochondria in neurons** (Liu et al., 2008).

Mitochondrial DNA copy number represents the number of mitochondrial genomes per cell. It is a measure of the mitochondrial health and functional status. The decline in the copy number of mtDNA in peripheral leukocytes in Huntington disease. Also, the transgenic mice model (R6/2 mice) (Petersen et al., 2014) of HD depicted reduction in the copy number of mtDNA.

Furthermore, mitochondrial disulfide relay system (oxidoreductase Mia40 and sulfhydryl oxidase Erv1/ALR) in the intermembrane space of mitochondria is disturbed concomitant to the expression of mutant huntingtin protein (Napoli et al., 2013).

In addition to ATP synthesis and generation of ROS, mitochondria are involved in the maintaining cytosolic calcium homeostasis (Giacomello et al., 2007). The calcium ions are sequestered into the calcium phosphate complexes within mitochondrial matrix after the rise in the cytosolic calcium levels. Contrarily, lowering of cytosolic calcium levels stimulate the release of calcium ions into the cytosol (Nicholls and Budd, 2000). Presence of sarcoplasmic reticulum ryanodine receptor (Ca^{2+} channel in ER) is essential in the calcium transport mediated by mitochondria. Nicholls and Budd (2000) revealed that compounds like dantrolene and Ryanodine (inhibitors of ryanodine receptor) reduced neuronal death caused by mutant huntingtin protein whereas concomitant expression of mHTT and ryanodine receptor led to increase in the mHTT mediated toxicity (Nicholls and Budd, 2000).

Furthermore, mutant huntingtin mediates enormous sequestration of calcium ions inside the mitochondrial matrix. The large influx of calcium ions occurs via activity of NMDA receptors leading to excessive generation of ROS, oxidative stress, and damage of mtDNA (Paul and Snyder, 2019).

Thus, disturbed calcium homeostasis in mitochondria in neurons is suggestive of key component in the etiopathogenesis of Huntington disease.

In vivo study (Suzuki et al., 2012) involved identification of role of ryanodine receptor in the striatal and cortical neuronal death in R6/2 transgenic mice model of HD.

Authors reported that mutant huntingtin induced release of calcium ions via ryanodine receptor from the mitochondrial matrix into the cytosol in the striatal and cortical neurons in R6/2 mice. The efflux of calcium was coupled with decline in mitochondrial calcium store leading to neuronal death in cortex and striatum in mice model of HD.

Thus, mutant huntingtin induced ryanodine receptor-mediated disturbance in calcium homeostasis in neuronal mitochondria is implicated in the neuronal death and progression of Huntington disease.

Furthermore, neuronal stability is dependent on normal calcium homeostasis. The calcium overload in the cytosol might be involved in opening of mitochondrial permeability transition pore (mPTP) located in the inner membrane of mitochondria. Thus, calcium efflux via the nonspecific pore leads to decline in oxidative phosphorylation and ATP synthesis. Hence, electrons are coupled with molecular oxygen to form surplus superoxide mediating oxidative stress and damage of mtDNA (Krieger and Duchen, 2002).

Further, aforementioned studies suggest that impaired calcium handling in mitochondria in neurons leads to etiopathogenesis of HD.

Additional studies (Choo et al., 2004; Green and Reed, 1998) described that mutant huntingtin due to its lipophilic nature, interacts with outer membrane of mitochondria and leads to opening of mitochondrial permeability transition pore in inner membrane that is manifested as release of cytochrome c into the cytosol and neuronal apoptosis.

Tumor suppressor protein (p53) in mitochondrial dysfunction and Huntington's disease

It is actively involved in mediating cytotoxicity of mutant huntingtin and mitochondrial dysfunction involved in the Huntington disease.

The toxic amino-terminal fragment of huntingtin (httex1p) was identified to aggregate with tumor suppressor protein (p53) in the inclusions in cell culture

(Steffan et al., 2000). In vitro study revealed the interaction potential of pathogenic fragment (httex1p) with p53 protein leading to repression of transcription factors namely **multidrug resistance protein 1 (MDR-1)** (Steffan et al., 2000) and **cyclin-dependent kinase inhibitor 1**, p21 (WAF1/CIP1) regulated by p-53 protein and suppression of transcription of genes (Steffan et al., 2000).

Thus, toxic amino-terminal fragment of huntingtin (httex1p) can mediate repression of transcription of genes regulated by p-53 proteins leading to neuronal dysfunction and neuronal death in the manifestation of HD.

Additionally, tumor suppressor gene (p53)activation led to rise in transcription of mutant HTT mRNA and stimulated expression of mutant huntingtin implicated in etiopathogenesis of HD (Feng et al., 2006), suppression of p53gene reduced the expression and transcription of mutant huntingtin leading to escape from the mitochondrial depolarization, calcium efflux and toxicity of mHTT in transgenic mice model of HD (Bae et al., 2005).

Thus, tumor suppressor protein (p-53) is closely involved in mediating cascade of events involved in mitochondrial dysfunction and neuronal death in response to pathogenic N-terminal fragment of mutant huntingtin.

Peroxisome proliferator-activated receptor γ coactivator 1-α (PGC-1α) in mitochondrial dysfunction and Huntington's disease

The **Peroxisome proliferator-activated receptor γ coactivator 1-α is the regulator of mitochondrial proliferation, ATP synthesis and antioxidant mechanism in mitochondria** (Johri et al., 2013).

Several studies described the potential of mutant huntingtin in regulating the expression of transcriptional coregulator named as **Peroxisome proliferator-activated receptor γ coactivator 1-α and generation of reactive oxygen species and oxidative stress** (Johri et al., 2013).

Mutant huntingtin binding with PGC-1α gene
First probably mechanism is the binding of mutant huntingtin to the PGC-1α gene inside the nucleus.

The mutant huntingtin decreases the transcription of **Peroxisome proliferator-activated receptor γ coactivator 1-α** (Chaturvedi et al., 2012).

Additionally, mutant huntingtin can bind with the PGC-1α protein directly and inhibits the expression of copper/zinc superoxide dismutase (SOD1), mitochondrial uncoupling proteins and manganese superoxide dismutase (SOD2), and glutathione peroxidase (Gpx-1) which serve as targets of PGC-1α protein (St-Pierre et al., 2006).

The reduction in the expression of elements of antioxidant system of body result into accumulation of reactive oxygen species and hyperactivity of prooxidant mechanism (Gandhi and Abramov, 2012a,b).

Thus, raised levels of ROS promote the mitochondrial dysfunction and neuronal death implicated in etiopathogenesis of HD.

Mutant HTT with the dynamin-related protein 1

Still additional mechanism is the interaction of mutant HTT with the dynamin-related protein 1 (Drp1) (Shirendeb et al., 2011a,b; Weydt et al., 2014).

The pathogenic N-terminal fragment of mutant huntingtin impairs the mitochondrial fission—fusion activities and generation of ATP leading to raised oxidative stress Tsunemi et al. (2012).

The aforementioned facts were supplemented by Lin et al. (2004) that identified improvement in the mitochondrial function in transgenic mice model with deleted PGC-1α.

Thus, it can be suggested that PGC-1α protein has a role in inducing oxidative damage and neuronal death after the pathological response by mHTT.

Advanced glycation end-products in pathology of Huntington disease

Introduction

Advanced glycation end-products (AGEs) constitute essential biomarkers of the oxidative stress and oxidative damage in the cellular components, thereby have close relation to the etiopathogenesis of several neurodegenerative diseases including the Huntington disease.

These are produced after exposure of macromolecules like proteins, nucleic acids and lipids to sugar molecules leading to oxidation of macromolecules via nonenzymatic process.

Therefore, AGEs can be termed as "glycotoxins" and exert strong potent "oxidant Effect." These have both endogenous as well as exogenous origins (Vistoli et al., 2013).

Classical Maillard reaction was proposed in the beginning of 20th century (Maillard, 1912) for the production of advanced glycation end-products yielded in response to the classical.

The "Maillard theory of Aging" was proposed by Monnier and Cerami to explain the perpetual accumulation of AGEs in tissues and associated degenerative changes in the affected body tissues and organs (Monnier, 1989).

Thus, continuous buildup and impaired clearance of AGEs are certainly involved in the protein denaturation, mutation in DNA, and lipid peroxidation owing to the prooxidant effect of AGEs.

These pathological changes in the biomacromolecules gradually contribute to the etiopathogenesis of onset of neurodegenerative diseases.

Brief overview of formation of advanced glycation end-products

The sugars like fructose and glucose contain free ketone (C=O) group and aldehyde (-CHO) group, respectively. Accordingly, these sugars molecules have high chemical reactivity and strong reducing potential (Njoroge and Monnier, 1989).

Thus, nonenzymatic and spontaneous glycation between carbonyl group of reducing sugar moiety and free amino group of a protein molecule occurs. The reaction takes place by formation of Schiff base via nucleophilic addition reaction (Schiff, 1866).

The Schiff base ($R_2C = NR'$) is unstable and reversible, thus it is subjected to Amadori rearrangement to form Amadori products namely ketoamine or fructosamine). These products have stable configuration (Baynes et al., 1989).

The Amadori products are the intermediates in the formation AGEs because Amadori products react with primary amines as the ε-amino-lysine and/or two molecules of Amadori product interact directly to form AGEs (Baynes et al., 1989).

Overall, series of nonenzymatic chemical reactions including the formation of covalent bond between sugar moiety and protein molecule is described as Maillard reactions that is, prominently involved in the formation of endogenous Advanced glycation end-products.

Furthermore, Maillard reaction results into synthesis of highly reactive dicarbonyls and Strecker aldehydes in the processed foods (Nursten, 2005) that becomes the basis of formation of exogenous Advanced glycation end-products.

The Strecker aldehydes include acetaldehyde, 2-methylbutanal, 2-methylpropanal, and 3-methylbutanal frequently found in the cooked and processed plant based food (Nursten, 2005).

The Strecker aldehydes interact with macromolecules including proteins and nucleic acids to produce N^ε-(carboxylmethyl)-L-lysine(Nursten, 2005), N^ε-(carboxylethyl)-L-lysine, N^δ-(5-hydro-5-methyl-4-imidazolon-2-yl)-L-ornithine like compounds.

These end products belong to exogenous AGEs (Nursten, 2005).

The dicarbonyls compounds are liberated from dietary carbohydrates after food processing. Predominant dicarbonyls can be 3-deoxygalactosone, 3-deoxyglucosone, glucosone, glyoxal, 3,4-dideoxyglucosone-3-ene and methylglyoxal (Nursten, 2005).

These high reactivity dicarbonyls are the precursors of dietary AGEs (exogenous AGEs) and possess high cytotoxicity.

For example, food cooking under dry conditions and high heat increases the probability of formation of AGEs.

Thus, both endogenous and exogenous factors contribute to formation of AGEs.

These glycation end products alter the normal physiology of living systems partly by their action on the AGE receptors and partly by covalent bonding with proteins and nucleic acids affecting their structure and functions.

Role of glycation and impaired carbohydrate metabolism in pathology of Huntington' disease

Contrary to Parkinson disease and Alzheimer disease, predominance of heredity prevailes as the causative factor in the pathogenesis of Huntington disease. Would it be appropriate to search association between advanced glycation end products and pathogenesis of Huntington disease?

Moreover, it can be hypothecated that factors involved in the impairment of carbohydrate metabolism, hyperglycemia and higher propensity for glycation of proteins and nucleic acids might be insinuated in the progression of Huntington disease.

Several studies provide supporting as well as contrary evidences in determining the potential of impaired glucose tolerance, hyperglycemia and AGEs in the etiopathogenesis of Huntington disease.

Supporting evidences

Longitudinal Studies (Podolsky et al., 1972; Farrer, 1985; Leibson et al., 1997) provided evidences supporting the association among Huntington disease, glucose intolerance, hyperglycemia and diabetes mellitus in patients.

Additional studies substantiated the interplay in causative factors as obesity, insulin resistance and type 2 diabetes mellitus with occurrence of Huntington disease (Kremer et al., 1989; Andreassen et al., 2002; Höybye et al., 2002; Ristow, 2004). These studies also explained the impaired endocrinal function of β cells in pancreas in patients with diabetes mellitus.

Moreover, population-based cohort study by Leibson et al. (1997) described the association between diabetes mellitus in patients with dementia.

In vitro cell culture study by Brás et al. (2019) utilizing mammalian cells and yeast reported that hyperglycemia for prolonged period resulted into higher probability of glycation of biomacromolecules that subsequently enhanced the neurotoxicity of mutant huntingtin leading to decline in its clearance and higher aggregation potential inside neurons.

The development of transgenic-mice model of Huntington disease with several CAG repeats (from 82 to 400 repeats) that exhibit several manifestations of HD phenotype, has revolutionized the research studies clinical studies in the field of neurology (Brás et al., 2019).

Josefsen et al. (2008) explained that R6/1 mice model showed several neurological manifestations at the age of 22–26 weeks of age coupled with the glucose intolerance and raised plasma insulin levels.

Josefsen et al. (2008) and Duan et al. (2003) described the raised plasma glucose levels in nondiabetic mice model of Huntington disease, thus linking association between impaired carbohydrate metabolism and Huntington disease.

Contradictory evidences

Moreover, contrary findings were provided indicating absence of impaired carbohydrate metabolism, hyperglycemia in patients with Huntington disease.

Longitudinal study of 2.6 years involving patients ($n = 10$) with Huntington disease and healthy controls ($n = 10$) were selected and were screened for diabetes mellitus by Oral glucose tolerance test, glycosylated Hb, C-peptide levels. patients ($n = 8$) with HD showed values of parameters not different from the control subjects after a duration of 2.6 years, thus study ruled out the association between impaired carbohydrate metabolism and Huntington disease (Kremer et al., 1989).

Thus, precise interaction among glycosylation of sugar moieties in patients with impaired carbohydrate metabolism, production of glycosylated end products and their effect on the progression of Huntington disease is still elusive.

Role of receptors for AGEs in pathology of Huntington's disease

The receptors for AGEs are prominently expressed in the astrocytes and medium spiny neurons in the subependymal layer and caudate nucleus in brains in HD (Anzilotti et al., 2012).

(Kim et al., 2015a,b) showed that receptors for AGEs are highly expressed in the astrocytes in the subependymal layer and striatum regions in brains in Huntington's disease. Further, the expression is concomitant with the rise in severity of Huntingtin disease.

The colocalization (Kim et al., 2015a,b) of receptor for advanced glycation end (RAGE) with the S100 calcium-binding protein B was predominant in the astrocytes in comparison to microglial cells and neurons.

Study (Kim et al., 2015a,b) revealed that astrocytes with receptors for AGEs contained nucleus translocated nuclear factor kappa-B (NF-kB) with the increase in the phosphorylation of Extracellular signal-related kinase (**ERK1/2**) (Kim et al., 2015a,b) in the brain regions in patients with HD.

Thus, it can be inferred that receptor for AGEs exhibit higher expression in astrocytes and involvement of astrocytes and RAGE in the etiopathogenesis of HD should be extensively studied to reach a uniform generalization.

Moreover, concluding evidences are still to be proposed to achieve a consensus whether degenerative diseases including Huntington disease are caused by the prooxidant effect exerted by excessive accumulation of AGEs in tissues or degenerative changes in tissues leads to poor scavenging of AGEs due to impaired and compromised antioxidant enzymes in the body.

Role of autophagy in pathology of Huntington's disease
Introduction

Autophagy is tightly regulated and coordinated process of self-degradation crucial in energy homeostasis in body.

It is helpful in scavenging the toxic misfolded proteins, protein aggregates, neuronal inclusion bodies, and damaged cytoplasmic organelles like endoplasmic reticulum, mitochondria, and damaged peroxisomes.

Hence, autophagy is essential for growth and development of tissues. moreover, dysregulation in autophagy is closely linked with neuronal cell death in several neurodegenerative disease.

The autophagy eliminates the mutant and pathogenic proteins with longer half-lives in comparison to ubiquitin proteasomal system that scavenges the damaged proteins with shorter half-lives (Klionsky, 2008).

Studies in yeast (*Saccharomyces cerevisiae*) (Nakatogawa et al., 2009) delineated the precise molecular mechanism in the regulation of autophagy.

Till date, 32 discrete autophagy-related genes (Atg) have been reported in the yeast that control autophagy.

Process of autophagosome formation

The "phagophore" is the isolation membrane and it starts the process of autophagosome formation. The phagophores is furnished by the phospholipid bilayer of membrane of the endoplasmic reticulum and trans-Golgi network (Axe et al., 2008).

The formation of phagophores is dependent on the signaling induced by several factors. The phagophores enclose the misfolded proteins including mutant HTT to be degraded (Walczak and Martens, 2013).

The Atg12−Atg5−Atg16L complex serves a key role in conversion of phagophores into autophagosomes (Walczak and Martens, 2013).

The Atg5 component can bind directly with phagophores membranes and its membrane binding ability is inversely controlled by conjugation of Atg12 (Walczak and Martens, 2013) component with the membranes of phagophores, while membrane binding potential of Atg5 component is enhanced by Atg16 (Walczak and Martens, 2013).

Before the formation of autophagosome, elongation of phagophore takes place that essential requires conjugation Atg12 with Atg5 and forms complex Atg12−Atg5 which in turn conjugates with ATG6 and finally formation of **Atg12−Atg5−Atg16L complex is established.**

The Atg12 element function is closely related to the activity microtubule-associated protein 1 light chain 3 (LC3) (Kabeya et al., 2000).

Protein LC3 is the mammalian protein and represents an ortholog of yeast Atg8. The LC3 protein specifically binds with autophagosome membrane (Kabeya et al., 2000).

LC3 protein is transformed into LC3-I as the cytosolic protein. After autophagy starts, LC3-I is attached to phosphatidylethanolamine component of the autophagosome membrane (Kabeya et al., 2000).

ATG8 lipidation in autophagosome formation

ATG8 protein belongs to the class of evolutionary conserved proteins. The ATG8 protein resembles structurally to ubiquitin. The ATG8 protein binds with the hydrophilic heads of the phospholipid molecules in the membranes of phagophores. This process is called as ATG8 lipidation.

ATG8 lipidation is crucial in the conversion of phagophores into autophagosomes and is the hallmark of autophagy (Martens and Fracchiolla, 2020).

The completion of autophagosome formation is followed by dissociation of the complex Atg12−Atg5−Atg16L form the membrane except LC3-II that remains bound with membrane.

The autophagosome fuses with the lysosome and forms autolysosome. The misfolded proteins and huntingtin proteins in the autophagosomes are degraded by the hydrolases of lysosomes.

Thus, autophagy is crucial in the clearance of unwanted proteins with tendency for aggregation namely mutant huntingtin (Huntington's disease), tau (dementias), ataxin-3 (spinocerebellar ataxia-3), and α-synuclein mutant (familial Parkinson's disease) (Berger et al., 2006).

Autophagy and mutant huntingtin

The N-terminal fragment of cleaved mutant huntingtin comprising is comprised of 100−150 amino acid residues with polyglutamine tract. This pathogenic cleaved product of mutant huntingtin is pathogenic and exhibits the neurotoxicity. Its aggregates are reported in inclusion bodies with questionable function whether protective or toxic, but latest studies (Takahashi et al., 2008) defined that soluble polyglutamine oligomers (preaggregate oligomers) and not the inclusion bodies, exhibits intense cytotoxic activity in the living cells. The cells having soluble oligomers suffered rapid neurodegeneration.

A potential association among impaired autophagy, accumulation of toxic misfolded proteins and neurodegeneration is suggestive in the etiopathogenesis of Huntington disease.

Study by Hara et al. (2006) involved transgenic mice model with knockout gene Atg5 (autophagy-related 5 protein) in the neural cells.

Hara et al. (2006) reported the presence of cytoplasmic inclusion bodies in affected neurons in the mice model (Atg5−/− cells) coupled with development of progressive motor function deficiency.

It is suggestive that basal autophagy is crucial in the removal of misfolded proteins, otherwise their gradual build up in the neurons disrupt neuronal activities leading to neuroinflammation and neurodegeneration implicated in the etiopathogenesis of Huntington disease.

Additional study related with deletion of ATG genes mice models (ATG5 and ATG7) essential in the formation of ATG proteins was conducted by Komatsu et al. (2006).

The Atg7 gene (autophagy-related 7) is essential in controlling synthesis of ATG 7 protein that is closely related with autophagosome formation and functioning.

Authors reported (Komatsu et al., 2006) deletion ATG7 gene mice model exhibited abnormal behavior including impaired coordinated muscle movements.

Possibly, Atg7 deletion resulted into impaired clearance of misfolded proteins in the neurons due to impaired autophagy. This led to massive neurodegeneration in cerebral and cerebellar cortices and aforementioned clinical manifestations.

Additional studies supplement the role of autophagy in the clearance of mutant huntingtin protein.

Compound 3-methyladenine is found to suppress formation of autophagosomes. It was reported to exhibit inhibitor of autophagy in concentration of (5 mM) in isolated hepatocytes of rat. The application of 3-methyladenine in mice model of HD and isolated rat hepatocytes resulted into inhibition of autophagy and impaired clearance and accumulation of soluble and aggregated mutant huntingtin (Ravikumar et al., 2002).

The findings are indicative of essentiality of autophagy in maintaining the health of neurons through scavenging the misfolded proteins from the neurons and thus, preventing their accumulation and minimizing their toxic effects on the neurons in brain regions.

Inducers and modulators of autophagy can affect the induced clearance of toxic elements. One such modulator is reported as Rapamycin which can induce clearance of mutant huntingtin by autophagy.

Thus, autophagy modulating and inducing drugs can be valuable therapeutic tool in the treatment of neurodegenerative disease owing to enhanced autophagy to remove the harmful misfolded proteins from the tissues that might halt the progression and or severity of neurodegenerative diseases including Huntington disease.

Ubiquitin proteasome functioning in Huntington's disease

Nascent proteins undergo precise folding in the endoplasmic reticulum to acquire the native three dimensional conformational states. Otherwise, misfolded proteins are efficiently removed from the tissues. These biological activities are crucial in determining cellular protein homeostasis.

The ubiquitin–proteasome system (UPS) is closely linked with the scavenging the misfolded and damaged proteins from the tissues.

The ubiquitin–proteasome has operational activity in cytoplasm and nucleus for degradation of misfolded and soluble proteins (Hershko and Ciechanover, 1998). The UPS represents the main scavenging system against the accumulation of misfolded proteins in tissues including the postmitotic neurons.

In routine physiological condition, the ubiquitin–proteasome system involves targeting of the misfolded or damaged protein by ubiquitination followed by proteolysis of ubiquitinated protein in the proteasome (Hershko and Ciechanover, 1998).

The target protein is covalently bound to ubiquitin called ubiquitination of target protein via activity of ubiquitin-activating enzyme (E1), ubiquitin-conjugating enzyme (E2) and ubiquitin-ligase (E3). The E1 enzyme splits the ATP molecule and catalyzes the adenylylation of ubiquitin molecule and thereafter adenylylated ubiquitin is transported to cysteine residue of E2 enzyme (Haas et al., 1982).

In the cascade of events, E3 enzyme catalyzes the transfer of ubiquitin protein from the ubiquitin-conjugating enzyme (E2) to the target protein for ubiquitination.

The target protein destined for degradation is conjugated with four ubiquitin monomers to form polyubiquitin chain and it is recognized by the proteasome system (Thrower et al., 2000).

The proteasome is the protein complex with ATP-dependent proteolytic activity. The complex has two 19S regulatory particles and 20S core particle. The 20S core particle is barrel shaped and consists of four heptagonal rings. These rings possess catalytic activity (Groll et al., 1997).

The 19S regulatory particles recognize the ubiquitinated protein via ATP-dependent pathway (Dong et al., 2018).

After the recognition of target protein, deubiquitilating enzyme separates the polyubiquitin chain from the target protein and it is broken down into constituent monomers for reutilization.

The target protein is further partially unfolded with the help of 19S regulatory particles (Smith et al., 2005) so that it can translocate through the narrow barrel-shaped interior of the 20S core particle. The target protein is attached to the proteolytic site in the 20S particle (Dong et al., 2018) for degradation.

Pathogenic N-terminal fragments of mutant huntingtin protein aggregate inside the neurons to form inclusion bodies that are also termed as neuronal intranuclear inclusions, or nuclear inclusions, or inclusion bodies in the neurons (Becher et al., 1998). **These are the hallmark of the Huntington pathology suggestive of failure of the ubiquitin—proteasome degradation system.**

Ubiquitin proteasome functioning is impaired in Huntington disease

The impaired functioning of **Ubiquitin Proteasome system is attributed to multiple factors. One of the several putative factors is the genetic mutation in the ubiquitin gene that controls the synthesis of ubiquitin.**

Mutation of UBB gene

The ubiquitin B gene (UBB gene) undergoes frame shift mutation to form mutant ubiquitin protein designated as ubiquitin B^+1 (UBB^+1).

The frame shift mutation results into dinucleotide deletions (ΔGA, ΔGU) at the level of mRNA transcript. These deletions are normally absent in the DNA. The dinucleotide deletions leads to carboxy-terminal extension in 19-amino-acid residue (van Leeuwen et al., 1998).

The frame shift mutation in ubiquitin B gene manifests as absence of amino acid residue, Gy76 so that the mutant ubiquitin B+1 cannot conjugate with target proteins in the ubiquitination process (van Leeuwen et al., 1998; Ortega and Lucas, 2014).

Thus, it can be asserted that basic genetic defect in the components of UPS is associated with impaired functioning and its implication in the neurodegeneration and etiopathogenesis of Huntington disease.

Proteasome sequestration into inclusion bodies

The aggregates of the pathogenic fragments of mutant huntingtin have been reported in the neuronal Inclusion bodies in striatum and cortex regions of brains in patients with HD (DiFiglia et al., 1997) as well as in brain regions in transgenic mice model of HD (Davies et al., 1997). I.

The inclusion bodies containing mutant huntingtin aggregates can be labeled with antibodies against the elements of UPS (ubiquitin and proteasome subunits as 19S caps and 20S core).

(Cummings et al., 1998; DiFiglia et al., 1997; Sherman and Goldberg, 2001; Goedert et al., 1998).

These findings are indicative of direct sequestration of the proteasome into inclusion bodies containing mutant huntingtin aggregates, thereby limiting the catalytic function of proteasome leading to reduction in proteolytic function of UPS.

Contrary findings were provided by study (Schipper-Krom et al., 2014) by positing that recruitment and sequestration of proteasome into the inclusion bodies is reversible activity. The proteasomes remain viable and functionally active in terms of their proteolytic potential.

Additional studies by Díaz-Hernández et al. (2003) and Bowman et al. (2005) involved the use of brain extracts obtained from transgenic mouse model of HD. The studies reported absence in reduction of proteolytic activity of proteasomes. Furthermore, study by Díaz-Hernández et al. (2003) provided additional evidence for the selective increase in the endopeptidase activity like chymotrypsin- and trypsin due to changes in the proteasomal subunits secondary to their sequestration in the inclusion bodies.

Thus, it can be concluded that nonuniformity in the aforementioned studies can be attributed to differences in the techniques used to determine the functioning of UPS, and differences in the samples used (whether in vitro culture of proteasomes with PolyQ or fresh frozen animal tissue or human frozen tissue after postmortem).

So it is difficult to reach a consensus over the modulation in functioning of UPS in association with Huntington disease.

Role of PARP-1 in pathology of Huntington disease

Poly (ADP-ribose) polymerase 1 is actively involved in ADP-ribosylation (addition of ADP-ribose residues) histone proteins. Thus, it is involved in several repair mechanisms of DNA including nucleotide exchange repair, DNA mismatch repair, and nonhomologous end joining repair (Pascal, 2018).

Thus, it helps to maintain stability of genome (Cardinale et al., 2015).

The NAD+ is the substrate of PARP-1 enzyme. It is converted into poly (ADP-ribose) (PAR) polymer.

It is the sensing molecule to detect DNA damage leading to binding of poly (ADP-ribose) polymer with histone protein of damaged DNA, thus mediates signaling to initiates DNA repair mechanisms.

Thus, several factors including ionizing radiation, and reactive oxygen species initiates activation of activation of PARP-1 that in turn activates the DNA repair machinery.

Several stimuli induce excessive activity of PARP-1 enzyme leads to surplus utilization of NAD+ and subsequent decline of cytosolic store of NAD+ and ATP depletion in the cell. These changes in the cells in turn aggravate DNA damage and further activates the PARP-1 enzyme in a cyclic event leading to necrotic cell death (Beneke et al., 2004; Andrabi et al., 2006).

Thus, overexpression of PARP-1 is implicated in the pathogenesis of several neurodegenerative diseases including Huntington disease.

The overactivation of poly(ADP-ribose) polymerase-1 mediates the liberation of apoptosis-inducing factor, also termed as poly(ADP-ribose) transferase or poly(-ADP-ribose) synthetase implicated in the necrotic cell death (Eliasson et al., 1997).

PARP-1 activates cascade of events leading to necrotic cell death in neurodegenerative diseases (Walker et al., 1988).

Additionally, PARP1 activity is linked with neuroinflammation.

PARp1 controls the transcription and nuclear translocation of NF-κB as well as activity of high mobility group box 1.

Overactivation of PARP1 leads to inhibition of SIRT1 followed by acetylation of HMGB1 and release of HMGB1 into extracellular environment (Mao et al., 2020).

Acetylated HMGB1 exerts as proinflammatory effect leading to activation of microglial cells and release of inflammatory cytokines in the brain. Thus, proinflammatory microenvironment mediated by HMGB1 might be involved in neurodegeneration (Mao et al., 2020).

In Huntington disease, raised immuno-reactivity of PARP-1 in microglial cells and neurons in brain regions was reported by Cardinale et al. (2015) suggesting the necrotic cell death of neurons in HD. PARP1 promotes the neuroinflammation implicated in the etiopathogenesis of Huntington disease.

In the study by Cardinale et al. (2015), neuro-protective role of PARP-1 inhibitor was tested in transgenic R6/2 mutant mice model. Study utilized **INO-1001** (strong and selective inhibitor of poly (ADP-ribose) polymerase (PARP) on the transgenic R6/2 mutant mice (Cardinale et al., 2015), study by Cardinale et al. (2015) identified that compound INO-1001 induced a rise in **cAMP-response element binding** protein and BDNF in the spiny neurons in striatum in brain in mice model of HD.

Overall, application of PARP-1 inhibitor in the mice model of HD resulted into decline in microglial activation, inclusion bodies formation with better morphology of neurons in striatum in brain leading to prolonged survival of mice model of HD (Cardinale et al., 2015).

Furthermore, gene knockout model of PARP-1 showed the protection of cells and neurons against the necrotic cell death in the brain regions (Endres et al., 1997).

Hence, drugs targeting PARP 1 inhibition might be possible tool in the management of Huntington disease.

Role of inflammasome and caspase 1 activation in Huntington's disease

Inflammasomes are cytosolic multiprotein complexes. These constitute essential element of the innate immune system. Multiple stimuli including pathogen-derived (bacteria, parasites, fungi, and viruses) chemicals and/or host-derived factors like reactive oxygen species, unfolded protein response, lipid peroxidation, damaged DNA, can mediate activation of inflammasomes.

Subsequently, the proteins NLRP3 or NLRP6 or NLRP7, and NLRC4 are involved in the inflammasome assembly by recruiting procaspase-1 with ASC adapter or without ASC adapter in canonical pathway (Zheng et al., 2020).

In noncanonical pathway, inflammasome assembly involves the recruitment of procaspase-8 (Zheng et al., 2020).

Later on, activated caspase 1 and caspase 8 are implicated in the cleavage of the pro-IL-1β and pro-IL-18 into IL-1β and pIL-18 that exert neuro-inflammatory effect in the brain regions (Zheng et al., 2020).

The activated capsase-1 promotes cleavage of Gasdermin D (executioner protein, substrate of caspase 1) leading to pyroptosis of neurons. Also, active caspase 1 can mediate the formation of NLRP3 inflammasome complex (Zheng et al., 2020).

Furthermore, postmortem studies of brain tissues in HD patients and tissues from the transgenic mouse model of HD provide experimental evidence for the activation of caspase 1 in brain regions in Huntington disease in Humans and mice model of HD (Ona et al., 1999).

Aforementioned findings were additionally supplemented by the study (Ona et al. 1999) that showed the role of caspase-1 inhibition in the decline of the progression of disease in mice model of HD.

Study (Ona et al. 1999) described that dominant-negative caspase-1 expression led to better survival of the transgenic mice model of HD with lesser neuronal inclusion bodies.

The study suggested the role of caspase 1 activation in the etiopathogenesis of Huntington disease.

Role of caspase 6 in Huntington's disease

Caspase-6 belongs to the family of cysteine-aspartate protease. It is encoded by CASP6 gene (Tiso et al., 1996).

Caspase 6 is the executioner caspase due to its involvement in the cleavage of nuclear lamina and changes in the cytoskeleton. Recent study showed that caspase 6 has a potential to activate caspase 3 in cerebellar granule cell apoptosis (Allsopp et al., 2000). Also, caspase 6 can activate the caspase 2. Thus studies delineated its potential to activate casp2 and casp3 and placed it in the category of initiator caspase Xanthoudakis et al. (1999).

Furthermore, enhanced activation of caspase 6 has been implicated in the etiopathogenesis of several degenerative diseases including Huntington disease (Graham et al., 2011; LeBlanc, 2013).

Surprisingly, aberrant rise in activity of caspase 6 was unrelated with the neuronal apoptosis. It was due to absence of marked apoptotic changes in the involved neurons with active caspase 6 (Graham et al., 2011).

It might be possible due to either incomplete activation of caspase 6 or partially suppressed activity of caspase 6 leading to gradual and progressive neuronal dysfunctioning and neuronal death (Unsain and Barker, 2015) and it is termed as sublethal activity of caspases.

Role of caspase 6 in the form of executioner caspase activity has linked to the neurodegeneration and etiopathogenesis of Huntington disease (Unsain and Barker, 2015).

The abnormal expansion of the CAG repeats in the Htt gene (Albin et al., 1990) leads to synthesis of mutant huntingtin protein (mHtt) with expansion of polyglutamine tract.

Furthermore, caspase induced cleavage of mutant huntingtin is crucial step in the etiopathogenesis of Huntington disease. The N-terminal fragments from mutant huntingtin are pathogenic (Mangiarini et al., 1996), while accumulation of these pathogenic N-terminal truncated fragments of mHTT have been reported in the neurons as inclusion bodies in patients with early Huntington disease.

The relation between caspase activation and caspase induced cleavage of mHtt were described in study by Graham et al. (2010).

The transgenic YAC mice model of HD expresses the caspase-6-resistant mutant huntingtin C6R mhtt) and additionally the mice lacks the HD phenotype (Absence of clinical manifestations of Huntington disease).

These facts demonstrate the role of caspase 6 mediated cleavage of huntingtin at 586 aa caspase-6 (casp6) site in the etiopathogenesis of Huntington disease (Graham et al., 2010).

Graham et al. (2010) reported that activation of casp6 precedes the clinical manifestations of motor abnormalities in the humans with HD and in mice model of HD. Furthermore, the levels of active caspase 6 in the HD humans and mice model of HD directly related with the length of polyglutamine tract.

But the expression of caspase-6-resistant mutant htt (C6R mhtt) was associated with diminished caspase activation and minimized injury to the neurons in the striatum in YAC model of HD.

Another study revealed that the C6R mice model is resistant to NMDA receptor-mediated toxicity (Milnerwood et al., 2010).

In the study by Graham et al. (2010) described the enhanced activation of caspase 6 and associated rise in neuronal apoptosis in the primary neurons in the striatum expressing caspase-cleavable mutant huntingtin but absence of caspase-6-resistant mutant htt after the exposure to NMDA.

The aforementioned findings provided by Graham et al. (2010) **in their study clearly suggest the increased activation of caspase 6 as biomarker and causative factor in the etiopathogenesis of Huntington disease.**

Furthermore, caspase-6-resistant mutant htt offers protection against the neuronal apoptosis in the Huntington disease.

Study by Graham et al. (2006) provided findings that aggregation of N-terminal pathogenic fragments in the neurons could be suppressed in the YAC mice model expressing C6R mHTT.

Thus, suppression of caspase 6 cleavage activity in turn limits the early-stage changes in the neurons in the striatum and cortex regions of brain in HD.

Exposure of neurons in striatum to staurosporine resulted into activation of caspase 6 and subsequent translocation to the nucleus. The active caspase 6 localizes with the 586 aa htt fragments (caspase 6 cleave product of mHTT) in nucleus indicating nuclear transport of caspase 6 crucial role in the neurotoxicity of N-terminal fragment of mutant huntingtin (Warby et al., 2008).

Tumor suppressor (p53) protein and caspase 6 in Huntington disease.

The TP53 gene in humans encodes the tumor suppressor p53 protein. The gene is regarded as the "guardian of genome" and protects the genome mutations of gene. it is termed as tumor suppressor gene (Matlashewski et al., 1984).

The tumor suppressor (p53) protein enhances the expression and activation of caspase 6 in skeletal muscles in patients with Huntington disease.

Important study was conducted by Ehrnhoefer et al. (2014) utilizing two separate transgenic mouse models of HD. Ehrnhoefer et al. (2014) study described the role of caspase 6 activation in the manifestations of peripheral phenotype like muscle wasting in the mouse model of HD.

The p53 is the transcriptional activator of caspase 6. Ehrnhoefer et al. (2014) reported increase in expression of tumor suppressor p53 in the neurons expressing mutant huntingtin leading to upregulated expression of casp6-mRNA (Ehrnhoefer et al., 2014), increased synthesis and activation of caspase-6 terminating (Ehrnhoefer et al., 2014) into raised proteolysis of lamin A (substrate of caspase 6) in the isolated skeletal muscle derived from patients with HD and in muscle tissues obtained from mice models of HD (Ehrnhoefer et al., 2014).

Thus, authors suggested that activation of caspase 6 mediated by increased expression of tumor suppressor protein (p53) has been reported in central nervous system as well as in the peripheral tissues as in skeletal muscles expressing mutant huntingtin.

Furthermore, expression and localization of pathogenic N-terminal fragment of mutant huntingtin mediates higher p53 activity.

References

Acheson, A., Conover, J.C., Fandl, J.P., DeChiara, T.M., Russell, M., Thadani, A., Squinto, S.P., Yancopoulos, G.D., Lindsay, R.M., 1995. A BDNF autocrine loop in adult sensory neurons prevents cell death. Nature 374 (6521), 450–453.

Albin, R.L., Young, A.B., Penney, J.B., Handelin, B., Balfour, R., Anderson, K.D., Markel, D.S., Tourtellotte, W.W., Reiner, A., 1990. Abnormalities of striatal projection neurons and N-methyl-D-aspartate receptors in presymptomatic Huntington's disease. N. Engl. J. Med. 322 (18), 1293–1298. https://doi.org/10.1056/NEJM199005033221807.

Allsopp, T.E., McLuckie, J., Kerr, L.E., Macleod, M., Sharkey, J., Kelly, J.S., 2000. Caspase 6 activity initiates caspase 3 activation in cerebellar granule cell apoptosis. Cell Death Differ. 7 (10), 984–993.

Andrabi, S.A., Kim, N.S., Yu, S.W., Wang, H., Koh, D.W., Sasaki, M., Klaus, J.A., Otsuka, T., Zhang, Z., Koehler, R.C., Hurn, P.D., Poirier, G.G., Dawson, V.L., Dawson, T.M., 2006. Poly(ADP-ribose) (PAR) polymer is a death signal. Proc. Natl. Acad. Sci. U. S. A. 103 (48), 18308–18313.

Andreassen, O.A., Dedeoglu, A., Stanojevic, V., Hughes, D.B., Browne, S.E., Leech, C.A., et al., 2002. Huntington's disease of the endocrine pancreas: insulin deficiency and diabetes mellitus due to impaired insulin gene expression. Neurobiol. Dis. 11, 410–424.

Andrew, S.E., Goldberg, Y.P., Kremer, B., Telenius, H., Theilmann, J., Adam, S., Starr, E., Squitieri, F., Lin, B., Kalchman, M.A., 1993. The relationship between trinucleotide (CAG) repeat length and clinical features of Huntington's disease. Nat. Genet. 4 (4), 398–403.

Anzilotti, S., Giampà, C., Laurenti, D., Perrone, L., Bernardi, G., Melone, M.A., Fusco, F.R., 2012. Immunohistochemical localization of receptor for advanced glycation end (RAGE) products in the R6/2 mouse model of Huntington's disease. Brain Res. Bull. 87 (2–3), 350–358.

Arrasate, M., Mitra, S., Schweitzer, E.S., Segal, M.R., Finkbeiner, S., 2004. Inclusion body formation reduces levels of mutant huntingtin and the risk of neuronal death. Nature 431 (7010), 805–810.

Axe, E.L., Walker, S.A., Manifava, M., Chandra, P., Roderick, H.L., Habermann, A., Griffiths, G., Ktistakis, N.T., 2008. Autophagosome formation from membrane compartments enriched in phosphatidylinositol 3-phosphate and dynamically connected to the endoplasmic reticulum. J. Cell Biol. 182 (4), 685–701.

Bae, B.I., Xu, H., Igarashi, S., Fujimuro, M., Agrawal, N., Taya, Y., Hayward, S.D., Moran, T.H., Montell, C., Ross, C.A., Snyder, S.H., Sawa, A., 2005. p53 mediates cellular dysfunction and behavioral abnormalities in Huntington's disease. Neuron 47 (1), 29–41.

Barbaro, B.A., et al., 2015. Comparative study of naturally occurring huntingtin fragments in Drosophila points to exon 1 as the most pathogenic species in Huntington's disease. Hum. Mol. Genet. 24, 913–925.

Baynes, J.W., Watkins, N.G., Fisher, C.I., Hull, C.J., Patrick, J.S., Ahmed, M.U., Dunn, J.A., Thorpe, S.R., 1989. The Amadori product on protein: structure and reactions. Prog. Clin. Biol. Res. 304, 43–67.

Becher, M.W., Kotzuk, J.A., Sharp, A.H., Davies, S.W., Bates, G.P., Price, D.L., Ross, C.A., 1998. Intranuclear neuronal inclusions in Huntington's disease and dentatorubral and pallidoluysian atrophy: correlation between the density of inclusions and IT15 CAG triplet repeat length. Neurobiol. Dis. 4 (6), 387–397.

Beneke, S., Diefenbach, J., Bürkle, A., 2004. Poly(ADP-ribosyl)ation inhibitors: promising drug candidates for a wide variety of pathophysiologic conditions. Int. J. Cancer 111 (6), 813–818.

Berger, Z., Ravikumar, B., Menzies, F.M., Oroz, L.G., Underwood, B.R., Pangalos, M.N., Schmitt, I., Wullner, U., Evert, B.O., O'Kane, C.J., et al., 2006. Rapamycin alleviates toxicity of different aggregate-prone proteins. Hum. Mol. Genet. 15, 433–442.

Björkqvist, M., Wild, E.J., Thiele, J., Silvestroni, A., Andre, R., Lahiri, N., Raibon, E., Lee, R.V., Benn, C.L., Soulet, D., Magnusson, A., Woodman, B., Landles, C., Pouladi, M.A., Hayden, M.R., Khalili-Shirazi, A., Lowdell, M.W., Brundin, P., Bates, G.P., Leavitt, B.R., Möller, T., Tabrizi, S.J., 2008. A novel pathogenic pathway of immune activation detectable before clinical onset in Huntington's disease. J. Exp. Med. 205 (8), 1869—1877.

Bowman, A.B., Yoo, S.Y., Dantuma, N.P., Zoghbi, H.Y., 2005. Neuronal dysfunction in a polyglutamine disease model occurs in the absence of ubiquitin-proteasome system impairment and inversely correlates with the degree of nuclear inclusion formation. Hum. Mol. Genet. 14 (5), 679—691.

Brás, I.C., König, A., Outeiro, T.F., 2019. Glycation in Huntington's disease: a possible modifier and target for intervention. J. Huntingtons Dis. 8 (3), 245—256.

Brinkman, R.R., Mezei, M.M., Theilmann, J., Almqvist, E., Hayden, M.R., 1997. The likelihood of being affected with Huntington disease by a particular age, for a specific CAG size. Am. J. Hum. Genet. 60 (5), 1202—1210.

Bushong, E.A., Martone, M.E., Jones, Y.Z., Ellisman, M.H., January, 2002. Protoplasmic astrocytes in CA1 stratum radiatum occupy separate anatomical domains. J. Neurosci. 22 (1), 183—192.

Cardinale, A., Paldino, E., Giampà, C., Bernardi, G., Fusco, F.R., 2015. PARP-1 inhibition is neuroprotective in the R6/2 mouse model of Huntington's disease. PLoS One 10 (8), e0134482.

Carroll, J.B., Southwell, A.L., Graham, R.K., Lerch, J.P., Ehrnhoefer, D.E., Cao, L.-P., Zhang, W.-N., Deng, Y., Bissada, N., Henkelman, R.M., Hayden, M.R., 2011. Mice lacking caspase-2 are protected from behavioral changes, but not pathology, in the YAC128 model of Huntington disease. Mol. Neurodegener. 6, 59.

Chan, D.C., 2006. Mitochondria: dynamic organelles in disease, aging, and development. Cell 125 (7), 1241—1252.

Chance, B., Sies, H., Boveris, A., 1979. Hydroperoxide metabolism in mammalian organs. Physiol. Rev. 59, 527—605.

Chang, R., Liu, X., Li, S., Li, X.J., 2015. Transgenic animal models for study of the pathogenesis of Huntington's disease and therapy. Drug Des. Dev. Ther. 9, 2179—2188.

Chaturvedi, R.K., Hennessey, T., Johri, A., Tiwari, S.K., Mishra, D., Agarwal, S., Kim, Y.S., Beal, M.F., 2012. Transducer of regulated CREB-binding proteins (TORCs) transcription and function is impaired in Huntington's disease. Hum. Mol. Genet. 21 (15), 3474—3488.

Chhor, V., Le Charpentier, T., Lebon, S., Oré, M.V., Celador, I.L., Josserand, J., Degos, V., Jacotot, E., Hagberg, H., Sävman, K., Mallard, C., Gressens, P., Fleiss, B., 2013. Characterization of phenotype markers and neuronotoxic potential of polarised primary microglia in vitro. Brain Behav. Immun. 32, 70—85.

Choi, Y.B., Kadakkuzha, B.M., Liu, X.A., Akhmedov, K., Kandel, E.R., Puthanveettil, S.V., 2014. Huntingtin is critical both pre- and postsynaptically for long-term learning-related synaptic plasticity in Aplysia. PLoS One 9 (7), e103004.

Choo, Y.S., Johnson, G.V., MacDonald, M., Detloff, P.J., Lesort, M., 2004. Mutant huntingtin directly increases susceptibility of mitochondria to the calcium-induced permeability transition and cytochrome c release. Hum. Mol. Genet. 13 (14), 1407—1420.

Cornett, J., Cao, F., Wang, C.E., Ross, C.A., Bates, G.P., Li, S.H., Li, X.J., 2005. Polyglutamine expansion of huntingtin impairs its nuclear export. Nat. Genet. 37 (2), 198—204.

Cummings, C.J., Mancini, M.A., Antalffy, B., DeFranco, D.B., Orr, H.T., Zoghbi, H.Y., 1998. Chaperone suppression of aggregation and altered subcellular proteasome localization imply protein misfolding in SCA1. Nat. Genet. 19 (2), 148—154.

Davies, S.W., Turmaine, M., Cozens, B.A., DiFiglia, M., Sharp, A.H., Ross, C.A., Scherzinger, E., Wanker, E.E., Mangiarini, L., Bates, G.P., 1997. Formation of neuronal intranuclear inclusions underlies the neurological dysfunction in mice transgenic for the HD mutation. Cell 90 (3), 537–548.

Desmond, C.R., Atwal, R.S., Xia, J., Truant, R., 2012. Identification of a karyopherin β1/β2 proline-tyrosine nuclear localization signal in huntingtin protein. J. Biol. Chem. 287 (47), 39626–39633.

Di Pardo, A., Alberti, S., Maglione, V., Amico, E., Cortes, E.P., Elifani, F., et al., 2013. Changes of peripheral TGF-beta1 depend on monocytes-derived macrophages in Huntington disease. Mol. Brain 6, 55. https://doi.org/10.1186/1756-6606-6-55.

Díaz-Hernández, M., Hernández, F., Martín-Aparicio, E., Gómez-Ramos, P., Morán, M.A., Castaño, J.G., Ferrer, I., Avila, J., Lucas, J.J., 2003. Neuronal induction of the immunoproteasome in Huntington's disease. J. Neurosci. 23 (37), 11653–11661.

DiFiglia, M., Sapp, E., Chase, K., Schwarz, C., Meloni, A., Young, C., Martin, E., Vonsattel, J.P., Carraway, R., Reeves, S.A., 1995. Huntingtin is a cytoplasmic protein associated with vesicles in human and rat brain neurons. Neuron 14 (5), 1075–1081.

DiFiglia, M., Sapp, E., Chase, K.O., Davies, S.W., Bates, G.P., Vonsattel, J.P., Aronin, N., 1997. Aggregation of huntingtin in neuronal intranuclear inclusions and dystrophic neurites in brain. Science 277 (5334), 1990–1993.

Dissing-Olesen, L., Ladeby, R., Nielsen, H.H., Toft-Hansen, H., Dalmau, I., Finsen, B., 2007. Axonal lesion-induced microglial proliferation and microglial cluster formation in the mouse. Neuroscience 149 (1), 112–122.

Dong, Y., Zhang, S., Wu, Z., Li, X., Wang, W.L., Zhu, Y., Stoilova-McPhie, S., Lu, Y., Finley, D., Mao, Y., 2018. Cryo-EM structures and dynamics of substrate-engaged human 26S proteasome. Nature 565 (7737), 49–55.

Donley, D.W., Olson, A.R., Raisbeck, M.F., Fox, J.H., Gigley, J.P., 2016. Huntingtons disease mice infected with toxoplasma gondii demonstrate early kynurenine pathway activation, altered CD8+ T-cell responses, and premature mortality. PLoS One 11 (9), e0162404.

Duan, W., Guo, Z., Jiang, H., Ware, M., Li, X.J., Mattson, M.P., 2003. Dietary restriction normalizes glucose metabolism and BDNF levels, slows disease progression, and increases survival in huntingtin mutant mice. Proc. Natl. Acad. Sci. U. S. A. 100 (5), 2911–2916.

Ehrnhoefer, D.E., Skotte, N.H., Ladha, S., Nguyen, Y.T., Qiu, X., Deng, Y., Huynh, K.T., Engemann, S., Nielsen, S.M., Becanovic, K., et al., 2014. p53 increases caspase-6 expression and activation in muscle tissue expressing mutant huntingtin. Hum. Mol. Genet. 23, 717–729.

Eliasson, M.J., Sampei, K., Mandir, A.S., et al., 1997. Poly(ADP-ribose) polymerase gene disruption renders mice resistant to cerebral ischemia. Nat. Med. 3 (10), 1089–1095.

Endres, M., Wang, Z.Q., Namura, S., et al., 1997. Ischemic brain injury is mediated by the activation of poly(ADP-ribose)polymerase. J. Cerebr. Blood Flow Metabol. 17 (11), 1143–1151.

Farrer, L.A., 1985. Diabetes mellitus in Huntington disease. Clin. Genet. 27, 62–67.

Feng, Z., Jin, S., Zupnick, A., Hoh, J., de Stanchina, E., Lowe, S., Prives, C., Levine, A.J., 2006. p53 tumor suppressor protein regulates the levels of huntingtin gene expression. Oncogene 25 (1), 1–7.

Fields, R.D., Araque, A., Johansen-Berg, H., Lim, S.-S., Lynch, G., Nave, K.-A., Nedergaard, M., Perez, R., Sejnowski, T., Wake, H., 2014. Glial biology in learning and cognition. Neuroscientist 20 (5), 426–431.

Franco, R., Fernandez-Suarez, D., 2015. Alternatively activated microglia and macrophages in the central nervous system. Prog. Neurobiol. 131, 65–86. https://doi.org/10.1016/j.pneurobio.2015.05.003.

Gandhi, S., Abramov, A.Y., 2012a. Mechanism of oxidative stress in neurodegeneration. Oxid. Med. Cell. Longev. 2012, 1–11.

Gandhi, S., Abramov, A.Y., 2012b. Mechanism of oxidative stress in neurodegeneration. Oxid. Med. Cell. Longev. 2012, 428010.

Gehrmann, J., Matsumoto, Y., Kreutzberg, G.W., 1995. Microglia: intrinsic immuneffector cell of the brain. Brain Res. Rev. 20 (3), 269–287.

Ghilan, M., Bostrom, C.A., Hryciw, B.N., Simpson, J.M., Christie, B.R., Gil-Mohapel, J., 2014. YAC128 Huntington's disease transgenic mice show enhanced short-term hippocampal synaptic plasticity early in the course of the disease. Brain Res. 1581, 117–128.

Ghosh, S., Karin, M., 2002. Missing pieces in the NF-kappaB puzzle. Cell 109 (Suppl. 1), S81–S96.

Ghosh, S., May, M.J., Kopp, E.B., 1998. NF-kappa B and Rel proteins: evolutionarily conserved mediators of immune responses. Annu. Rev. Immunol. 16, 225–260.

Giacomello, M., Drago, I., Pizzo, P., Pozzan, T., 2007. Mitochondrial Ca^{2+} as a key regulator of cell life and death. Cell Death Differ. 14, 1267–1274.

Goedert, M., Spillantini, M.G., Davies, S.W., 1998. Filamentous nerve cell inclusions in neurodegenerative diseases. Curr. Opin. Neurobiol. 8 (5), 619–632.

Gómez-Márquez, J., Dosil, M., Segade, F., Bustelo, X.R., Pichel, J.G., Dominguez, F., Freire, M., 1989. Thymosin-beta 4 gene. Preliminary characterization and expression in tissues, thymic cells, and lymphocytes. J. Immunol. 143 (8), 2740–2744.

Graham, R.K., Deng, Y., Slow, E.J., Haigh, B., Bissada, N., Lu, G., Pearson, J., Shehadeh, J., Bertram, L., Murphy, Z., Warby, S.C., et al., 2006. Cleavage at the caspase-6 site is required for neuronal dysfunction and degeneration due to mutant huntingtin. Cell 125 (6), 1179–1191. https://doi.org/10.1016/j.cell.2006.04.026.

Graham, R.K., Deng, Y., Carroll, J., Vaid, K., Cowan, C., Pouladi, M.A., Metzler, M., Bissada, N., Wang, L., Faull, R.L., et al., 2010. Cleavage at the 586 amino acid caspase-6 site in mutant huntingtin influences caspase-6 activation in vivo. J. Neurosci. 30 (45), 15019–15029. https://doi.org/10.1523/JNEUROSCI.2071-10.2010.

Graham, R.K., Ehrnhoefer, D.E., Hayden, M.R., 2011. Caspase-6 and neurodegeneration. Trends Neurosci. 34, 646–656.

Green, D.R., Reed, J.C., 1998. Mitochondria and apoptosis. Science 281 (5381), 1309–1312.

Groll, M., Ditzel, L., Löwe, J., Stock, D., Bochtler, M., Bartunik, H.D., Huber, R., 1997. Structure of 20S proteasome from yeast at 2.4 A resolution. Nature 386 (6624), 463–471.

Haas, A.L., Warms, J.V., Hershko, A., Rose, I.A., 1982. Ubiquitin-activating enzyme. Mechanism and role in protein-ubiquitin conjugation. J. Biol. Chem. 257 (5), 2543–2548.

Hanisch, U.K., Kettenmann, H., 2007. Microglia: active sensor and versatile effector cells in the normal and pathologic brain. Nat. Neurosci. 10 (11), 1387–1394.

Hara, T., Nakamura, K., Matsui, M., Yamamoto, A., Nakahara, Y., Suzuki-Migishima, R., Yokoyama, M., Mishima, K., Saito, I., Okano, H., et al., 2006. Suppression of basal autophagy in neural cells causes neurodegenerative disease in mice. Nature 441, 885–889.

Hermel, E., Gafni, J., Propp, S.S., Leavitt, B.R., Wellington, C.L., Young, J.E., Hackam, A.S., Logvinova, A.V., Peel, A.L., Chen, S.F., Hook, V., Singaraja, R., Krajewski, S., Goldsmith, P.C., Ellerby, H.M., Hayden, M.R., Bredesen, D.E., Ellerby, L.M., 2004. Specific caspase interactions and amplification are involved in selective neuronal vulnerability in Huntington's disease. Cell Death Differ. 11 (4), 424–438.

Hershko, A., Ciechanover, A., 1998. The ubiquitin system. Annu. Rev. Biochem. 67, 425–479.

Heyes, M.P., Achim, C.L., Wiley, C.A., Major, E.O., Saito, K., Markey, S.P., 1996. Human microglia convert l-tryptophan into the neurotoxin quinolinic acid. Biochem. J. 320 (Pt 2), 595–597.

Höybye, C., Hilding, A., Jacobsson, H., Thorén, M., 2002. Metabolic profile and body composition in adults with Prader-Willi syndrome and severe obesity. J. Clin. Endocrinol. Metab. 87, 3590–3597.

Hsiao, H.Y., Chiu, F.L., Chen, C.M., Wu, Y.R., Chen, H.M., Chen, Y.C., Kuo, H.C., Chern, Y., 2014. Inhibition of soluble tumor necrosis factor is therapeutic in Huntington's disease. Mol. Genet. 23 (16), 4328–4344.

Jack, C.S., Arbour, N., Manusow, J., Montgrain, V., Blain, M., McCrea, E., Shapiro, A., Antel, J.P., 2005. TLR signaling tailors innate immune responses in human microglia and astrocytes. J. Immunol. 175 (7), 4320–4330.

Jenkins, B.G., Koroshetz, W.J., Beal, M.F., Rosen, B.R., 1993. Evidence for impairment of energy metabolism in vivo in Huntington's disease using localized 1H NMR spectroscopy. Neurology 43 (12), 2689–2695.

Jha, M.K., Lee, W.H., Suk, K., 2016. Functional polarization of neuroglia: implications in neuroinflammation and neurological disorders. Biochem. Pharmacol. 103, 1–16. https://doi.org/10.1016/j.bcp.2015.11.003.

Johri, A., Chandra, A., Beal, M.F., 2013. PGC-1α, mitochondrial dysfunction and Huntington's disease. Free Radic. Biol. Med. 62, 37–46.

Josefsen, K., Nielsen, M.D., Jørgensen, K.H., Bock, T., Nørremølle, A., Sørensen, S.A., Naver, B., Hasholt, L., 2008. Mpaired glucose tolerance in the R6/1 transgenic mouse model of Huntington's disease. J. Neuroendocrinol. 20 (2), 165–172.

Kabeya, Y., Mizushima, N., Ueno, T., Yamamoto, A., Kirisako, T., Noda, T., Kominami, E., Ohsumi, Y., Yoshimori, T., 2000. LC3, a mammalian homologue of yeast Apg8p, is localized in autophagosome membranes after processing. EMBO J. 19, 5720–5728.

Khoshnan, A., Ko, J., Watkin, E.E., Paige, L.A., Reinhart, P.H., Patterson, P.H., 2004. Activation of the IkappaB kinase complex and nuclear factor-kappaB contributes to mutant huntingtin neurotoxicity. J. Neurosci. 24 (37), 7999–8008.

Kim, G.H., Kim, J.E., Rhie, S.J., Yoon, S., 2015a. The role of oxidative stress in neurodegenerative diseases. Exp. Neurobiol. 24, 325–340.

Kim, J., Waldvogel, H.J., Faull, R.L., Curtis, M.A., Nicholson, L.F., 2015b. The RAGE receptor and its ligands are highly expressed in astrocytes in a grade-dependant manner in the striatum and subependymal layer in Huntington's disease. J. Neurochem. 134 (5), 927–942.

Klionsky, D.J., 2008. Autophagy revisited: a conversation with Christian de Duve. Autophagy 4 (6), 740–743.

Komatsu, M., Waguri, S., Chiba, T., Murata, S., Iwata, J.I., Tanida, I., Ueno, T., Koike, M., Uchiyama, Y., Kominami, E., et al., 2006. Loss of autophagy in the central nervous system causes neurodegeneration in mice. Nature 441, 880–884.

Kremer, H.P., Roos, R.A., Frölich, M., Radder, J.K., Nieuwenhuijzen Kruseman, A.C., Van der Velde, A., Buruma, O.J., 1989. Endocrine functions in Huntington's disease. A two-and-a-half years follow-up study. J. Neurol. Sci. 90 (3), 335–344.

Krieger, C., Duchen, M.R., 2002. Mitochondria, Ca^{2+} and neurodegenerative disease. Eur. J. Pharmacol. 447, 177–188. https://doi.org/10.1016/S0014-2999(02)01842-3.

Kroner, A., Greenhalgh, A.D., Zarruk, J.G., Passos Dos Santos, R., Gaestel, M., David, S., 2014. TNF and increased intracellular iron alter macrophage polarization to a detrimental M1 phenotype in the injured spinal cord. Neuron 83 (5), 1098–1116.

Krumschnabel, G., Sohm, B., Bock, F., Manzl, C., Villunger, A., February, 2009. The enigma of caspase-2: the laymen's view. Cell Death Differ. 16 (2), 195−207.

Landwehrmeyer, G.B., McNeil, S.M., Dure 4th, L.S., Ge, P., Aizawa, H., Huang, Q., Ambrose, C.M., Duyao, M.P., Bird, E.D., Bonilla, E., 1995. Huntington's disease gene: regional and cellular expression in brain of normal and affected individuals. Ann. Neurol. 37 (2), 218−230.

Lawson, L.J., Perry, V.H., Gordon, S., 1992. Turnover of resident microglia in the normal adult mouse brain. Neuroscience 48 (2), 405−415.

LeBlanc, A.C., 2013. Caspase-6 as a novel early target in the treatment of Alzheimer's disease. Eur. J. Neurosci. 37, 2005−2018.

Leibson, C.L., Rocca, W.A., Hanson, V.A., Cha, R., Kokmen, E., O'Brien, P.C., et al., 1997. Risk of dementia among persons with diabetes mellitus: a population-based cohort study. Am. J. Epidemiol. 145, 301−308.

Li, S.H., Schilling, G., Young 3rd, W.S., Li, X.J., Margolis, R.L., Stine, O.C., Wagster, M.V., Abbott, M.H., Franz, M.L., Ranen, N.G., 1993. Huntington's disease gene (IT15) is widely expressed in human and rat tissues. Neuron 11 (5), 985−993.

Li, H., Li, S.H., Johnston, H., Shelbourne, P.F., Li, X.J., 2000. Amino-terminal fragments of mutant huntingtin show selective accumulation in striatal neurons and synaptic toxicity. Nat. Genet. 25 (4), 385−389.

Li, W., Serpell, L.C., Carter, W.J., Rubinsztein, D.C., Huntington, J.A., 2006. Expression and characterization of full-length human huntingtin, an elongated HEAT repeat protein. J. Biol. Chem. 281 (23), 15916−15922.

Lilla, H., Nikoletta, D., Gábor, K., Zsolt, K., Dénes, P., Zoltán, S., Gábor, H., Lóránt, S., 2018. The role of indoleamine-2,3-dioxygenase in cancer development, diagnostics, and therapy. Front. Immunol. 9, 151.

Lin, B., Rommens, J.M., Graham, R.K., Kalchman, M., MacDonald, H., Nasir, J., Delaney, A., Goldberg, Y.P., Hayden, M.R., 1993. Differential 3′ polyadenylation of the Huntington disease gene results in two mRNA species with variable tissue expression. Hum. Mol. Genet. 2, 1541−1545.

Lin, J., Wu, P.H., Tarr, P.T., Lindenberg, K.S., St-Pierre, J., Zhang, C.Y., Mootha, V.K., Jäger, S., Vianna, C.R., Reznick, R.M., Cui, L., Manieri, M., Donovan, M.X., Wu, Z., Cooper, M.P., Fan, M.C., Rohas, L.M., Zavacki, A.M., Cinti, S., Shulman, G.I., Lowell, B.B., Krainc, D., Spiegelman, B.M., 2004. Defects in adaptive energy metabolism with CNS-linked hyperactivity in PGC-1alpha null mice. Cell 119 (1), 121−135.

Liu, C.-S., Cheng, W.-L., Kuo, S.-J., Li, J.-Y., Soong, B.-W., Wei, Y.-H., 2008. Depletion of mitochondrial DNA in leukocytes of patients with poly-Q diseases. J. Neurol. Sci. 264, 18−21.

Lunkes, A., Lindenberg, K.S., Ben-Haïem, L., Weber, C., Devys, D., Landwehrmeyer, G.B., Mandel, J.L., Trottier, Y., 2002. Proteases acting on mutant huntingtin generate cleaved products that differentially build up cytoplasmic and nuclear inclusions. Mol. Cell. 10 (2), 259−269.

MacDonald, M.E., Ambrose, C.M., Duyao, M.P., Myers, R.H., Lin, C., Srinidhi, L., et al., 1993. A novel gene containing a trinucleotide repeat that is expanded and unstable on Huntington's disease chromosomes. Cell 72, 971−983.

Maillard, L.C., 1912. Action des acides amines sur les sucres; formation de melanoidines par voie méthodique [Action of amino acids on sugars. Formation of melanoidins in a methodical way]. Comptes Rendus 154, 66−68 (in French).

Mangiarini, L., et al., 1996. Exon 1 of the HD gene with an expanded CAG repeat is sufficient to cause a progressive neurological phenotype in transgenic mice. Cell 87, 493—5069.

Mao, K., Chen, J., Yu, H., Li, H., Ren, Y., Wu, X., Wen, Y., Zou, F., Li, W., 2020. Poly (ADP-ribose) polymerase 1 inhibition prevents neurodegeneration and promotes α-synuclein degradation via transcription factor EB-dependent autophagy in mutant α-synuclein A53T model of Parkinson's disease. Aging Cell 19, e13163.

Marques, C., Sousa, S., 2013. Humbert Huntingtin: here, there, everywhere! J. Huntingtons Dis. 2, 395—403.

Martens, S., Fracchiolla, D., 2020. Activation and targeting of ATG8 protein lipidation. Cell Discov 6, 23.

Matlashewski, G., Lamb, P., Pim, D., Peacock, J., Crawford, L., Benchimol, S., 1984. Isolation and characterization of a human p53 cDNA clone: expression of the human p53 gene. EMBO J. 3 (13), 3257—3262.

Milnerwood, A.J., Gladding, C.M., Pouladi, M.A., Kaufman, A.M., Hines, R.M., Boyd, J.D., Ko, R.W., Vasuta, O.C., Graham, R.K., Hayden, M.R., Murphy, T.H., Raymond, L.A., 2010. Early increase in extrasynaptic NMDA receptor signaling and expression contributes to phenotype onset in Huntington's disease mice. Neuron 65 (2), 178—190.

Miron, V.E., Boyd, A., Zhao, J.W., Yuen, T.J., Ruckh, J.M., Shadrach, J.L., van Wijngaarden, P., Wagers, A.J., Williams, A., Franklin, R.J.M., Ffrench-Constant, C., 2013. M2 microglia and macrophages drive oligodendrocyte differentiation during CNS remyelination. Nat. Neurosci. 16 (9), 1211—1218.

Möller, T., 2010. Neuroinflammation in Huntington's disease. J. Neural. Transm. 117 (8), 1001—1008.

Monnier, V.M., 1989. Toward a Maillard reaction theory of aging. Prog. Clin. Biol. Res. 304, 1—22.

Myers, R.H., 2004. Huntington's disease genetics. NeuroRx 1 (2), 255—262.

Nakatogawa, H., Suzuki, K., Kamada, Y., Ohsumi, Y., 2009. Dynamics and diversity in autophagy mechanisms: lessons from yeast. Nat. Rev. Mol. Cell Biol. 10 (7), 458—467.

Napoli, E., Wong, S., Hung, C., et al., 2013. Defective mitochondrial disulfide relay system, altered mitochondrial morphology and function in Huntington's disease. Hum. Mol. Genet. 22 (5), 989—1004.

National Center for Biotechnology Information, 2021. HTT Huntingtin. https://www.ncbi.nlm.nih.gov/gene/3064.

Neueder, A., Landles, C., Ghosh, R., Howland, D., Myers, R.H., Faull, R.L.M., Tabrizi, S.J., Bates, G.P., 2017. The pathogenic exon 1 HTT protein is produced by incomplete splicing in Huntington's disease patients. Sci. Rep. 7, 1307. https://doi.org/10.1038/s41598-017-01510-z.

Neveklovska, M., Clabough, E.B., Steffan, J.S., Zeitlin, S.O., 2012. Deletion of the huntingtin proline-rich region does not significantly affect normal huntingtin function in mice. J. Huntingtons Dis. 1 (1), 71—87. https://doi.org/10.3233/JHD-2012-120016.

Nicholls, D.G., Budd, S.L., 2000. Mitochondria and neuronal survival. Physiol. Rev. 80, 315—360. https://doi.org/10.1152/physrev.2000.80.1.315.

Njoroge, F.G., Monnier, V.M., 1989. The chemistry of the Maillard reaction under physiological conditions: a review. Prog. Clin. Biol. Res. 304, 85—107.

Nursten, H.E., 2005. The Maillard Reaction. Chemistry, Biochemistry, and Implications. Royal Society of Chemistry, Cambridge.

Ona, V.O., Li, M., Vonsattel, J.P., Andrews, L.J., Khan, S.Q., Chung, W.M., Frey, A.S., Menon, A.S., Li, X.-J., Stieg, P.E., et al., 1999. Inhibition of caspase-1 slows disease progression in a mouse model of Huntington's disease. Nature 399, 263—267.

Orre, M., Kamphuis, W., Dooves, S., Kooijman, L., Chan, E.T., Kirk, C.J., Dimayuga Smith, V., Koot, S., Mamber, C., Jansen, A.H., Ovaa, H., Hol, E.M., 2013. Reactive glia show increased immunoproteasome activity in Alzheimer's disease. Rain 136 (Pt 5), 1415–1431.

Ortega, Z., Lucas, J.J., 2014. Ubiquitin–proteasome system involvement in Huntington's disease. Front. Mol. Neurosci. 7, 77.

Palidwor, G.A., Shcherbinin, S., Huska, M.R., Rasko, T., Stelzl, U., Arumughan, A., Foulle, R., Porras, P., Sanchez-Pulido, L., Wanker, E.E., Andrade-Navarro, M.A., 2009. Detection of alpha-rod protein repeats using a neural network and application to huntingtin PLoS Comput. Biol. 5, e1000304.

Pascal, J.M., 2018. The comings and goings of PARP-1 in response to DNA damage. DNA Repair 71, 177–182.

Paul, B.D., Snyder, S.H., 2019. Impaired redox signaling in Huntington's disease: therapeutic implications. Front. Mol. Neurosci. 12, 68.

Pavese, N., Gerhard, A., Tai, Y.F., Ho, A.K., Turkheimer, F., Barker, R.A., et al., 2006. Microglial activation correlates with severity in Huntington disease: a clinical and PET study. Neurology 66, 1638–1643.

Penney Jr., J.B., Vonsattel, J.P., Macdonald, M.E., Gusella, J.F., Myers, R.H., 1997. CAG repeat number governs the development rate of pathology in Huntington's disease. Ann. Neurol. 41, 689–692.

Petersen, M.H., Budtz-Jørgensen, E., Sørensen, S.A., Nielsen, J.E., Hjermind, L.E., Vinther-Jensen, T., Nielsen, S.M.B., Nørremølle, A., 2014. Reduction in mitochondrial DNA copy number in peripheral leukocytes after onset of Huntington's disease. Mitochondrion 17, 14–21. https://doi.org/10.1016/j.mito.2014.05.001.

Perutz, M.F., Johnson, T., Suzuki, M., Finch, J.T., 1994. Glutamine repeats as polar zippers: their possible role in inherited neurodegenerative diseases. Proc. Natl. Acad. Sci. U. S. A. 91 (12), 5355–5358.

Petersén, A., Mani, K., Brundin, P., 1999. Recent advances on the pathogenesis of Huntington's disease. Exp. Neurol. 157 (1), 1–18.

Podolsky, S., Leopold, N.A., Sax, D.S., 1972. Increased frequency of diabetes mellitus in patients with Huntington's chorea. Lancet 1 (7765), 1356–1358.

Polidori, M.C., Mecocci, P., Browne, S.E., Senin, U., Beal, M.F., 1999. Oxidative damage to mitochondrial DNA in Huntington's disease parietal cortex. Neurosci. Lett. 272 (1), 53–56.

Politis, M., Pavese, N., Tai, Y.F., Kiferle, L., Mason, S.L., Brooks, D.J., et al., 2011. Microglial activation in regions related to cognitive function predicts disease onset in Huntington's disease: a multimodal imaging study. Hum. Brain. Mapp. 32, 258–270. https://doi.org/10.1002/hbm.21008.

Ravikumar, B., Duden, R., Rubinsztein, D.C., 2002. Aggregate-prone proteins with polyglutamine and polyalanine expansions are degraded by autophagy. Hum. Mol. Genet. 11, 1107–1117.

Reiner, A., Albin, R.L., Anderson, K.D., D'Amato, C.J., Penney, J.B., Young, A.B., 1988. Differential loss of striatal projection neurons in Huntington disease. Proc. Natl. Acad. Sci. U. S. A. 85 (15), 5733–5737.

Ristow, M., 2004. Neurodegenerative disorders associated with diabetes mellitus. J. Mol. Med. 82, 510–529.

Rosas, H.D., Liu, A.K., Hersch, S., Glessner, M., Ferrante, R.J., Salat, D.H., van der Kouwe, A., Jenkins, B.G., Dale, A.M., Fischl, B., 2002. Regional and progressive thinning of the cortical ribbon in Huntington's disease. Neurology 58 (5), 695–701.

Rosenblatt, A., 2007. Neuropsychiatry of Huntington's disease. Dialogues Clin. Neurosci. 9 (2), 191–197.

Rubinsztein, D.C., Leggo, J., Coles, R., Almqvist, E., Biancalana, V., Cassiman, J.J., Chotai, K., Connarty, M., Crauford, D., Curtis, A., Curtis, D., Davidson, M.J., Differ, A.M., Dode, C., Dodge, A., Frontali, M., Ranen, N.G., Stine, O.C., Sherr, M., Abbott, M.H., Franz, M.L., Graham, C.A., Harper, P.S., Hedreen, J.C., Hayden, M.R., 1996. Phenotypic characterization of individuals with 30-40 CAG repeats in the Huntington disease (HD) gene reveals HD cases with 36 repeats and apparently normal elderly individuals with 36-39 repeats. Am. J. Hum. Genet. 59 (1), 16–22.

Sapp, E., Kegel, K.B., Aronin, N., Hashikawa, T., Uchiyama, Y., Tohyama, K., Bhide, P.G., Vonsattel, J.P., DiFiglia, M., 2001. Early and progressive accumulation of reactive microglia in the Huntington disease brain. J. Neuropathol. Exp. Neurol. 60 (2), 161–172.

Sathasivam, K., et al., 2013. Aberrant splicing of HTT generates the pathogenic exon 1 protein in Huntington disease. Proc. Natl. Acad. Sci. U. S. A. 110, 2366–2370.

Saudou, F., Finkbeiner, S., Devys, D., Greenberg, M.E., 1998. Huntingtin acts in the nucleus to induce apoptosis but death does not correlate with the formation of intranuclear inclusions. Cell 95 (1), 55–66. https://doi.org/10.1016/s0092-8674(00)81782-1.

Saudou, F., Humbert, S., 2016. The biology of Huntingtin. Neuron 89 (Issue 5), 910–926.

Savitz, J., 2020. The kynurenine pathway: a finger in every pie. Mol. Psychiatr. 25 (1).

Schiff, H., 1866. Eine neue Reihe organischer Diamine" [A new series of organic diamines]. Annalen der Chemie und Pharmacie 3, 343–370. Supplementband (in German).

Schipper-Krom, S., Juenemann, K., Jansen, A.H., Wiemhoefer, A., van den Nieuwendijk, R., Smith, D.L., Hink, M.A., Bates, G.P., Overkleeft, H., Ovaa, H., Reits, E., 2014. Dynamic recruitment of active proteasomes into polyglutamine initiated inclusion bodies. FEBS Lett. 588 (1), 151–159.

Sherman, M.Y., Goldberg, A.L., 2001. Cellular defenses against unfolded proteins: a cell biologist thinks about neurodegenerative diseases. Neuron 29 (1), 15–32.

Shirendeb, U., Reddy, A.P., Manczak, M., Calkins, M.J., Mao, P., Tagle, D.A., et al., 2011a. Abnormal mitochondrial dynamics, mitochondrial loss and mutant huntingtin oligomers in Huntington's disease: implications for selective neuronal damage. Hum. Mol. Genet. 20, 1438–1455.

Shirendeb, U.P., Calkins, M.J., Manczak, M., Anekonda, V., Dufour, B., McBride, J.L., et al., 2011b. Mutant huntingtin's interaction with mitochondrial protein Drp1 impairs mitochondrial biogenesis and causes defective axonal transport and synaptic degeneration in Huntington's disease. Hum. Mol. Genet. 21, 406–420.

Siddiqui, A., Rivera-Sánchez, S., Castro, M.d.R., Acevedo-Torres, K., Rane, A., Torres-Ramos, C.A., et al., 2012. Mitochondrial DNA damage is associated with reduced mitochondrial bioenergetics in Huntington's disease. Free Radic. Biol. Med. 53, 1478–1488. https://doi.org/10.1016/j.freeradbiomed.2012.06.008.

Sies, H., 1991. Oxidative stress: from basic research to clinical application. Am. J. Med. 91, S31–S38. https://doi.org/10.1016/0002-9343(91)90281-2.

Singhrao, S.K., Neal, J.W., Morgan, B.P., Gasque, P., 1999. Increased complement biosynthesis by microglia and complement activation on neurons in Huntington's disease. Exp. Neurol. 159 (2), 362–376.

Smith, D.M., Kafri, G., Cheng, Y., Ng, D., Walz, T., Goldberg, A.L., 2005. ATP binding to PAN or the 26S ATPases causes association with the 20S proteasome, gate opening, and translocation of unfolded proteins. Mol. Cell 20 (5), 687–698.

Squitieri, F., Cannella, M., Simonelli, M., 2002. CAG mutation effect on rate of progression in Huntington's disease. Neurol. Sci. 23 (Suppl. 2), S107—S108.

St-Pierre, J., Drori, S., Uldry, M., Silvaggi, J.M., Rhee, J., Jäger, S., Handschin, C., Zheng, K., Lin, J., Yang, W., Simon, D.K., Bachoo, R., Spiegelman, B.M., 2006. Suppression of reactive oxygen species and neurodegeneration by the PGC-1 transcriptional coactivators. Cell 127 (2), 397—408.

Steffan, J.S., Kazantsev, A., Spasic-Boskovic, O., Greenwald, M., Zhu, Y.Z., Gohler, H., Wanker, E.E., Bates, G.P., Housman, D.E., Thompson, L.M.I., 2000. The Huntington's disease protein interacts with p53 and CREB-binding protein and represses transcription. Proc. Natl. Acad. Sci. U. S. A. 97 (12), 6763—6768.

Su, P., Zhang, J., Wang, D., Zhao, F., Cao, Z., Aschner, M., Luo, W., 2016. The role of autophagy in modulation of neuroinflammation in microglia. Neuroscience 319, 155—167.

Suzuki, M., Nagai, Y., Wada, K., Koike, T., 2012. Calcium leak through ryanodine receptor is involved in neuronal death induced by mutant huntingtin. Biochem. Biophys. Res. Commun. 429 (1—2), 18—23.

Tai, Y.F., Pavese, N., Gerhard, A., Tabrizi, S.J., Barker, R.A., Brooks, D.J., et al., 2007. Imaging microglial activation in Huntington's disease. Brain Res. Bull. 72, 148—151.

Takahashi, T., Kikuchi, S., Katada, S., Nagai, Y., Nishizawa, M., Onodera, O., 2008. Soluble polyglutamine oligomers formed prior to inclusion body formation are cytotoxic. Hum. Mol. Genet. 17, 345—356.

Tartari, M., Gissi, C., Lo Sardo, V., Zuccato, C., Picardi, E., Pesole, G., Cattaneo, E., 2008. Phylogenetic comparison of huntingtin homologues reveals the appearance of a primitive polyQ in sea urchin. Mol. Biol. Evol. 25, 330—338.

The Huntington's Disease Collaborative Research Group (HDCRG), 1993. A novel gene containing a trinucleotide repeat that is expanded and unstable on Huntington's disease chromosomes. The Huntington's Disease Collaborative Research Group. Cell 72 (6), 971—983.

Thrower, J.S., Hoffman, L., Rechsteiner, M., Pickart, C.M., 2000. Recognition of the polyubiquitin proteolytic signal. EMBO J. 19 (1), 94—102.

Tiso, N., Pallavicini, A., Muraro, T., Zimbello, R., Apolloni, E., Valle, G., Lanfranchi, G., Danieli, G.A., 1996. Chromosomal localization of the human genes, CPP32, Mch2, Mch3, and Ich-1, involved in cellular apoptosis. Biochem. Biophys. Res. Commun. 2.

Träger, U., Andre, R., Lahiri, N., Magnusson-Lind, A., Weiss, A., Grueninger, S., McKinnon, C., Sirinathsinghji, E., Kahlon, S., Pfister, E.L., Moser, R., Hummerich, H., Antoniou, M., Bates, G.P., Luthi-Carter, R., Lowdell, M.W., Björkqvist, M., Ostroff, G.R., Aronin, N., Tabrizi, S.J., 2014. HTT-lowering reverses Huntington's disease immune dysfunction caused by NFκB pathway dysregulation. Brain 137 (Pt 3), 819—833.

Tschampa, H.J., Kallenberg, K., Kretzschmar, H.A., Meissner, B., Knauth, M., Urbach, H., et al., 2007. Pattern of cortical changes in sporadic Creutzfeldt-Jakob disease. Am J Neuroradiol 28, 1114—1118.

Tsunemi, T., Ashe, T.D., Morrison, B.E., Soriano, K.R., Au, J., Roque, R.A., Lazarowski, E.R., Damian, V.A., Masliah, E., La Spada, A.R., 2012. PGC-1α rescues Huntington's disease proteotoxicity by preventing oxidative stress and promoting TFEB function. Sci. Transl. Med. 4 (142), 142ra97.

Unsain, N., Barker, P.A., 2015. New views on the misconstrued: executioner caspases and their diverse non-apoptotic roles. Neuron 88, 461—474.

Valekova, I., Jarkovska, K., Kotrcova, E., Bucci, J., Ellederova, Z., Juhas, S., Motlik, J., Gadher, S.J., Kovarova, H., 2016. Revelation of the IFNα, IL-10, IL-8 and IL-1β as promising biomarkers reflecting immuno-pathological mechanisms in porcine Huntington's disease model. J. Neuroimmunol. 293, 71–81.

van Leeuwen, F.W., de Kleijn, D.P., van den Hurk, H.H., Neubauer, A., Sonnemans, M.A., Sluijs, J.A., et al., 1998. Frameshift mutants of beta amyloid precursor protein and ubiquitin-B in Alzheimer's and Down patients. Science 279, 242–247. https://doi.org/10.1126/science.279.5348.242.

Vistoli, G., De Maddis, D., Cipak, A., Zarkovic, N., Carini, M., Aldini, G., 2013. Advanced glycoxidation and lipoxidation end products (AGEs and ALEs): an overview of their mechanisms of formation. Free Radic. Res. 47 (Suppl. 1), 3–27.

Walczak, M., Martens, S., 2013. Dissecting the role of the Atg12-Atg5-Atg16 complex during autophagosome formation. Autophagy 9 (3), 424–425.

Walker, N.I., Harmon, B.V., Gobe, G.C., et al., 1988. Patterns of cell death. Methods Achiev. Exp. Pathol. 13, 18–54.

Wang, Y., Lin, F., Qin, Z.H., 2010. The role of post-translational modifications of huntingtin in the pathogenesis of Huntington's disease. Neurosci. Bull. 26 (2), 153–162.

Warby, S.C., Doty, C.N., Graham, R.K., Carroll, J.B., Yang, Y.Z., Singaraja, R.R., Overall, C.M., Hayden, M.R., 2008. Activated caspase-6 and caspase-6-cleaved fragments of huntingtin specifically colocalize in the nucleus. Hum. Mol. Genet. 17 (15), 2390–2404.

Weidinger, A., Kozlov, A.V., 2015. Biological activities of reactive oxygen and nitrogen species: oxidative stress versus signal transduction. Biomolecules 5, 472–484. https://doi.org/10.3390/biom5020472.

Weydt, P., Soyal, S.M., Landwehrmeyer, G.B., Patsch, W., 2014. A single nucleotide polymorphism in the coding region of PGC-1α is a male-specific modifier of Huntington disease age-at-onset in a large European cohort. BMC Neurol. 14, 1. https://doi.org/10.1186/1471-2377-14-1.

Wu, L.L., Zhou, X.F., 2009. Huntingtin associated protein 1 and its functions. Cell Adhes. Migrat. 3 (1), 71–76.

Xanthoudakis, S., Roy, S., Rasper, D., Hennessey, T., Aubin, Y., Cassady, R., Tawa, P., Ruel, R., Rosen, A., Nicholson, D.W., 1999. Hsp60 accelerates the maturation of procaspase-3 by upstream activator proteases during apoptosis. EMBO J. 18 (8), 2049–2056.

Xu, M., Wong, A., 2018. GABAergic inhibitory neurons as therapeutic targets for cognitive impairment in schizophrenia. Acta Pharmacol. Sin. 39, 733–753.

Yang, H.-M., Yang, S., Huang, S.-S., Tang, B.-S., Guo, J.-F., 2017. Microglial activation in the pathogenesis of Huntington's disease. Front. Aging Neurosci. 9, 193.

Yu, B.P., 1994. Cellular defenses against damage from reactive oxygen species. Physiol. Rev. 74, 139–162. https://doi.org/10.1152/physrev.1994.74.1.139.

Zheng, S., Clabough, E.B.D., Sarkar, S., Futter, M., Rubinsztein, D.C., Zeitlin, S.O., 2010. Deletion of the huntingtin polyglutamine stretch enhances neuronal autophagy and longevity in mice. PLoS Genet. https://doi.org/10.1371/journal.pgen.1000838.

Zheng, J., Winderickx, J., Vanessa, F., Liu, B., 2018. A mitochondria-associated oxidative stress perspective on Huntington's disease. Front. Mol. Neurosci. 11, 329.

Zheng, D., Liwinski, T., Elinav, E., 2020. Inflammasome activation and regulation: toward a better understanding of complex mechanisms. Cell Discov. 6, 36.

Index

Printed in the United States
by Baker & Taylor Publisher Services